Estimations AND Tests IN Change-Point Models

Estimations AND Tests IN Change-Point Models

Odile Pons

INRA, National Institute for Agronomical Research, France

NEW JERSEY · LONDON · SINGAPORE · BEIJING · SHANGHAI · HONG KONG · TAIPEI · CHENNAI · TOKYO

Published by

World Scientific Publishing Co. Pte. Ltd.
5 Toh Tuck Link, Singapore 596224
USA office: 27 Warren Street, Suite 401-402, Hackensack, NJ 07601
UK office: 57 Shelton Street, Covent Garden, London WC2H 9HE

Library of Congress Cataloging-in-Publication Data
Names: Pons, Odile, author.
Title: Estimations and tests in change-point models / by Odile Pons
(INRA, National Institute for Agronomical Research, France).
Description: New Jersey : World Scientific, 2018. | Includes bibliographical references and index.
Identifiers: LCCN 2017060417 | ISBN 9789813231764 (hardcover : alk. paper)
Subjects: LCSH: Change-point problems. | Mathematical statistics. | Mathematical analysis.
Classification: LCC QA274.42 .P66 2018 | DDC 519.2/3--dc23
LC record available at https://lccn.loc.gov/2017060417

British Library Cataloguing-in-Publication Data
A catalogue record for this book is available from the British Library.

For any available supplementary material, please visit
http://www.worldscientific.com/worldscibooks/10.1142/10757#t=suppl

Printed in Singapore

Preface

The interest and usefulness of an automatic detection of changes have been proved by an increasing number of applications to a wide range of technical domains. The failure of a system may be expressed by an abrupt change in the mean of sequential measurements of its features, this is a frequent question in the analysis of data in quality control, in medical or ecological follow-up. Chemical models are most often described by nonlinear parametric models. The deviation from a required pattern is detected by tests of constant models against models with a change which compares the parameters of each model through a statistical criterion.

The methods used for estimating the parameters are the minimization of the empirical variance of the variable of interest and the maximization of their likelihood. The first chapters consider samples of independent and identically distributed variables on several phases determined by the change-points and the asymptotic behaviour are related to the sampling sizes in each phase. We study changes in the mean of a variable, changes in the parameters of its density and changes in the parameters of regression functions. The samples of independent observations are extended to series of dependent observations. The following chapters consider counting processes and their intensity, the asymptotic study relies on observations in increasing time intervals or in an increasing number of independent observations on the same interval.

In models with a change in the mean or in the density, including Poisson and point processes, and in linear regression models, we consider changes of the parameters according to the level of variables of the model and chronological changes at a sampling index. The choice of parametric models with a few number of parameters may be restrictive though it can be statistically assessed, so they are generalized to models where unknown functions are replaced by nonparametric estimators.

Introducing change-points and discontinuities in regular models of differentiable functions modifies deeply the behaviour of their maximum likelihood and least squares estimators so their weak convergence and the limit of the test statistics are not standard. This is the motivation of this book and it contains many new results.

Odile M.-T. Pons
July 2017

Contents

Chapter 1

Introduction

1.1 Detection of changes in the distribution of a sample

A model with change-point in the mean at an unknown sample index k is defined as

$$EY_i = \mu_1, \, i = 1, \ldots, k,$$
$$EY_i = \mu_2, \, i = k+1, \ldots, n,$$

it can be detected recursively with a sequential calculus of the mean of the observations. Let $N_j = n_1 + \cdots + n_j$ be the total size of a sample with j independent sub-samples having respective sizes n_j, the empirical mean of Y is calculated recursively as

$$\bar{Y}_{N_j} = \frac{N_{j-1}}{N_j} \bar{Y}_{N_{j-1}} + \frac{n_j}{N_j} \bar{Y}_{n_j}$$

where $\bar{Y}_{N_{j-1}}$ and $\bar{Y}_{n_j}$ are the empirical mean of $Y_1, \ldots, Y_{N_{j-1}}$ and, respectively $Y_{N_{j-1}+1}, \ldots, Y_{N_j}$. A recursive formula for the empirical variance of Y is

$$\widehat{\sigma}^2_{N_j} = \frac{N_{j-1}}{N_j} \widehat{\sigma}^2_{N_{j-1}} + \frac{n_j}{N_j} \widehat{\sigma}^2_{n_j}$$

with the empirical variances $\widehat{\sigma}^2_{N_{j-1}}$ of $Y_1, \ldots, Y_{N_{j-1}}$ and $\widehat{\sigma}^2_{n_j}$ of the $(j-1)$th sub-sample $Y_{N_{j-1}+1}, \ldots, Y_{N_j}$. Sequential Student's tests of a constant mean are performed with the statistics

$$S_{N_j} = \frac{N_{j-1} n_j (\bar{Y}_{N_{j-1}} - \bar{Y}_{n_j})}{N_j \widehat{\sigma}_{N_j}}$$

which detect a change at an index between N_{j-1} and N_j, when the mean of the first N_{j-1} observations is constant.

If the variable $Y_i - \mu_j$ has a common distribution function F, the distribution function of Y_i is

$$P(Y_i \le y) = F(y - \mu_1),\ i = 1, \dots, k,$$
$$P(Y_i \le y) = F(y - \mu_2),\ i = k+1, \dots, n.$$

The empirical estimators of the partial means in the model with a change of mean at k are

$$\widehat{\mu}_{1n,k} = k^{-1} \sum_{i=1}^{k} Y_i,$$

$$\widehat{\mu}_{2n,k} = (n-k)^{-1} \sum_{i=k+1}^{n} Y_i$$

they are independent and have the variances $\sigma_{1k}^2 = k^{-1}\sigma_0^2$ and respectively $\sigma_{2k}^2 = (n-k)^{-1}\sigma_0^2$ where σ_0^2 is the variance of Y. Then σ_0^2 is estimated by

$$\widehat{\sigma}_{n,k}^2 = k^{-1} \sum_{i=1}^{k} (Y_i - \widehat{\mu}_{1n,k})^2 + (n-k)^{-1} \sum_{i=k+1}^{n} (Y_i - \widehat{\mu}_{2n,k})^2.$$

Student's test statistic for the comparison of the means μ_1 and μ_2 is

$$S_{n,k} = \frac{\widehat{\mu}_{1n,k} - \widehat{\mu}_{2n,k}}{\widehat{\sigma}_{n,k}(k^{-1} + (n-k)^{-1})^{\frac{1}{2}}},$$

the asymptotic distribution of the variables $S_{n,k}$ is normal, the discrete process $(S_{n,i})_{i=1,\dots,n}$ is Gaussian centered but their covariances are not standard. A test for the detection of a change between 1 and $n-1$ is performed with the statistic

$$S_n = \max_{1<k\le n-1} S_{n,k}.$$

The asymptotic of sequential tests have been widely studied, in particular by Jackson and Bradley (1961) for χ^2 tests.

With a random sampling time T, the observations Y_i, at times T_i, belong to consecutively sampled sub-populations with the probabilities λ_τ and $1 - \lambda_\tau$, when $T \le \tau$ and $T > \tau$ respectively. The threshold τ of the sampling times corresponds to a change at k/n for a sampling index k. The sampling times are not necessarily uniform or regular as supposed in most linear models with change-points. The distribution function of Y is then

$$\lambda_\tau F(\cdot - \mu_1) + (1 - \lambda_\tau) F(\cdot - \mu_2).$$

A change-point model is therefore related to a mixture model but they differ from a statistical point of view. Whereas the estimator of a mixture probability λ_τ in $]0,1[$ has a convergence rate $n^{\frac{1}{2}}$, the estimator of the threshold parameter τ converges with rate n. For the tests of a single mean, the logarithm of the likelihood ratio has the property

$$E_{\mu_1} \log \frac{f(Y-\mu_1)}{f(Y-\mu_2)} > 0,$$
$$E_{\mu_2} \log \frac{f(Y-\mu_1)}{f(Y-\mu_2)} < 0$$

and it provides an optimal test statistic. With a Gaussian density f having the same variance before and after the change-point, the log-likelihood ratio test statistic for a n-sample is

$$T_n = \frac{\widehat{\mu}_{1n} - \widehat{\mu}_{2n}}{\widehat{\sigma}_n} \sum_{i=\widehat{k}_n}^{n} \Big(Y_i - \frac{\widehat{\mu}_{1n} + \widehat{\mu}_{2n}}{2}\Big),$$

where $\widehat{\mu}_{1n}$ and $\widehat{\mu}_{2n}$ are the empirical estimators of the parameters of the Gaussian distributions and k is the maximum likelihood estimator of the change-point. Let $\mu_0 = \frac{1}{2}(\mu_1 + \mu_2)$ and let $\mu_1 = \mu_0 + \mu$ and $\mu_2 = \mu_0 - \mu$ where μ is their difference $\frac{1}{2}(\mu_1 - \mu_2)$, the test statistic is proportional to $T_n = \widehat{\sigma}_n^{-1} \widehat{\mu}_n \sum_{i=\widehat{k}_n}^{n} (Y_i - \widehat{\mu}_{0n})$. The statistic is negative if μ is different from zero and it tends to zero if μ is zero. The sum

$$S_n = \widehat{\sigma}_n^{-2} \sum_{i=\widehat{k}_n}^{n} \Big(Y_i - \frac{\widehat{\mu}_{1n} + \widehat{\mu}_{2n}}{2}\Big),$$

is positive if μ is negative and negative if μ is positive so a test which rejects the hypothesis of a model with a constant mean is defined by the probability of error under the hypothesis

$$P_0(|S_n| > c_\alpha) = \alpha$$

where the threshold c_α determined by the asymptotic distribution of $|S_n|$ under the hypothesis. The hypothesis is therefore rejected if $|S_n| > c_\alpha$. A sequential test of the hypothesis of a constant mean may be performed without the estimation of the change-point k using the process $S_n(k) = \widehat{\sigma}_n^{-2} \sum_{i=k}^{n} \{Y_i - \frac{1}{2}(\widehat{\mu}_{1n} + \widehat{\mu}_{2n})\}$ until the stopping point

$$t_\alpha = \min\{k : |S_n(k)| > c_\alpha\}.$$

Under the hypothesis, $S_n(k)$ converges weakly to a normal variable, for every k and as n tends to infinity, the constant c_α is therefore the $(1-\alpha)$th quantile of the normal distribution, however the estimator $\widehat{k}_n$ tends to infinity.

In a model with a change-point in the mean of a variable Y at an unknown threshold of the variable and such that $Y - EY$ has a distribution function F, the distribution of Y is characterized by

$$EY = \mu_1 P(Y \leq \gamma) + \mu_2 P(Y > \gamma)$$
$$P(Y \leq y) = F(y - \mu_1)P(Y \leq \gamma) + F(y - \mu_2)P(Y > \gamma).$$

It is extended to a change of distribution

$$F(y) = P(Y \leq y) = p_\gamma F_1(y) + (1 - p_\gamma)F_2(y)$$

with $F_1(y) = P(Y \leq y \mid Y \leq \gamma)$ and $F_2(y) = P(Y \leq y \mid Y > \gamma)$. The estimators of the partial means determined by a change at a threshold γ are

$$\widehat{\mu}_{1n,\gamma} = n^{-1} \sum_{i=1}^{n} Y_i 1_{\{Y_i \leq \gamma\}},$$
$$\widehat{\mu}_{2n,\gamma} = n^{-1} \sum_{i=1}^{n} Y_i 1_{\{Y_i > \gamma\}}$$

they are independent and have the variances $\sigma^2_{1n,\gamma} = n^{-1} Var(Y 1_{\{Y \leq \gamma\}})$ and respectively $\sigma^2_{2n,\gamma} = n^{-1} Var(Y 1_{\{Y > \gamma\}})$. The variance of Y assuming a change of mean as Y reaches γ is estimated by

$$\widehat{\sigma}^2_{n,\gamma} = n^{-1} \sum_{i=1}^{n} \{(Y_i 1_{\{Y_i \leq \gamma\}} - \widehat{\mu}_{1n,\gamma})^2 + (Y_i 1_{\{Y_i > \gamma\}} - \widehat{\mu}_{2n,\gamma})^2\}.$$

A Student's test statistic for the comparison of the means μ_1 and μ_2 is

$$S_{n,k} = \frac{n^{\frac{1}{2}}(\widehat{\mu}_{1n,\gamma} - \widehat{\mu}_{2n,\gamma})}{\widehat{\sigma}_{n,\gamma}}$$

then a test for the detection of a change of the mean between $Y_{n:1}$ and $Y_{n:n}$ is performed with the statistic

$$S_n = \sup_{Y_{n:1} < \gamma < Y_{n:n}} S_{n,\gamma}.$$

Under the hypothesis of no change, the process $S_{n,\gamma}$ converges to a centered Gaussian process with variance 1 and the asymptotic distribution of the maximum Student's statistics is the supremum of this process. Other estimators have been considered (Chernoff and Zacks 1964, Hinkley 1970).

The asymptotic distribution of the maximum likelihood estimators for a parametric density with change-points have been studied by Ibragimov and Has'minskii (1981), they established that the convergence rate of the estimator of the change-point is n^{-1}, the estimators of the other parameters of the density converge with the rate. The log-likelihood ratio test statistic for a n-sample $n^{-\frac{1}{2}}$, like the estimators of twice continuously differentiable densities.

The log-likelihood ratio test for a change of mean in a Gaussian distribution is

$$S_n = \widehat{\sigma}_n^{-1} \sum_{k=1}^{n} \Big(Y_k - \frac{\widehat{\mu}_{1n} + \widehat{\mu}_{2n}}{2} \Big) 1_{\{Y_k > \widehat{\gamma}_n\}},$$

with the empirical estimators of the parameters of the Gaussian distributions and the maximum likelihood estimator of γ. A test is defined as in the first model. A sequential test of the hypothesis of a constant mean may be performed without the estimation of the change-point γ using the process $|S_n(t)|$ until the stopping point

$$t_\alpha = \inf\{s : |S_n(s)| > c_\alpha\},$$

or a sequence of statistics $\{|S_n(t_i)|\}_{i=1,\dots,T}$, at a sequence of points, until the stopping point $t_\alpha = \min\{i : |S_n(t_i)| > c_\alpha\}$ and t_α is also an estimator of the change-point parameter γ. Under the hypothesis, $S_n(s)$ converges weakly to a normal variable, for every s and as n tends to infinity, the constant c_α may be chosen as the $(1-\alpha)$th quantile of the normal distribution though the estimator $\widehat{\gamma}_n$ tends to infinity.

If the distribution function of a real random sample $(Y_k)_{k=1,\dots,n}$ is unknown, a nonparametric test of homogeneity, with a single distribution function F_0, against an alternative of a change of distribution function at an unknown sampling index $k_0 = [nt_0]$, is performed with the two-samples test Kolmogorov–Smirnov statistic which compares the estimators $\widehat{F}_{1k}(y) = k^{-1} \sum_{i=1}^{k} 1_{\{Y_i \le y\}}$ of the F_{1k} from the sub-sample $Y_1, \dots, Y_k$ and $\widehat{F}_{2,k,n}(y) = (n-k)^{-1} \sum_{i=k+1}^{n} 1_{\{Y_i \le y\}}$ of $F_{2k,n}$ from the sub-sample $Y_{k+1}, \dots, Y_n$

$$T_n = \min_{1 \le k \le n} \sup_y \Big(\frac{k(n-k)}{n} \Big)^{\frac{1}{2}} |\widehat{F}_{1k}(y) - \widehat{F}_{2,k,n}(y)|.$$

Under the hypothesis of homogeneity, if n and k tend to infinity, the Kolmogorov–Smirnov statistic

$$\sup_y \Big(\frac{k(n-k)}{n} \Big)^{\frac{1}{2}} |\widehat{F}_{1k}(y) - \widehat{F}_{2,k,n}(y)|$$

converges weakly to the supremum of the transformed Brownian motion $T_0 = \sup_t |W \circ F_0|$ and it is free of the ratio $n^{-1}k$. This convergence implies that the statistic T_n converges weakly to the same limit, it diverges under a fixed alternative.

The unknown index $k_0 = [n\tau_0]$, $0 < \tau_0 < 1$, where a change of distribution function occurs is estimated empirically by

$$\widehat{k}_n = \arg\min\Big\{1 < k < n : \sup_y \Big(\frac{k(n-k)}{n}\Big)^{\frac{1}{2}} |\widehat{F}_n(y) - \widehat{F}_{k,n}(y)| > c_\alpha\Big\}$$

where c_α is the $1-\alpha$ quantile of the distribution of $\sup_{0 \le x \le 1} |W(x)|$.

1.2 Change-points in regression models

A model with change-point in a regression model of a real variable Y on a vector X of explanatory variables at an unknown sample index k is defined for an n-sample under a probability $P_{\eta,t}$ as

$$E_{\eta,t}(Y_i \mid X_i) = r_1(X_i, \eta_1),\ i = 1, \dots, k,$$
$$E_{\eta,t}Y_i \mid X_i) = r_2(X_i, \eta_2),\ i = k+1, \dots, n,$$

where t is the integer part of $n^{-1}k$. The distribution function of (X, Y) is determined under $P_{\eta,t}$ by the marginal distribution function F_X of X and by the conditional distribution function F of $Y - E_{\eta,t}(Y \mid X)$ given X

$$P_{\eta,t}(Y_i \le y \mid X_i) = F(y - r_1(X_i, \eta_1)),\ i = 1, \dots, k,$$
$$P_{\eta,t}(Y_i \le y) = F(y - r_2(X_i, \eta_2)),\ i = k+1, \dots, n,$$

then

$$P_{\eta,t}(Y \le y \mid X) = \lambda_t \int F(y - r_1(x, \eta_1))\, dF_X(x) + (1 - \lambda_t) \int F(y - r_2(x, \eta_2))\, dF_X(x).$$

The parameters are estimated by minimization of the quadratic error of the regression or by maximum likelihood.

The regression model of (X, Y) with change-point at a threshold γ of the variable X, occurring with a probability

$$\Pr(X \le \gamma) = \lambda_\gamma,$$

is defined by unknown functions

$$r_{1,\gamma}(X) = E(Y|X, 1_{\{X \le \gamma\}}),$$
$$r_{2,\gamma}(X) = E(Y|X, 1_{\{X > \gamma\}}),$$

where the parameter γ is unknown. A nonparametric change-point for a family of regression functions with derivatives of order p means a modification of the order of differentiability or a break-point for some derivative of $E(Y|X)$ when $r_1(\gamma_0) = r_2(\gamma_0)$ at the change-point γ_0, or simply a discontinuity between the functions r_1 and r_2 at γ_0.

The variable Y has then a mixture distribution function F_γ with the probabilities λ_γ and $1 - \lambda_\gamma$

$$F_\gamma(y) = \lambda_\gamma \Pr(Y \le y|X \le \gamma) + (1 - \lambda_\gamma) \Pr(Y \le y|X > \gamma).$$

When the regression model is only conditional on X, the conditional distribution function of Y is always discontinuous with respect to X

$$F_{Y|X}(y \mid x) = 1_{\{x \le \gamma\}} \Pr(Y \le y|X = x) + 1_{\{X > \gamma\}} \Pr(Y \le y|X = x),$$

and the conditional expectation of Y is

$$E(Y|X) = 1_{\{X \le \gamma\}} r_{1,\gamma}(X) + 1_{\{X > \gamma\}} r_{2,\gamma}(X).$$

The earliest results on the estimation in regression models have been published by Quant (1958, 1960), and Hinkley (1970). In linear regression, the parameters are estimated independently from the sub-samples defined by the change-points which are estimated by minimization of the empirical variance with unknown change-points. In parametric regression models, estimating the parameters by the minimization of the estimated error provides estimators with the smallest residual variance. Maximum likelihood estimators are defined when the density is known.

The same methods apply to series of dependent data in autoregressive models. They are Markov chains and their ergodicity ensures the convergence of the empirical estimators in each phase of the models and of statistical tests of homogeneity.

1.3 Models for processes

The hazard function λ of a variable X with density function f and survival function $\bar{F} = 1 - F$ is defined as

$$\lambda(t) = \frac{f(t)}{\bar{F}(t)},$$

it extends the notion of intensity for a point process where the duration between the jumps of the process has an exponential distribution function $F(x) = 1 - \exp\{-\int_0^x \lambda(t)\,dt\}$. The detection of changes in a hazard function or a Poisson process, with a piece-wise constant intensity, has been studied by Matthews, Farewell and Pyke (1985), Loader (1991, 1992) among others. They generalize to point processes with dependent random times under convergence properties and to conditional hazard functions depending on explanatory variables or processes. In proportional hazard regression model (Cox, 1969), the conditional hazard function has the form

$$\lambda(t,z) = \lambda_0(t)\exp\{r_\theta(Z)\}$$

with an unknown baseline hazard function λ_0 and a linear regression function. Changes occur at discontinuities of the functions λ_0 or r_θ.

For a point process with independent increments, the likelihood of a sample of the observations is expressed using the intensity function λ so the estimation of parametric functions λ and likelihood ratio tests of homogeneity may be considered.

Nelson's nonparametric estimator of the cumulative intensity function $\Lambda(x) = \int_0^x \lambda(t)\,dt$ does not require the knowledge of a parametric function λ, estimating the variance of the estimators provides a mean squares criterion for the detection of changes. Maximum likelihood methods are developed with a kernel estimator of the intensity function λ.

Let α and β be functions on a metric space $(\mathbb{X}, \|\cdot\|)$, and let B be the standard Brownian motion. A diffusion process is defined by a stochastic differential equation

$$dX_t = \alpha(X_t)dt + \beta(X_t)dB_t,$$

for t in a time interval $[0,T]$. By a discretization on n sub-intervals of length Δ_n which tends to zero as n tends to infinity, the equation has the approximation

$$Y_i = X_{t_{i+1}} - X_{t_i} = \Delta_n\alpha(X_{t_i}) + \beta(X_{t_i})\Delta B_{t_i}$$

where $\varepsilon_i = \Delta B_{t_i} = B_{t_{i+1}} - B_{t_i}$ is a Gaussian variable with mean zero and variance Δ_n conditionally on the σ-algebra $\mathcal{F}_{t_i}$ generated by the sample-paths of X up to t_i, then

$$E\{\alpha(X_{t_i})\varepsilon_i\} = 0, \quad Var(Y_i|X_{t_i}) = \beta^2(X_{t_i})\Delta_n.$$

Kernel estimators of nonparametric functions α and β^2 of $C_2(\mathbb{X})$ were defined and studied by Pons (2008) and the estimators where extended to the diffusion equation without discretization and to a model where the drift function α has jumps, the estimator is continuous except at the change-point. The detection of jumps in the drift function of a diffusion is performed with this estimator.

1.4 Empirical processes on small intervals

On a probability space $(\Omega, \mathcal{A}, P)$, we consider a sequence $(Y_k)_{i=k,\dots,n}$ of independent and identically distributed random variables with values in an interval I of $\mathbb{R}$ and a sequence of indicator (I_{nk}), $k = 1, \dots, n$ with length of order n^{-1} in a neighborhood of γ_0 belonging to I. Let f be the density of the variables Y_k, we assume that there exists $m \geq 2$ such that the mth moment $\int_I y^m f(y)\,dy$ is finite, and we consider the random intervals $I_{nk}(u) = 1_{]\gamma_0, \gamma_0+n^{-1}u]}(Y_k)$. For measurable function g, the process

$$S_n(u) = \sum_{k=1}^{n} g(Y_k) I_{nk}(u) \tag{1.1}$$

is defined as a random variable on the space $D(I)$ of the right-continuous processes with left-hand limits on I. For every integer j, its jth moment

$$\mu_{jn}(u) = n \int_{\gamma_0}^{\gamma_0+n^{-1}u} g^j(y)\, f(y)\, dy$$

has the approximation $\mu_j(u) = u\, g^j(\gamma_0) f(\gamma_0) + o(1)$ and it is finite.

According to Prohorov's theorem, the weak convergence of the process S_n is equivalent to its tightness in $D(I)$ and to the weak convergence of its finite dimensional distributions $(S_n(u_1), \dots, S_n(u_j))$, for every integer j and for all $u_1, \dots, u_j$ in I. The convergence of all moments μ_{jn} of the process S_n implies the weak convergence of its finite dimensional distributions. Billingsley (1968) proved that a sufficient condition of tightness for a process X_n of $C[0,1]$ is the existence of two exponents $\alpha > 0$ and $\beta > 1$, and an increasing continuous function F such that

$$E|X_n(t_1) - X_n(t_2)|^\alpha \leq |F(t_1) - F(t_2)|^\beta.$$

Billingsley's criterion (15.21) for the tightness of a process $X_n(t)$ of $D[0,1]$ is a second order moment condition for its variations on the intervals $]t_2, t_1]$ and $]t_3, t_2]$, for all $t_1 > t_2 > t_3$

$$E|X_n(t_1) - X_n(t_2)|\,|X_n(t_2) - X_n(t_3)| \leq c(t_1 - t_3)^2,$$

with a constant c.

The empirical process $\nu_n = n^{-\frac{1}{2}}(\widehat{F}_n - F)$ of a n-sample uniformly distributed on $[0,1]$ has a second order moment

$$E|\nu_n(t_1) - \nu_n(t_2)|^2 = (t_2 - t_1)(1 - t_1 + t_2),$$

for $t_1 < t_2$, and all its moments are $O(|t_2 - t_1|)$, it satisfies the tightness criterion (15.21) and it converges weakly to the Brownian motion W with mean zero and covariance function $EW_sW_t = s \wedge t - st$.

As the intervals $I_{n1} =]\gamma_0 + n^{-1}t_2, \gamma_0 + n^{-1}t_1]$ and $I_{n2} =]\gamma_0 + n^{-1}t_3, \gamma_0 + n^{-1}t_2]$ are disjoint, the expectation of the variations of the process X_n are bounded, for every n, as

$$\begin{aligned} E\{|S_n(t_1) - S_n(t_2)|\,|S_n(t_2) - S_n(t_3)|\} \\ \leq \sum_{k \neq k'} E\{1_{I_{n1}}(Y_k)1_{I_{n2}}(Y_{k'})g(Y_k)g(Y_{k'})\} \\ \leq \{(t_1 - t_3)g(\gamma_0)f(\gamma_0)\}^2 \end{aligned} \tag{1.2}$$

and the process S_n satisfies Billingsley's criterion (15.21), it converge weakly to a process S of $C(I)$. The same method is used for the process defined on $[-A, A]$, $A > 0$ finite, by a measurable and integrable function g, as

$$S_n(u) = \sum_{k=1}^{n} g(Y_k)\{1_{]\gamma_0, \gamma_0 + n^{-1}u]}(Y_k) + 1_{]\gamma_0 + n^{-1}u, \gamma_0]}(Y_k)\}.$$

Its mean and its variance are $O(|u|)$ and they converge to finite limits.

Proposition 1.1. *The process S_n converges weakly in $D([-A, A])$ to an uncentered Gaussian process.*

In models with a change-point at a sampling index k_n of a n-sample, we assume that the integer part t_n of $n^{-1}k_n$ converges to a limit t belonging to $]0, 1[$. The weak convergence of the estimators of the mean of a variable relies on the behaviour of a process of partial sums

$$S_{[nt]} = \sum_{i=1}^{[nt]} \xi_i$$

of independent and identically distributed centered variables ξ_i, with variance σ^2. Donsker's theorem states that the process $X_n(t) = \sigma^{-1}n^{-\frac{1}{2}}S_{[nt]}$ converges weakly to the Brownian motion W in the space $C[0, 1]$. The empirical process $\nu_{[nt]} = n^{\frac{1}{2}}(\widehat{F}_{[nt]} - F)$ of a variable Y with distribution function F converges weakly in $C[0, 1]$ to the Brownian motion $W \circ F$.

The estimator of the change-point t_0 depends on the variation of the process of partial sums on small intervals $]nt, nt_0]$ and $]nt_0, nt]$, for t in a neighborhood of t_0. Under a probability P_0 such that the variables ξ_i have the distribution function F_1 if $i \leq k_0$ and F_2 if $i > k_0$, the processes

$$S_{1,[nt]}(y) = [nt]^{-1} \sum_{i=1}^{[nt]} \xi_i, \tag{1.3}$$

$$S_{2,n-[nt]}(y) = (n - [nt])^{-1} \sum_{i=[nt]+1}^{n} \xi_i$$

have means asymptotically equivalent, as n tends to infinity, to

$$\mu_{1,t} = \mu_1 1_{\{t \leq t_0\}} + \left\{ \frac{t_0}{t} \mu_1 + \frac{t - t_0}{t} \mu_2 \right\} 1_{\{t > t_0\}},$$

$$\mu_{2,t} = \mu_2 1_{\{t > t_0\}} + \left\{ \frac{1 - t_0}{1 - t} \mu_2 + \frac{t_0 - t}{1 - t} \mu_1 \right\} 1_{\{t \leq t_0\}},$$

where μ_1 is the mean of ξ_i for $i \leq k_0$ and μ_2 is their mean for $i > k_0$. They are biased, with

$$\mu_{1,t} - \mu_1 = \frac{t - t_0}{t} (\mu_2 - \mu_1) 1_{\{t > t_0\}},$$

$$\mu_{2,t} - \mu_2 = \frac{t_0 - t}{1 - t} (\mu_1 - \mu_2) 1_{\{t \leq t_0\}}$$

and their variance is proportional to $|t - t_0|$. Donsker's theorem applies to the empirical processes

$$\nu_{1,[nt]} = [nt]^{\frac{1}{2}} (S_{1,[nt]} - \mu_{1,t}), \tag{1.4}$$

$$\nu_{2,n-[nt]} = (n - [nt])^{\frac{1}{2}} (S_{2,n-[nt]} - \mu_{2,t})$$

of variables ξ_i with distribution functions F_1 or F_2, according to $i \leq [nt_0]$ or $i > [nt_0]$, for t in $]0, 1[$.

Proposition 1.2. *The empirical processes $\nu_{1,[nt]}$ and $\nu_{2,n-[nt]}$ converge weakly in $D[0, 1]$ to the composed Brownian motions*

$$\nu_{1,t} = t^{-\frac{1}{2}} [W \circ F_1 1_{\{t \leq t_0\}} + \{t_0 W \circ F_1 + (t - t_0) W \circ F_2\} 1_{\{t > t_0\}}],$$

$$\nu_{2,t} = (1 - t)^{-\frac{1}{2}} [W \circ F_2 1_{\{t > t_0\}} + \{(1 - t_0) F_2 + (t_0 - t) F_1\} 1_{\{t \leq t_0\}}].$$

The asymptotic behaviour of the estimator $\widehat{t}_n$ of the change-point t_0 is related to the variation of the processes $S_{1,[nt]} - S_{1,[nt_0]}$ and $S_{2,n-[nt]} - S_{2,n-[nt_0]}$, for t in a neighborhood of t_0. Their means are the biases of the processes $S_{1,[nt]}$, and respectively $S_{2,n-[nt]}$, and all their moments are $O[|t - t_0|)$. Let $t_3 < t_2 < t_1$, the variations of the process $S_{1,[nt]}$ on the intervals $]t_{j-1}, t_j]$, for $j = 2, 3$, are

$$S_{1,[nt_j]} - S_{1,[nt_{j-1}]} = \frac{1}{[nt_j]} \sum_{i=[nt_{j-1}]+1}^{[nt_j]} \xi_i - \left(\frac{1}{[nt_{j-1}]} - \frac{1}{[nt_j]} \right) \sum_{i=1}^{[nt_{j-1}]} \xi_i,$$

and $E|S_{1,[nt_1]}(y) - S_{1,[nt_2]}|\,|S_{1,[nt_2]} - S_{1,[nt_3]} = O(|t_1 - t_2|\,|t_2 - t_3|)$ which proves Billingsley's tightness criterion (15.21) for the processes $S_{1,[nt]} - S_{1,[nt_0]}$, and the proof is similar for the process $S_{2,n-[nt]} - S_{2,n-[nt_0]}$. By the same argument, at $t_{n,u} = t_0 + n^{-1}u$, the processes $n(S_{j,n-[nt_{n,u}]} - S_{j,n-[nt_0]})$, $j = 1, 2$, are tight and they converge weakly to uncentered Gaussian processes with finite mean and variance functions.

The tightness criterion (15.21) is still satisfied for the processes $nS_{1,[nt_{n,u}]}$ and $nS_{2,n-[nt_{n,u}]}$, with a sequence $t_{n,u} = t_0 + n^{-1}u$, $|u| \leq A$, for $A > 0$. Their means and their variances converge to finite limits. The same properties are satisfied at $t_{n,u}$ by the processes

$$X_{1,n,t} = \sum_{i=[nt]+1}^{[nt_0]} \xi_i,$$

$$X_{2,n,t} = \sum_{i=[nt_0]+1}^{[nt]} \xi_i,$$

with t_0 and $t_{n,u}$ in $]0, 1[$.

Proposition 1.3. *The processes $nS_{1,[nt_{n,u}]}$ and $nS_{2,n-[nt_{n,u}]}$ converge weakly in $D([-A, A])$ to uncentered Gaussian processes.*

Proposition 1.4. *Let t_0 and $t_{n,u} = t_0 + n^{-1}u$ in $]0, 1[$, with $|u| \leq A$, the processes $X_{1,n,t_{n,u}}$ and $X_{2,n,t}$ converge weakly in $D([-A, A])$ to uncentered Gaussian processes with finite variances.*

1.5 Content of the book

The chapters are organized according to the methods: minimization of the empirical variance of a variable of interest or maximization of the likelihood in parametric models. We study the limits of estimators and tests performed with the same criterion for the hypothesis of models without change. They are powerful in the sense that the test statistics diverge under fixed alternatives and they detect local alternatives converging to the hypothesis. A sequential detection of consecutive changes is performed by the same methods starting from the preceding change or after a delay of repair of the observed system. This procedure applies to all models for the estimation of consecutive change-points.

The first chapters consider samples of independent and identically distributed variables on several phases determined by the change-points and the asymptotic behaviour are related to the sampling sizes in each phase. They concern changes in the mean of a variable, in the parameters of its density and changes in the parameters of a regression function. The following chapters consider counting processes and their intensity, the asymptotic study relies on the behaviour of local martingales as the observation interval increases or an increasing number of independent samples of observations on the same interval. For auto-regressive series, the limits of discrete martingales determine the asymptotic behaviour of the sample-paths of the series and the estimators of their parameters. In models with a change in the mean and in linear regression models, we consider additive changes of the parameters according to the level of variables of the model and chronological changes at a sampling index. The methods extend to changes in the variance and multiplicative changes in proportional hazard models for counting processes, and to nonparametric models.

Chapter 2 presents estimators and tests for changes in the mean of a n-sample at an unknown threshold or at an unknown sampling index. The empirical estimation of the partial means and variance of the variable enables to estimate the change-point by minimization of the empirical variance and a test of homogeneity is performed with the difference of the estimated variance under the models. The convergence rate of the change-point estimators is n^{-1} in models with a change of mean, it is unknown under the hypothesis of homogeneity and depends on the behaviour of the density at infinity. Furthermore the estimators of the mean of the sub-samples after the change have the same convergence rate as the change-point estimators under the hypothesis.

Change-points of the density of a n-sample are estimated in Chapter 3 by maximum likelihood with a known density or with a nonparametric estimator of the density. The change-point estimators and the likelihood ratio tests of homogeneity have a behaviour similar to the estimators of means by maximum likelihood, the test statistics converge in all cases to χ^2 variables under the hypothesis.

In regression models of a variable Y on an explanatory vector of variables X, the parameters are estimated by minimization of the empirical variance of the variable $Y - E(Y|X)$ and by maximization of the likelihood of sample conditionally on the regressor in parametric models. With a n-sample, the regression parameters have convergence rate $n^{-\frac{1}{2}}$ and the estimators of the change-points have the convergence rate n^{-1}. Tests for the detection of change-points rely on the same criteria as the estimation, they are tests for a chronological change at the index n or at the end of the support of regressors. The asymptotic behaviour of the test statistics differ from those of tests of sub-hypotheses in regular parametric models, their asymptotic expansions are similar to those for densities. Models of regressions with change-points according to thresholds of the explanatory variable or the observation time are studied in Chapters 4 for parametric models and 8 for nonparametric models, with samples of independent and identically distributed variables.

Auto-regressive series are observed at a discrete time sequence, their domain of convergence depend on the values of the parameters. They are markovian models and the convergence of their transition probabilities to an invariant measure relies on properties of the mean duration time in recurrence intervals of the series. These properties provide convergence of functionals of the dependent observations of the series. Parametric models of series with change-points are considered in Chapter 7, the convergence rates of the estimators of their parameters depend on the domain of convergence of the series. After a change, the series and their estimators behave as in new series with new initial random value depending on the change-point. Estimators and tests in nonparametric models of series are studied in Chapter 8.

In models for Poisson and counting processes (Chapters 5 and 6), the maximum likelihood estimators of the change-point have the same convergence rates as in parametric models for densities or regressions. According to the model, the likelihood ratio tests or the tests based on differences of the residual variances converge weakly to transformed Gaussian limits. In Chapters 8 and 9, the changes in nonparametric models are located at discontinuities of piece-wise C^2 densities, regression functions, conditional distribution functions and drift of diffusions. Using d-dimensional kernel estimators which converge with the rate $(nh^d)^{-\frac{1}{2}}$, the convergence rate of the change-points is $(nh^d)^{-1}$. We determine the asymptotic behaviour of the estimators and of tests statistics for the hypothesis of continuous functions.

Chapter 2

Change-points in the mean

Abstract. This chapter studies models with change-point in the mean of a real variable at an unknown threshold or at an unknown sampling index. We first consider the empirical estimators and the least squares estimator of the change-point then the maximum likelihood with a known or unknown density, their convergence rates are established. Tests for the hypothesis of a constant mean are defined by the same methods. The weak convergence of the estimators and the test statistics under the hypothesis and alternatives are proved.

2.1 Empirical estimators

On a probability space $(\Omega, \mathcal{A}, P_0)$, let Y be a real variable with a continuous distribution function F_0 on $\mathbb{R}$ such that $E_0(Y^2)$ is finite and let $Y_1, \ldots, Y_n$ be a sample of Y under P_0. In a model with a change-point in the mean of Y at an unknown value γ_0, we denote

$$E_0 Y = \begin{cases} \mu_{01} & \text{if } Y \leq \gamma_0, \\ \mu_{02} & \text{if } Y > \gamma_0. \end{cases} \tag{2.1}$$

Let F_{01} be the distribution function of Y up to γ_0 and let F_{02} be its distribution function on the set after γ_0, under P_0.

Considering γ as a real parameter, the distribution function of Y under $P_\gamma(y) = P_\gamma(Y \leq y)$ is a distribution mixture

$$F_\gamma(y) = p_\gamma P_\gamma(Y \leq y \mid Y \leq \gamma) 1_{\{y \leq \gamma\}} + (1 - p_\gamma) P_\gamma(Y \leq y \mid Y > \gamma) 1_{\{y > \gamma\}}$$

with the probability $p_\gamma = F_\gamma(\gamma)$. It is denoted

$$F_\gamma(y) = F_{1\gamma}(y) 1_{\{y \leq \gamma\}} + F_{2\gamma}(y) 1_{\{y > \gamma\}}$$

and Y has the partial means under P_γ

$$\mu_{1\gamma} = E_\gamma(Y1_{\{Y\leq\gamma\}}),$$
$$\mu_{2\gamma} = E_\gamma(Y1_{\{Y>\gamma\}}),$$

they are denoted μ_{01}, μ_{02} and p_0 at γ_0, under P_0. The distribution function F_γ splits according to the change of conditional mean as

$$F_\gamma(y) = F(y-\mu_{1\gamma})1_{\{y\leq\gamma\}} + F(y-\mu_{2\gamma})1_{\{y>\gamma\}}$$

where F is the distribution function of $Y - E_\gamma Y$ and, under P_0, the sub-distributions functions F_{01} and F_{02} are

$$F_{01}(y) = F(y-\mu_{01})1_{\{y\leq\gamma_0\}}, \quad F_{02}(y) = F(y-\mu_{02})1_{\{y>\gamma_0\}}.$$

We assume that the distribution function F has a continuous density f so that the change of mean is the unique singularity of the model.

The mean of the variable Y under a probability P_γ is a mixture of the partial means

$$\mu_\gamma = \mu_{1\gamma} + \mu_{2\gamma}.$$

Let $(Y_1, \ldots, Y_n)$ be a sample of the variable Y under P_0 and let

$$\delta_{k\gamma} = 1_{\{Y_k>\gamma\}},$$

the empirical ratios of the sub-samples determined by a change at γ, with partial means $\mu_{1\gamma}$ and $\mu_{2\gamma}$, are

$$\widehat{p}_{n\gamma} = n^{-1}\sum_{k=1}^{n}(1-\delta_{k\gamma})$$

and, respectively $1-\widehat{p}_{n\gamma}$, the estimator $\widehat{p}_{n\gamma}$ converges a.s. under P_γ to p_γ, as n tends to infinity. Under P_0, $\widehat{p}_{n\gamma}$ converges a.s. to

$$\begin{aligned} E_0\widehat{p}_{n\gamma} &= P_0(Y\leq\gamma) \\ &= P_0(Y\leq\gamma\wedge\gamma_0) + 1_{\{\gamma_0\leq\gamma\}}P_0(\gamma_0<Y\leq\gamma) \\ &= F_{01}(\gamma\wedge\gamma_0) + 1_{\{\gamma_0\leq\gamma\}}\{F_{02}(\gamma) - F_{02}(\gamma_0)\}. \end{aligned}$$

The probability $P_0(Y>\gamma)$ is expressed in the same way according to γ_0

$$\begin{aligned} P_0(Y>\gamma) &= P_0(Y>\gamma\vee\gamma_0) + 1_{\{\gamma\leq\gamma_0\}}P_0(\gamma<Y\leq\gamma_0) \\ &= F_{02}(\gamma\vee\gamma_0) + 1_{\{\gamma\leq\gamma_0\}}\{F_{01}(\gamma_0) - F_{01}(\gamma)\}. \end{aligned}$$

The partial means $\mu_{1\gamma}$ and $\mu_{2\gamma}$ are estimated from the sub-samples determined by γ as

$$\widehat{\mu}_{1n,\gamma} = n^{-1} \sum_{k=1}^{n} Y_k(1 - \delta_{k,\gamma}),$$
$$\widehat{\mu}_{2n,\gamma} = n^{-1} \sum_{k=1}^{n} Y_k \delta_{k,\gamma},$$

they are independent and they converge a.s. under P_0 to $\mu_{01,\gamma}$ and, respectively $\mu_{02,\gamma}$

$$\mu_{01,\gamma} = \mu_{1,\gamma\wedge\gamma_0} + 1_{\{\gamma>\gamma_0\}}\{\mu_{01} - \mu_{1,\gamma}\},$$
$$\mu_{02,\gamma} = \mu_{2,\gamma\vee\gamma_0} + 1_{\{\gamma\le\gamma_0\}}\{\mu_{02} - \mu_{2,\gamma}\}.$$

The variable $n^{\frac{1}{2}}(\widehat{\mu}_{in,\gamma} - \mu_{i0,\gamma})$ converges weakly to a centered Gaussian variable with finite variance, for $i = 1, 2$. For every γ, the sum and the difference of the estimators of the partial means are

$$\widehat{\mu}_{1n,\gamma} + \widehat{\mu}_{2n,\gamma} = \bar{Y}_n,$$
$$\widehat{\mu}_{1n,\gamma} - \widehat{\mu}_{2n,\gamma} = \bar{Y}_n - 2\widehat{\mu}_{2n,\gamma},$$

where $\bar{Y}_n$ is the empirical mean of the sample. Let $\widehat{b}_{in,\gamma} = \widehat{\mu}_{in,\gamma} - \widehat{\mu}_{in,\gamma_0}$, for $i = 1, 2$, we have

$$\widehat{b}_{1n,\gamma} = n^{-1} \sum_{k=1}^{n} Y_k(\delta_{k,\gamma_0} - \delta_{k,\gamma}) = \int_{\gamma_0}^{\gamma} y\, d\widehat{F}_{Y,n}(y)$$
$$\widehat{b}_{2n,\gamma} = -\widehat{b}_{1n,\gamma},$$

they converge a.s. under P_0 to non-zero biases $b_{i,\gamma}$, $i = 1, 2$, where

$$b_{1,\gamma} = 1_{\{\gamma<\gamma_0\}} \int_{\gamma}^{\gamma_0} y\, dF_{01}(y) + 1_{\{\gamma_0<\gamma\}} \int_{\gamma_0}^{\gamma} y\, dF_{02}(y).$$

The limits of the variances of $\widehat{\mu}_{1n,\gamma}$ and $\widehat{\mu}_{2n,\gamma}$ under P_0 are

$$v_{01,\gamma} = n^{-1}\{E_0(Y^2 1_{\{Y\le\gamma\}}) - \mu_{01,\gamma}^2\},$$
$$v_{02,\gamma} = n^{-1}\{E_0(Y^2 1_{\{Y>\gamma\}}) - \mu_{02,\gamma}^2\}.$$

On a bounded interval Γ that contains γ_0 and where the density of Y is C^1, the bias and variance functions satisfy a Lipschitz inequality, there exist constants c_1 and c_2 such that for every γ in Γ

$$|b_{1,\gamma} - b_{1,\gamma_0}| \le c_1|\gamma - \gamma_0|,$$
$$|v_{i0,\gamma} - v_{i0,\gamma_0}| \le c_2|\gamma - \gamma_0|.$$

Assuming that a change occurs at an arbitrary γ, the variance of the sample under P_γ is a function of $(\mu_{1,\gamma}, \mu_{2,\gamma}, \gamma)$ and we denote

$$\begin{aligned}\sigma^2_{n,\gamma} &= n^{-1}\sum_{k=1}^{n}\{(Y_k 1_{\{Y_k\leq\gamma\}} - \mu_{1\gamma})^2 + (Y_k 1_{\{Y_k>\gamma\}} - \mu_{2\gamma})^2\}\\ &= n^{-1}\sum_{k=1}^{n}(Y_k^2 1_{\{Y_k\leq\gamma\}} + Y_k^2 1_{\{Y_k>\gamma\}}) - \mu^2_{1\gamma} - \mu^2_{2\gamma},\end{aligned}$$

it has the mean $\sigma^2_\gamma = \sigma^2_{1\gamma} + \sigma^2_{2\gamma}$ where $\sigma^2_{1\gamma} = E_\gamma(Y^2 1_{\{Y\leq\gamma\}}) - \mu^2_{1\gamma}$ and $\sigma^2_{2\gamma} = E_\gamma(Y^2 1_{\{Y>\gamma\}}) - \mu^2_{2\gamma}$. Under P_0, we denote

$$\sigma^2_{n,\mu_{0\gamma},\gamma} = n^{-1}\sum_{k=1}^{n}(Y_k^2 1_{\{Y_k\leq\gamma\}} + Y_k^2 1_{\{Y_k>\gamma\}}) - \mu^2_{01,\gamma} - \mu^2_{02,\gamma},$$

its expectation is $\sigma^2_{0\gamma} = \sigma^2_{01\gamma} + \sigma^2_{02\gamma}$ where

$$\sigma^2_{01\gamma} = E_0(Y^2 1_{\{Y\leq\gamma\}}) - \mu^2_{01,\gamma}, \quad \sigma^2_{02\gamma} = E_0(Y^2 1_{\{Y>\gamma\}}) - \mu^2_{02,\gamma}.$$

Under P_0, $\sigma^2_{n,\gamma}$ converges a.s. to the sum σ^2_γ of the partial variances

$$\begin{aligned}\sigma^2_{1\gamma} &= \int 1_{\{y\leq\gamma\wedge\gamma_0\}} y^2\, dF_{01}(y) + 1_{\{\gamma>\gamma_0\}}\int_{\gamma_0}^{\gamma} y^2\, dF_{02}(y) - \mu^2_{1\gamma}\\ \sigma^2_{2\gamma} &= \int 1_{\{y>\gamma\vee\gamma_0\}} y^2\, dF_{02}(y) + 1_{\{\gamma\leq\gamma_0\}}\int_{\gamma}^{\gamma_0} y^2\, dF_{01}(y) - \mu^2_{2\gamma},\end{aligned}$$

and σ^2_γ is minimum at γ_0 where its value is σ^2_0. For an arbitrary γ belonging to the open interval determined by the first and the last order statistics of the sample $]Y_{n:1}, Y_{n:n}[$, the empirical variance function $\sigma^2_{n,\mu_\gamma,\gamma}$ is minimum with respect to $\mu_\gamma = (\mu_{1\gamma}, \mu_{2\gamma})$, at $\widehat{\mu}_{n,\gamma}$ which is the minimum variance estimator of μ_γ. Let

$$\widehat{\sigma}^2_{n,\gamma} = \sigma^2_{n,\widehat{\mu}_{n,\gamma},\gamma} = n^{-1}\sum_{k=1}^{n}\Big\{(Y_k 1_{\{Y_k\leq\gamma\}} - \widehat{\mu}_{1n,\gamma})^2 + (Y_k 1_{\{Y_k>\gamma\}} - \widehat{\mu}_{2n,\gamma})^2\Big\}.$$

Under P_0, the empirical variance $\widehat{\sigma}^2_{n,\gamma}$ converges to the variance $\sigma^2_{0\gamma}$ which has a unique minimum at γ_0. Then γ_0 is estimated by the first value that minimizes the empirical variance

$$\widehat{\gamma}_n = \arg\inf_{Y_{n:1}<\gamma<Y_{n:n}} \widehat{\sigma}^2_{n,\gamma}.$$

The conditional means and the variance are estimated as $\widehat{\mu}_{in} = \widehat{\mu}_{in,\widehat{\gamma}_n}$, for $i = 1, 2$, and $\widehat{\sigma}^2_n = \widehat{\sigma}^2_{n,\widehat{\gamma}_n}$.

For every γ, the process $\widehat{\sigma}^2_{n,\gamma}$ converges a.s. under P_0 to the expectation $\sigma^2_{\mu_{0\gamma},\gamma}$ of $\sigma^2_{n,\mu_{0\gamma},\gamma}$. The process

$$X_n(\gamma) = \widehat{\sigma}^2_{n,\gamma} - \widehat{\sigma}^2_{n,\gamma_0} \tag{2.2}$$

is positive and it converges a.s. uniformly under P_0 to the function

$$X(\gamma) = \sigma^2_{\mu_{0\gamma},\gamma} - \sigma^2_0, \tag{2.3}$$

the function $X(\gamma)$ is positive and minimum at γ_0 where it is zero.

Theorem 2.1. *If Y has a density in an open interval J_0 such that γ_0 and $\widehat{\gamma}_n$ belongs to J_0, then $\widehat{\gamma}_n$ converges a.s. under P_0 to γ_0 as n tends to infinity.*

Proof. By the uniform convergence of X_n to X and the properties of X and X_n, we have

$$0 \le X_n(\gamma_0) \le \sup_\Gamma |X_n(\gamma) - X(\gamma)| + X(\gamma_0)$$

moreover X is minimum at γ_0 with $X(\gamma_0) = 0$. If $\widehat{\gamma}_n$ did not converge a.s. to γ_0, we could extract a sub-sequence converging to a value γ distinct from γ_0 and such that $X(\gamma) = 0$ but this is impossible for X having a unique minimum at γ_0. □

The process $X_n(\gamma)$ is expressed with the empirical means as

$$\begin{aligned} X_n(\gamma) &= \widehat{\mu}^2_{1n,\gamma_0} + \widehat{\mu}^2_{2n,\gamma_0} - \widehat{\mu}^2_{1n,\gamma} - \widehat{\mu}^2_{2n,\gamma} \\ &= (\widehat{\mu}_{1n,\gamma_0} - \widehat{\mu}_{1n,\gamma})(\widehat{\mu}_{1n,\gamma_0} + \widehat{\mu}_{1n,\gamma}) \\ &\quad + (\widehat{\mu}_{2n,\gamma_0} - \widehat{\mu}_{2n,\gamma})(\widehat{\mu}_{2n,\gamma_0} + \widehat{\mu}_{2n,\gamma}). \end{aligned} \tag{2.4}$$

Proposition 2.1. *For $\varepsilon > 0$ sufficiently small, there exists a constant κ_0 such that*

$$\inf_{|\gamma-\gamma_0|\le\varepsilon} X(\gamma) \ge \kappa_0|\gamma - \gamma_0|. \tag{2.5}$$

Proof. The function $X(\gamma)$ depends on integrals on the interval $]\gamma, \gamma_0]$ or $]\gamma_0, \gamma]$ according to the sign of $\gamma_0 - \gamma$ and on integrals on the intervals defined by γ_0

$$\begin{aligned} X(\gamma) &= \mu^2_{01,\gamma_0} + \mu^2_{02,\gamma_0} - \mu^2_{01,\gamma} - \mu^2_{02,\gamma} \\ &= (\mu_{01,\gamma_0} - \mu_{01,\gamma})(2\mu_{01,\gamma_0} + \mu_{01,\gamma} - \mu_{01,\gamma_0}) \\ &\quad + (\mu_{02,\gamma_0} - \mu_{02,\gamma})(2\mu_{02,\gamma_0} + \mu_{02,\gamma} - \mu_{02,\gamma_0}). \end{aligned} \tag{2.6}$$

As $|\gamma - \gamma_0|$ tends to zero, the differences $\mu_{0j,\gamma_0} - \mu_{0j,\gamma}$ are $O(|\gamma - \gamma_0|)$, for $j = 1, 2$. At γ_0, μ_{01,γ_0} is an integral with respect to F_{01} up to γ_0 and μ_{02,γ_0} is an integral with respect to F_{02} on the interval starting at γ_0. □

The functions p_γ and $\mu_{1,\gamma}$, and respectively $\mu_{2,\gamma}$, have right, and respectively left, derivatives. The function X has then left and right derivatives with respect to γ.

In the space D of right-continuous functions with left-hand limits, we consider the centered process

$$W_n(\gamma) = n^{\frac{1}{2}}\{X_n(\gamma) - X(\gamma)\},$$

it has a finite variance v_γ if $E_0(Y^4)$ is finite.

Proposition 2.2. *For Y such that $E_0(Y^4)$ is finite and for every $\varepsilon > 0$, there exists a constant κ_1 such that for n sufficiently large*

$$E_0 \sup_{|\gamma-\gamma_0|\leq\varepsilon} |W_n(\gamma)| \leq \kappa_1 \varepsilon^{\frac{1}{2}}. \tag{2.7}$$

Proof. Like the function X, the process X_n is expressed as a function of integrals with respect to the empirical distribution of the sample $\widehat{F}_{n,Y}$ on intervals $]\gamma, \gamma_0]$ or $]\gamma_0, \gamma]$ according to the sign of $\gamma_0 - \gamma$ and on the intervals defined by γ_0. The process W_n is defined in the same way by integrals with respect to the empirical processes $\nu_{jn} = n^{\frac{1}{2}}(\widehat{F}_{jn,Y} - F_{0j})$, $j = 1, 2$

$$\begin{aligned}
W_n(\gamma) &= (\widehat{\mu}_{1n,\gamma_0} + \widehat{\mu}_{1n,\gamma}) \int_\gamma^{\gamma_0} y\, d\nu_{1n} + (\widehat{\mu}_{2n,\gamma_0} + \widehat{\mu}_{2n,\gamma}) \int_\gamma^{\gamma_0} y\, d\nu_{2n} \\
&\quad + \int_\gamma^{\gamma_0} y\, dF_{01}\Big\{2\int_{-\infty}^{\gamma_0} y\, d\nu_{1n} + \int_{\gamma_0}^{\gamma} y\, d\nu_{1n}\Big\} \\
&\quad + \int_\gamma^{\gamma_0} y\, dF_{02}\Big\{2\int_{\gamma_0}^{\infty} y\, d\nu_{2n} + \int_{\gamma_0}^{\gamma} y\, d\nu_{2n}\Big\},
\end{aligned}$$

and the variance of $\int_\gamma^{\gamma_0} y\, d\nu_{jn}$ is a $O(|\gamma - \gamma_0|)$.

Let $\varepsilon > 0$ and let n tend to infinity, by monotonicity we have

$$E_0 \sup_{|\gamma-\gamma_0|\leq\varepsilon} \Big|\int_\gamma^{\gamma_0} y\, d\nu_{jn}\Big| \leq \Big\{E_0 \sup_{|\gamma-\gamma_0|\leq\varepsilon} \Big|\int_\gamma^{\gamma_0} y\, d\nu_{jn}\Big|^2\Big\}^{\frac{1}{2}}$$

and it is a $O(\varepsilon^{\frac{1}{2}})$. If $E_0(Y^4)$ is finite, by the same argument for W_n, we have

$$E_0 \sup_{|\gamma-\gamma_0|\leq\varepsilon} |W_n(\gamma)|^2 = O(\varepsilon)$$

and the result follows by the Cauchy–Schwarz inequality. □

Let $\mathcal{U}_n = \{u = n(\gamma - \gamma_0), \gamma \in \mathbb{R}\}$ and for $\varepsilon > 0$, let

$$\mathcal{U}_{n,\varepsilon} = \{u \in \mathcal{U}_n : |u| \leq n\varepsilon\}.$$

For γ in $V_\varepsilon(\gamma_0)$, let $u_{n,\gamma} = n(\gamma - \gamma_0)$.

Theorem 2.2. *If $E_0(Y^4)$ is finite, then*

$$\overline{\lim}_{n,A\to\infty} P_0(n|\widehat{\gamma}_n - \gamma_0| > A) = 0.$$

Proof. Let $\widehat{u}_n = n(\widehat{\gamma}_n - \gamma_0)$, let $\eta > 0$ and let $\varepsilon > 0$ be sufficiently small to ensure the inequalities (2.5) and (2.7). From Theorem 2.1, for n larger than some integer n_0, $P_0\{\widehat{u}_n \in \mathcal{U}_{n,\varepsilon}\} = P_0\{\widehat{\gamma}_n \in V_\varepsilon(\gamma_0)\} > 1 - \eta$ and

$$\begin{aligned} P_0(|\widehat{u}_n| > A) &\leq P_0\Big\{\inf_{u\in\mathcal{U}_{n,\varepsilon}, |u|>A} \widehat{\sigma}^2_{n,\gamma_{n,u}} \leq \widehat{\sigma}^2_{n,\gamma_0}\Big\} + \eta \\ &= P_0\Big\{\inf_{u\in\mathcal{U}_{n,\varepsilon}, |u|>A} X_n(\gamma_{n,u}) \leq 0\Big\} + \eta. \end{aligned}$$

This probability is bounded following the arguments of Theorem 5.1 in Ibragimov and Has'minskii (1981), where $\mathcal{U}_{n,\varepsilon}$ is split into subsets $H_{n,j}$ defined by its intersection with the sets $\{g(j) < |u| \leq g(j+1)\}$, for the integers j such that $g(j) > A$ and $g(j+1) \leq n\varepsilon$, where g is an increasing function g such that $\sum_{g(j)>A} g(j+1)^{\frac{1}{2}} g^{-1}(j)$ tends to zero as A tends to infinity. The inequalities (2.7) and (2.5) and the Bienaymé–Chebyshev inequality imply that as n tends to infinity

$$\begin{aligned} P_0\Big\{\inf_{u\in\mathcal{U}_{n,\varepsilon}, |u|>A} X_n(\gamma_{n,u}) \leq 0\Big\} &\leq \sum_{g(j)>A} P_0\Big\{\sup_{H_{n,j}} |W_n(\gamma_{n,u})| \geq \kappa_0 n^{-\frac{1}{2}} g(j)\Big\} \\ &\leq \kappa_0^{-1}\kappa_1 \sum_{g(j)>A} g^{\frac{1}{2}}(j+1) g^{-1}(j) \end{aligned}$$

and this bound tends to zero as A tends to infinity. □

For $A > 0$, let $\mathcal{U}_n^A = \{u \in \mathcal{U}_n; |u| < A\}$. By Theorem 2.2, the asymptotic distribution of $n(\widehat{\gamma}_n - \gamma_0)$ may be deduced from the asymptotic distribution of the process nX_n in the bounded set $\mathcal{U}_n^A$, as A tends to infinity. The proof of the weak convergence of the process nX_n follows the arguments of Section 1.4 for weighted distribution functions.

Theorem 2.3. *Under the conditions $E_0(Y^4)$ finite and $f(\gamma_0)$ different from zero, the variable $n(\widehat{\gamma}_n - \gamma_0)$ is bounded in probability and it converges weakly to the limit U of the location of the minimum U_A of an uncentered Gaussian process L_X in $\mathcal{U}_n^A$, as n and A tend to infinity.*

Proof. Let $\widehat{\mu}_{1n,0} = \widehat{\mu}_{1n,\gamma_0}$, $\mu_{01,\gamma} = E_0(Y1_{\{Y\leq\gamma\}})$ and $\mu_{01} = \mu_{01,\gamma_0}$. By Theorem 2.2, the differences

$$\widehat{\mu}_{1n,\gamma_{n,u}} - \widehat{\mu}_{1n,0} = n^{-1}\sum_{i=1}^{n} Y_i 1_{\{\gamma_0<Y_i\leq\gamma_{n,u}\}},$$

$$\widehat{\mu}_{2n,\gamma_{n,u}} - \widehat{\mu}_{2n,0} = n^{-1}\sum_{i=1}^{n} Y_i 1_{\{\gamma_{n,u}<Y_i\leq\gamma_0\}},$$

are $O_p(n^{-1})$. The estimator $\widehat{u}_n = n(\widehat{\gamma}_n - \gamma_0)$ of $u = n(\gamma_{n,u} - \gamma_0)$ in $\mathcal{U}_n^A$ minimizes the process $nX_n(\gamma_{n,u})$. For $u > 0$, the expectation of $n(\widehat{\mu}_{1n,\gamma_{n,u}} - \widehat{\mu}_{1n,0})$ is approximated by $u_\gamma\gamma_0 f(\gamma_0)$ and its variance by $\gamma_0^2 u_\gamma f(\gamma_0)\{1 - u_\gamma f(\gamma_0)\}$. The mean of $n(\widehat{\mu}_{2n,\gamma_{n,u}} - \widehat{\mu}_{2n,0})$ is approximated by $-u_\gamma\gamma_0 f(\gamma_0)$ and its variance by

$$\gamma_0^2 u_\gamma f(\gamma_0)\{1 - u_\gamma f(\gamma_0)\},$$

as n tends to infinity. They are integrals with respect to the empirical distribution function $\widehat{F}_{n,Y}$ on the intervals $]\gamma, \gamma_0]$ or $]\gamma_0, \gamma]$ and these integrals satisfy Billingsley's tightness criterion (15.21) on disjoint intervals $]\gamma_0 + n^{-1}t_2, \gamma_0 + n^{-1}t_1]$ and $]\gamma_0 + n^{-1}t_3, \gamma_0 + n^{-1}t_2]$, for all $t_3 < t_2 < t_1$. It follows that the processes $n(\widehat{\mu}_{kn,\gamma_{n,u}} - \widehat{\mu}_{kn,0})$, $k = 1, 2$, converge weakly to uncentered Gaussian processes L_{μ_k}, $k = 1, 2$, in $\mathcal{U}_n^A$. From (2.4), $nX_n(\gamma_{n,u})$ converges weakly in $\mathcal{U}_n^A$ to the uncentered Gaussian process

$$L_X = -(2\mu_{01} + L_{\mu_1})L_{\mu_1} - (2\mu_{02} + L_{\mu_2})L_{\mu_2}.$$

The variable $\widehat{u}_n$ converges weakly in $\mathcal{U}_n^A$ to the location of the minimum U_A of this process. By Theorem 2.2, the limit of $\widehat{u}_n$ is bounded in probability and it converges as A tends to infinity to this limit. □

Theorem 2.4. *Under the conditions $E_0(Y^4)$ finite and $f(\gamma_0)$ different from zero, the variable $n^{\frac{1}{2}}(\widehat{\mu}_n - \mu_0)$ converges weakly to a centered Gaussian variable on $\mathbb{R}^2$ with respective variances $v_{10\gamma_0}$ and $v_{20\gamma_0}$ and with covariance zero, as n tends to infinity.*

Proof. The bias of the estimator satisfies a Lipschitz inequality and the convergence rate of the estimator $\widehat{\gamma}_n$ entails that the limit is centered. It is Gaussian as a normalized sum of variables independent conditionally on $\widehat{\gamma}_n$. The variances converge from their Lipschitz inequality. □

2.2 Test for a change of mean

Consider the hypothesis H_0 of a constant mean μ and the alternative K of a change at an unknown threshold γ_0 belonging to the interval $]Y_{n:1}, Y_{n:n}[$ and μ_{01} different from μ_{02}. Under H_0, $\mu_{01} = \mu_{02}$ and γ_0 is arbitrary or $\mu_{01} = \mu_0$, $\mu_{02} = 0$ and γ_0 is at the end point of the support of the variable Y. A test of H_0 against the alternative is defined by a statistic comparing the process X_n at the estimators of the parameter γ under the hypothesis and the alternative

$$T_n = n\{X_n(\widehat{\gamma}_n) - X_n(\widehat{\gamma}_{0n})\}.$$

Under H_0, the estimator of the mean is $\widehat{\mu}_n = \bar{Y}_n$, all Y_k being smaller than γ_0, the empirical variance $\widehat{\sigma}^2_{n,\gamma_0}$ is the empirical variance of the sample $\widehat{\sigma}^2_n = n^{-1}\sum_{k=1}^n Y_k^2 - \bar{Y}_n^2$ and the test statistic is

$$T_n = \inf_{Y_{n:1}<\gamma\leq Y_{n:n}} n(\widehat{\sigma}^2_{n,\gamma} - \widehat{\sigma}^2_{n,\gamma_0}) = n(\widehat{\sigma}^2_{n,\widehat{\gamma}_n} - \widehat{\sigma}^2_{n,\gamma_0}).$$

The process $X_n(\gamma) = \widehat{\sigma}^2_{n,\gamma} - \widehat{\sigma}^2_{n,\gamma_0}$ is the difference (2.4) and it converges a.s. uniformly under P_0 to the function X given by (2.6). The function X is minimum with value zero as

$$p_0 = 1, \quad \mu_{01} = \mu_0, \tag{2.8}$$

and $\mu_{02} = 0$ for a general mean $\mu_0 = \mu_{01} + \mu_{02}$, γ_0 being the first point where μ_{01} and μ_{02} are distinct, so that γ_0 is at the end-point of the support of Y. Under H_0, the process X_n and the function X reduce to zero at γ_0. The process $X_n(\gamma)$ is expressed in (2.2) with $\mu_{02} = 0$

$$\begin{aligned} X_n(\gamma) &= \widehat{\mu}^2_{0n} - \widehat{\mu}^2_{1n,\gamma} - \widehat{\mu}^2_{2n,\gamma} \\ &= (\widehat{\mu}_{0n} - \widehat{\mu}_{1n,\gamma})(\widehat{\mu}_{0n} + \widehat{\mu}_{1n,\gamma}) - \widehat{\mu}^2_{2n,\gamma}. \end{aligned}$$

By Theorem 2.1, the estimator $\widehat{\gamma}_n$ converges a.s. to γ_0 under H_0 and $\widehat{p}_n = \widehat{P}_n(Y \leq \widehat{\gamma}_n)$ converges a.s. to one. The inequalities (2.5) and (2.7) are still true under H_0 and Theorem 2.2 is still valid, the convergence rate of $\widehat{\gamma}_n$ is not larger then n^{-1}.

If the limit of $\gamma^2 f(\gamma)$, as γ tends to γ_0, is finite, then the asymptotic behaviour of $\widehat{\gamma}_n$ described by Theorem 2.4 is still valid under H_0. Let U be the limit of $n(\gamma_0 - \widehat{\gamma}_n)$ and let ϕ_{01} be the limit of $\gamma f(\gamma)$ as γ tends to γ_0. If ϕ_{01} is finite, let G_U be the variable $U\phi_{01}$.

Proposition 2.3. *If $E_0(Y^4)$ and the limit of $\gamma^2 f(\gamma)$, as γ tends to γ_0, are finite then the statistic T_n converges weakly under H_0 to $T_0 = 2\mu_0 G_U$.*

Proof. Under H_0, p_0 is known and the limit of the process $X_n(\gamma)$ is the function

$$X(\gamma) = -(\mu_{01\gamma}^2 + \mu_{02\gamma}^2 - \mu_0^2).$$

Let $\gamma_n = \gamma_0 - n^{-1}u_n$ with $u_n > 0$ in $\mathcal{U}_n^A$, the difference of the estimators at γ_0 and γ_n is

$$1 - \widehat{p}_{n,\gamma_n} = n^{-1}\sum_{i=1}^{n} 1_{\{Y_i > \gamma_n\}},$$

$$\widehat{\mu}_{n,0} - \widehat{\mu}_{1n,\gamma_n} = n^{-1}\sum_{i=1}^{n} Y_i 1_{\{Y_i > \gamma_n\}} = \widehat{\mu}_{2n,\gamma_n},$$

therefore $\widehat{\mu}_{n,0} - \widehat{\mu}_{1n,\gamma_{n,u}}$ and $\widehat{\mu}_{2n,\gamma_{n,u}}$ have the same convergence rate as $1 - \widehat{p}_{n,\gamma_{n,u}}$ and

$$\widehat{\mu}_{1n,\gamma}^2 + \widehat{\mu}_{2n,\gamma}^2 - \widehat{\mu}_{n,0}^2 = -2\widehat{\mu}_{1n,\gamma}\widehat{\mu}_{2n,\gamma} + o_p(n^{-1}).$$

Then the process nX_n has the approximation

$$nX_n(\widehat{\gamma}_n) = 2n\widehat{\mu}_{2n}\mu_0 + o_p(1)$$

and it converges weakly to T_0.

The expectation of $n\widehat{\mu}_{2n,\gamma_n}$ is approximated under H_0 by $n(\gamma_0 - \widehat{\gamma}_n)\gamma_0 f(\gamma_0)$ and its variance by $\gamma_0^2 u_n f(\gamma_0)\{1 - u_n f(\gamma_0)\}$, then $n\widehat{\mu}_{2n}$ converges weakly to an uncentered Gaussian variable G_U under the conditions. The limit of the test statistic T_n follows. □

Under the alternative K of a probability measure P_γ for the observations space, with a change point at γ between $Y_{n:1}$ and $Y_{n:n} < \gamma_0$, the empirical mean $\bar{Y}_n$ converges a.s. under P_γ to μ_γ and the estimator $\widehat{\mu}_{kn}$ converges a.s. under P_γ to $\mu_{k\gamma}$, for $k = 1, 2$. The limit of $n(\widehat{\gamma}_n - \gamma)$ under P_γ is given by Theorem 2.3.

Local alternatives P_{θ_n} are defined by parameters $\gamma_{n,u} = \gamma_0 - n^{-1}u_{n,\gamma}$, $\mu_{1n,u} = \mu_0 + n^{-\frac{1}{2}}u_{1n,\mu}$ and $\mu_{2n,u} = n^{-1}u_{2n,\mu}$ such that u_n converges to a limit u with strictly positive components.

Proposition 2.4. *Let the conditions of Proposition 2.3 be satisfied. Under a fixed alternative P_θ, the statistic T_n tends to infinity. Under local alternatives P_{θ_n}, the statistic T_n converges weakly to $T_0 + c$, with a non null constant c.*

Proof. Under P_γ, the empirical mean $\bar{Y}_n$ converges a.s. to the expectation $\mu_\gamma = \mu_{1\gamma} + \mu_{2\gamma}$ of Y and the process

$$X_{n\gamma}(\gamma') = \widehat{\mu}^2_{1n,\gamma} - \widehat{\mu}^2_{1n,\gamma'} + \widehat{\mu}^2_{2n,\gamma} - \widehat{\mu}^2_{2n,\gamma'}$$

converges a.s. uniformly to the function

$$X_\gamma(\gamma') = \mu^2_{1\gamma} - \mu^2_{1\gamma'} + \mu^2_{2\gamma} - \mu^2_{2\gamma'}.$$

With p_γ in $]0,1[$, the variable $n(\widehat{p}_{n,\gamma} - p_\gamma)$ converges weakly to the location of the minimum U_γ of a Gaussian process and $n(1-p_\gamma)$ diverges. In the same way, the differences of the estimators $\widehat{\mu}_{kn,\gamma} - \widehat{\mu}^2_{1n,\gamma'}$ are $O(|\gamma - \gamma'|)$, and the variables $n(\widehat{\mu}_{kn} - \mu_{k\gamma})$ converge weakly to non-degenerated Gaussian variables with means and variances proportional to $f(\gamma)$, for $k = 1, 2$. But $n(\mu_{1\gamma} - \mu_0)$ and therefore T_n diverge.

Under the local alternatives K_n, the probability is $p_n = 1 - n^{-1}u_{n,p}$ with a sequence $(u_{n,p})_n$ converging to a non-zero limit u_p, the variable $n(1-\widehat{p}_n)$ converges weakly to $u_p + U$ where the variable U is the limit of $n(p_n - \widehat{p}_n)$ under P_n, it is identical to the limit of $n(1-\widehat{p}_n)$ under P_0. The variables $n^{\frac{1}{2}}(\widehat{\mu}_{kn} - \mu_{kn,u})$ and $n(\widehat{\mu}_{kn,\gamma} - \widehat{\mu}_{kn,0})$ converge weakly to bounded limits, $n(\mu_0 - \mu_{1n,u})$ converges to $u_{1,\mu}$ and $n\mu_{2n}$ converges to $u_{2,\mu}$. These convergences entail the weak convergence of

$$nX_n(\widehat{\gamma}_n) = 2n\widehat{\mu}_{1n}\widehat{\mu}_{2n} + o_p(1)$$

to $2\mu_0(G_{\mu_2} - u_{2,\mu})$ and the constant is $c = -2\mu_0 u_{2,\mu}$. □

The limits of T_n under H_0 and local alternatives are well separated and a test at the level α has the rejection domain $\{T_n > c_\alpha\}$ such that $P_0(|T_n| > c_\alpha) = \alpha$. Its asymptotic power $\lim_{n\to\infty} P_{\gamma_{n,u}}(|T_n| > c_\alpha)$ is larger than α.

2.3 Maximum likelihood estimation

A change in the mean of a real variable Y is a change in the location of its density

$$f_{\mu,\gamma}(y) = f(y)1_{\{y\leq\gamma\}} + f(y-\mu)1_{\{y>\gamma\}} \tag{2.9}$$

with a non-zero difference of the partial means $\mu = \mu_{2,\gamma} - \mu_{1,\gamma}$ on $\{y > \gamma\}$ and $\{y \leq \gamma\}$. The parameter space is $\Theta = \{\theta = (\mu,\gamma) : \mu, \gamma \in \mathbb{R}\}$ and

θ_0 is the true parameter value. The empirical estimator of the unknown difference of means μ_γ, at fixed γ, is

$$\widehat{\mu}_{n,\gamma} = n^{-1} \sum_{k=1}^{n} Y_k (1_{\{Y_k > \gamma\}} - 1_{\{Y_k \leq \gamma\}}).$$

Under P_θ, the mean of Y is $E_\theta Y = \int y \, dF_\theta(y)$ the variance $\sigma_\theta^2 = \sigma_{F_\theta}^2$ of Y is estimated by the empirical variance $\widehat{\sigma}_n^2$. Under P_0, the density f_0 has the parameters γ_0 and $\mu_0 \neq 0$, and $f_0 = f_{\theta_0}$ has a variance σ_0^2.

The logarithm of the likelihood ratio of the sample under P_θ and P_0 defines a process

$$X_n(\theta) = n^{-1} \sum_{k=1}^{n} \Big\{ \log \frac{f(Y_k)}{f(Y_k - \mu_0)} 1_{]\gamma_0, \gamma]}(Y_k) + \log \frac{f(Y_k - \mu)}{f(Y_k)} 1_{]\gamma, \gamma_0]}(Y_k) + \log \frac{f(Y_k - \mu)}{f(Y_k - \mu_0)} 1_{\{Y_k > \gamma \vee \gamma_0\}} \Big\},$$

the parameters are estimated by maximization of

$$l_n(\theta) = \sum_{k=1}^{n} \Big\{ \log f(Y_k) 1_{\{Y_k \leq \gamma\}} + \log f(Y_k - \mu) 1_{\{Y_k > \gamma\}} \Big\}, \tag{2.10}$$

and $X_n(\theta) = l_n(\theta) - l_n(\theta_0)$. Under P_0, the process X_n converges a.s. uniformly on Θ to the function

$$\begin{aligned} X(\theta) = {} & 1_{\{\gamma > \gamma_0\}} \Big\{ \int_{\gamma_0}^{\gamma} \log \frac{f(y)}{f(y - \mu_0)} \, dF(y - \mu_0) \\ & + \int_{\gamma}^{\infty} \log \frac{f(y - \mu)}{f(y - \mu_0)} \, dF(y - \mu_0) \Big\} \\ & + 1_{\{\gamma < \gamma_0\}} \Big\{ \int_{\gamma}^{\gamma_0} \log \frac{f(y - \mu)}{f(y)} \, dF(y) \\ & + \int_{\gamma_0}^{\infty} \log \frac{f(y - \mu)}{f(y - \mu_0)} \, dF(y - \mu_0) \Big\}. \end{aligned}$$

The function X is also written as

$$X(\theta) = \int_{\gamma_0}^{\gamma} \log \frac{f(y)}{f(y - \mu)} \, dF_0(y) + \int_{\gamma_0}^{\infty} \log \frac{f(y - \mu)}{f(y - \mu_0)} \, dF(y - \mu_0).$$

Theorem 2.5. *If f is continuous and bounded on Θ, the maximum likelihood estimator $\widehat{\gamma}_n$ of γ_0 is a.s. consistent.*

Proof. The function X is C_b^1 on $\{\gamma < \gamma_0\}$ and on $\{\gamma_0 < \gamma\}$, its derivative with respect to γ is

$$X'_\gamma(\theta) = \{1_{\{\gamma<\gamma_0\}} f(\gamma) + 1_{\{\gamma_0<\gamma\}} f(\gamma - \mu_0)\} \log \frac{f(\gamma)}{f(\gamma - \mu)}$$

and $X'_\gamma(\theta)$ is zero if and only if $\gamma_0 = \gamma$, with $\mu \neq 0$. If $\gamma < \gamma_0$, $X'_\gamma(\theta_0) > 0$ and if $\gamma_0 < \gamma$, $X'_\gamma(\theta_0) < 0$ therefore the function X is concave at γ_0 where X reaches its maximum zero with respect to the parameter γ.

By the uniform convergence of X_n to X, we have

$$0 \leq X_n(\widehat{\gamma}_n) \leq \sup_\Gamma |X_n(\gamma) - X(\gamma)| + X(\widehat{\gamma}_n).$$

As $X(\widehat{\gamma}_n) \leq 0$, it is a $o_p(1)$ and $\widehat{\gamma}_n$ converges a.s. to γ_0. □

In the following, we assume that $\int \log^2 f(y)\, dF_0(y)$ and $\int (f'^2 f^{-1})$ are finite. By concavity of X, if F is C_b^2, the left-derivative $X''_\gamma(\theta_0)$ is negative, and its derivatives with respect to μ are zero at γ_0. The derivatives of $X(\theta)$ with respect to μ satisfy

$$\begin{aligned}
X'_\mu(\theta) &= \int_{\gamma_0}^{\gamma} \frac{f'}{f}(y - \mu)\, dF_0(y) - \int_{\gamma_0}^{\infty} \frac{f'}{f}(y - \mu)\, dF_0(y) \\
&= - \int_{\gamma}^{\infty} \frac{f'}{f}(y - \mu)\; dF_0(y), \\
X''_\mu(\theta) &= \int_{\gamma}^{\infty} \frac{ff'' - f'^2}{f^2}(y - \mu)\, dF_0(y), \\
X''_{\mu,\gamma}(\theta) &= \frac{f'}{f}(\gamma - \mu)\, f_0(\gamma),
\end{aligned}$$

$X'_\mu(\theta_0) = 0$ and $I_0 = -X''_\mu(\theta_0)$ reduces to $I_0 = \int_{\gamma_0}^\infty f_0'^2(y-\mu_0) f_0^{-1}(y-\mu_0)\, dy$, with $\int_{\gamma_0}^\infty f_0' = 0 = \int_{\gamma_0}^\infty f_0''$, and I_0 is positive. Let

$$\rho(\theta, \theta_0) = \{(\mu - \mu_0)^2 + |\gamma - \gamma_0|\}^{\frac{1}{2}},$$

ρ is semi-norm. For every $\varepsilon > 0$, let $V_\varepsilon(\theta_0)$ be a neighborhood of θ_0 for ρ.

Lemma 2.1. *For ε small enough, there exists a constant $\kappa_0 > 0$ such that for every θ in Θ*

$$X(\theta) \geq -\kappa_0 \rho^2(\theta, \theta_0). \tag{2.11}$$

Proof. By a second order expansion of $X(\theta)$ for θ in a neighborhood $V_\varepsilon(\theta_0)$ of θ_0 and by the equality $X'_\mu(\theta_0) = 0$, we have $X(\theta) = -\frac{1}{2}(\mu - \mu_0)^2 I_0\{1 + o(1)\}$ and the result follows for every positive and sufficiently small ε. □

The process $W_n = n^{\frac{1}{2}}(X_n - X)$ is written as

$$W_n(\theta) = 1_{\{\gamma>\gamma_0\}} \int_{\gamma_0}^{\gamma} \log \frac{f(y)}{f(y-\mu_0)} \, d\nu_{0n}(y)$$
$$+1_{\{\gamma<\gamma_0\}} \int_{\gamma}^{\gamma_0} \log \frac{f(y-\mu)}{f(y)} \, d\nu_{0n}(y)$$
$$+ \int_{\gamma\vee\gamma_0}^{\infty} \log \frac{f(y-\mu)}{f(y-\mu_0)} \, d\nu_{0n}(y),$$

where $\nu_{0n} = n^{\frac{1}{2}}(\widehat{F}_n - F_0)$ is the empirical process on each sub-interval under P_0, $\nu_{0n}(y) = 1_{\{\gamma_0<\gamma_{n,u}\}}\nu_n(y-\mu_0) + 1_{\{\gamma_{n,u}<\gamma_0\}}\nu_n(y)$.

For every $\varepsilon > 0$ and under the condition $E_0 \sup_{\theta\in V_\varepsilon(\theta_0)} \log^2 f(\theta)(Y)$ finite, there exists a constant $\kappa_1 > 0$ such that for n large enough

$$E_0 \sup_{\theta\in V_\varepsilon(\theta_0)} W_n^2(\theta) \leq \kappa_1^2\varepsilon,$$
$$E_0 \sup_{\theta\in V_\varepsilon(\theta_0)} |W_n(\theta)| \leq \kappa_1\varepsilon^{\frac{1}{2}}. \tag{2.12}$$

We first consider the empirical estimator $\widehat{\mu}_{n,\gamma}$ for the parameter μ_γ, at a fixed value γ. By Theorem 2.5, $\widehat{\mu}_{n,\widehat{\gamma}_n}$ is a.s. consistent and it converges to μ_0 with the rate $n^{-\frac{1}{2}}$. Let

$$\widetilde{X}_n(\gamma) = X_n(\widehat{\mu}_{n,\gamma}, \gamma),$$

we now define the estimator of the change-point by maximization of the estimated log-likelihood ratio

$$\widehat{\gamma}_n = \arg \max_{Y_{n:1}<\gamma<Y_{n:n}} \widetilde{X}_n(\gamma).$$

The uniform a.s. consistency of $\widehat{\mu}_{n,\gamma}$ implies $\widetilde{X}_n$ is a.s. uniformly consistent and it is concave at γ_0, like X_n. Since $\mu_\gamma - \mu_0 = \int_{\gamma_0}^{\gamma} dF = 0(|\gamma - \gamma_0|)$ in a neighborhood of γ_0, for ε sufficiently small there exists a constant κ_1 such that the process

$$\widetilde{W}_n = n^{\frac{1}{2}}(\widetilde{X}_n - X)$$

satisfies the same inequality as W_n.

Lemma 2.2. *For $\varepsilon > 0$ sufficiently small, there exists a constant κ_1 such that*

$$E_0 \sup_{\rho(\gamma,\gamma_0)\leq\varepsilon} |\widetilde{W}_n(\gamma)| \leq \kappa_1\varepsilon. \tag{2.13}$$

Proof. The inequality is true on the interval $]\gamma \vee \gamma_0, \infty[$ by an expansion of $f(y-\mu)$ for μ in a neighborhood $V_\varepsilon(\mu_0)$ of μ_0. For γ in a neighborhood of γ_0, we have $\mu_\gamma - \mu_0 = \int_{\gamma_0}^{\gamma} dF = 0(|\gamma - \gamma_0|)$ and on the interval $]\gamma, \gamma_0]$

$$E_0 \sup_{\rho(\theta,\theta_0)\leq\varepsilon} \log \frac{f(Y_k-\mu)}{f(Y_k)} 1_{]\gamma,\gamma_0]}(Y_k) \leq \int_{\gamma}^{\gamma_0} \sup_{|\mu-\mu_0|\leq\varepsilon} \left|\log \frac{f(y-\mu)}{f(y)}\right| dF(y)$$

and the inequality is true for $f(y-\mu)$ bounded in $V_\varepsilon(\mu_0)$, with $\varepsilon < 1$. □

Theorem 2.6. *For $\varepsilon > 0$ sufficiently small*

$$\overline{\lim}_{n,A\to\infty} P_0(\sup_{|\gamma-\gamma_0|\leq\varepsilon, |u_{n,\gamma}|>A} \widetilde{X}_n(\gamma) \geq 0) = 0,$$
$$\overline{\lim}_{n,A\to\infty} P_0(n|\widehat{\gamma}_n - \gamma_0| > A) = 0.$$

The proof for the process $\widetilde{X}_n$ is similar to the proof of Theorem 2.2, using the inequalities (2.11) and (2.13). The process X_n and the maximum likelihood estimator $\widehat{\theta}_n$ have the same asymptotic behaviour from the inequalities (2.11) and (2.12).

Theorem 2.7. *For $\varepsilon > 0$ sufficiently small*

$$P_0(\sup_{\rho(\theta,\theta_0)\leq\varepsilon, |u_{n,\theta}|>A} X_n(\theta) \geq 0),$$
$$\overline{\lim}_{n,A\to\infty} P_0(n|\widehat{\gamma}_n - \gamma_0| > A^2, n^{\frac{1}{2}}|\widehat{\mu}_n - \mu_0| > A,) = 0.$$

Proof. Let $\|\widehat{u}_n\| = n^{\frac{1}{2}}\rho(\widehat{\theta}_n, \theta_0)$, and let $\varepsilon > 0$ be sufficiently small to entail the inequalities (2.11) and (2.12). From Theorem 2.5, for $\eta > 0$ there exists n_0 such that n larger than n_0 implies $P_0\{\widehat{u}_n \in \mathcal{U}_{n,\varepsilon}\} = P_0\{\widehat{\theta}_n \in V_\varepsilon(\theta_0)\} > 1 - \eta$ and

$$P_0(\|\widehat{u}_n\| > A) \leq P_0\Big\{\sup_{u\in\mathcal{U}_{n,\varepsilon}, |u|>A} X_n(\theta_{n,u}) \geq 0\Big\} + \eta.$$

This probability is bounded in the subsets $H_{n,j}$ of $\mathcal{U}_{n,\varepsilon}$ defined by its intersection with the sets $\{g(j) < \|u\| \leq g(j+1)\}$, for the integers j such that $g(j) > A$ and $g(j+1) \leq n\varepsilon$, where g is a function g increasing at an exponential rate and $\sum_{g(j)>A} g(j+1)g^{-1}(j)$ tends to zero as A tends to infinity. The inequalities (2.11) and (2.12) imply that as n tends to infinity

$$|X(\theta_{n,u})| \leq \kappa_0 n^{-\frac{1}{2}}\|u\| \leq \kappa_0 n^{-\frac{1}{2}} g(j+1)$$

and $P_0\{\sup_{H_{n,j}} W_n(\theta_{n,u}) \geq c\} \leq c^{-1}\kappa_1 n^{\frac{1}{2}} g(j+1)$, therefore

$$P_0\Big\{\sup_{u\in\mathcal{U}_{n,\varepsilon}, |u|>A} X_{n,h}(\theta_{n,u}) \geq 0\Big\} \leq \sum_{g(j)>A} P_0\Big\{\sup_{H_{n,j}} |W_n(\theta_{n,u})| \geq \kappa_0 n^{-\frac{1}{2}} g(j)\Big\}$$
$$\leq \kappa_0^{-1}\kappa_1 \sum_{g(j)>A} g(j+1)g^{-1}(j)$$

and it tends to zero as A tends to infinity. □

Let $A > 0$, we can now restrict the asymptotic study of $\widetilde{W}_n(\gamma_{n,u})$ and $W_n(\gamma_{n,u})$ to $\gamma_{n,u} = \gamma_0 + n^{-1}u$ with u in $\mathcal{U}_n^A = \{u \in \mathcal{U}_n : |u| \leq A\}$.

Proposition 2.5. *On $\mathcal{U}_n^A$, $A > 0$, the processes $\widehat{\mu}_{n,\gamma_{n,u}}$ and $\widetilde{W}_n(\gamma_{n,u})$ satisfy $n^{\frac{1}{2}}(\widehat{\mu}_{n,\gamma_{n,u}} - \mu_0) = n^{\frac{1}{2}}(\widehat{\mu}_{n,\gamma_0} - \mu_0) + o_p(1)$ and*

$$\widetilde{W}_n(\gamma_{n,u}) = \int_{\gamma_0}^{\gamma_{n,u}} \log \frac{f(y)}{f(y-\mu_0)} \, d\nu_{0n}(y) - (\widehat{\mu}_{n,\gamma_0} - \mu_0) \int_{\gamma_0}^{\infty} \frac{f'(y-\mu_0)}{f(y-\mu_0)} \, d\nu_{0n}(y) + o_p(n^{-\frac{1}{2}}).$$

Proof. The process $\widehat{\mu}_{n,\gamma_{n,u}} = \int_{\gamma_{n,u}}^{\infty} y \, d\widehat{F}_{Y,n} - \int_{-\infty}^{\gamma_{n,u}} y \, d\widehat{F}_n$ is expanded as

$$\begin{aligned} n^{\frac{1}{2}}(\widehat{\mu}_{n,\gamma_{n,u}} - \mu_0) &= n^{\frac{1}{2}}(\widehat{\mu}_{n,\gamma_{n,u}} - \widehat{\mu}_{n,\gamma_0}) + n^{\frac{1}{2}}(\widehat{\mu}_{n,\gamma_0} - \mu_0) \\ &= n^{-\frac{1}{2}} \sum_{k=1}^{n} Y_k \{1_{]\gamma_{n,u},\gamma_0]}(Y_k) - 1_{]\gamma_0,\gamma_{n,u}]}(Y_k)\} \\ &\quad + n^{\frac{1}{2}}(\widehat{\mu}_{n,\gamma_0} - \mu_0), \end{aligned}$$

where the mean of $Y_k\{1_{]\gamma_{n,u},\gamma_0]}(Y_k)$ is $\int_{\gamma_{n,u}}^{\gamma_0} y \, dF(y) = -n^{-1}u\gamma_0 f(\gamma_0) + o(n^{-1})$, it follows that for u in $\mathcal{U}_n^A$, the mean of $n^{\frac{1}{2}}(\widehat{\mu}_{n,\gamma_{n,u}} - \widehat{\mu}_{n,\gamma_0})$ is a $o(1)$ uniformly on $\mathcal{U}_n^A$, like its variance, hence $n^{\frac{1}{2}}(\widehat{\mu}_{n,\gamma_{n,u}} - \widehat{\mu}_{n,\gamma_0})$ converges uniformly in probability to zero. Finally, by an expansion of $f(y - \widehat{\mu}_{n,\gamma_{n,u}})$, the process $\widetilde{W}_n(\gamma_{n,u}) = n^{\frac{1}{2}}\{\widetilde{X}_n(\gamma_{n,u}) - X(\widehat{\mu}_{n,\gamma_{n,u}}, \gamma_{n,u})\}$ is expanded as

$$\begin{aligned} \widetilde{W}_n(\gamma_{n,u}) &= 1_{\{\gamma_0 < \gamma_{n,u}\}} \int_{\gamma_0}^{\gamma_{n,u}} \log \frac{f(y)}{f(y-\mu_0)} \, d\nu_{0n}(y) \\ &\quad + 1_{\{\gamma_{n,u} < \gamma_0\}} \int_{\gamma_{n,u}}^{\gamma_0} \log \frac{f(y - \widehat{\mu}_{n,\gamma_{n,u}})}{f(y)} \, d\nu_{0n}(y) \\ &\quad + \int_{\gamma_{n,u} \vee \gamma_0}^{\infty} \log \frac{f(y - \widehat{\mu}_{n,\gamma_{n,u}})}{f(y-\mu_0)} \, d\nu_{0n}(y), \\ &= \int_{\gamma_0}^{\gamma_{n,u}} \log \frac{f(y)}{f(y-\mu_0)} \, d\nu_{0n}(y) \\ &\quad - (\widehat{\mu}_{n,\gamma_0} - \mu_0) \Big\{ \int_{\gamma_0}^{\infty} \frac{f'(y-\mu_0)}{f(y-\mu_0)} \, d\nu_{0n}(y) + o_p(1) \Big\}, \end{aligned}$$

the convergence of the empirical process ν_{0n} to $W \circ F_0$, where W is a Brownian bridge, and the weak convergence of $n^{\frac{1}{2}}(\widehat{\mu}_{n,\gamma_0} - \mu_0)$ to a centered Gaussian variable imply that the last term is a $O(n^{-\frac{1}{2}})$. □

By the approximation of Proposition 2.5 for the empirical process $\widehat{\mu}_{n,\gamma_{n,u}}$ on $\mathcal{U}_u^A$, it has the asymptotic distribution of the variable $n^{\frac{1}{2}}(\widehat{\mu}_{n,\gamma_0} - \mu_0)$. The process

$$S_n(u) = \sum_{k=1}^{n} \log \frac{f(Y_k - \mu_0)}{f(Y_k)} 1_{]\gamma_0,\gamma_{n,u}]}(Y_k) + \sum_{k=1}^{n} \log \frac{f(Y_k)}{f(Y_k - \mu_0)} 1_{]\gamma_{n,u},\gamma_0]}(Y_k)\}$$

satisfies the same bound (1.2) in $\mathcal{U}_u^A$ as the process defined by (1.1) therefore it is tight in $D([-A,A])$ and it converges weakly to an uncentered Gaussian process S with finite variance function.

Theorem 2.8. *Under P_0, the maximum likelihood estimators $\widehat{\gamma}_n$ and $\widehat{\mu}_n$ are asymptotically independent, $n^{\frac{1}{2}}(\widehat{\mu}_n - \mu_0)$ converges weakly to a centered Gaussian variable with variance I_0^{-1}, $n(\widehat{\gamma}_n - \gamma_0)$ is bounded in probability and it converges weakly to the location of the maximum of the Gaussian process S.*

Proof. The process $X'_{n,\mu}(\mu_0, \widehat{\gamma}_n) - X'_{n,\mu}(\theta_0)$ is asymptotically equivalent under P_0 to $\int_{\widehat{\gamma}_n}^{\gamma_0} f'(y - \mu_0)\,dy = 0_p(n^{-1})$, therefore the variable $n^{\frac{1}{2}}\{X'_{n,\mu}(\mu_0, \widehat{\gamma}_n) - X'_{n,\mu}(\theta_0)\}$ converges in probability to zero. Expansions of the process $X'_{n,\mu}$ in neighbourhoods of θ_0 yield

$$\begin{aligned} n^{\frac{1}{2}} X'_{n,\mu}(\mu_0, \widehat{\gamma}_n) &= -n^{\frac{1}{2}}(\widehat{\mu}_n - \mu_0) X''_{n,\mu}(\theta_0)\{1 + o_p(1)\} \\ n^{\frac{1}{2}}(\widehat{\mu}_n - \mu_0) &= n^{\frac{1}{2}} X'_{n,\mu}(\theta_0) I_0^{-1} + o_p(1). \end{aligned} \quad (2.14)$$

By the properties of I_0, the variable $n^{\frac{1}{2}}(\widehat{\mu}_n - \mu_0)$ converges weakly under P_0 to a centered Gaussian variable with variance I_0^{-1}. Let θ_{n,u_n} with u_n a vector with components $u_{1n} = n^{\frac{1}{2}}(\mu_{n,u} - \mu_0)$ and $u_{2n} = n(\gamma_{n,u} - \gamma_0)$, such that $n^{\frac{1}{2}}\rho(\theta_{n,u_n}, \theta_0) < A$. An expansion of $nX_n(\theta_{n,u})$ as n tends to infinity is written as

$$\begin{aligned} nX_n(\theta_{n,u}) = &\sum_{k=1}^{n} \log \frac{f(Y_k)}{f(Y_k - \mu_0)} 1_{]\gamma_0,\gamma_{n,u}]}(Y_k) \\ &+ \sum_{k=1}^{n} \log \frac{f(Y_k - \mu_0)}{f(Y_k)} 1_{]\gamma_{n,u},\gamma_0]}(Y_k) \\ &- u_{1n} \int_{\gamma_0}^{\infty} f'_0\, d\nu_{0n} - \frac{1}{2} u_{1n}^T I_0 u_{1n} + o_p(1), \end{aligned}$$

it is a the sum of processes depending on distinct components of u_n. As $Y_{n:1} < \gamma_0 < Y_{n:n}$, the sum of the last two terms at $\widehat{u}_n$ is asymptotically equivalent to $\frac{1}{2}\widehat{u}_{1n}^2 I_0^{-1}$ and $\widehat{u}_{1n}$ converges weakly to a centered Gaussian variable with variance I_0^{-1}. By the convergence rate of $\widehat{\gamma}_n$, the variable

$$\sum_{k=1}^{n}\{\log f(Y_k - \widehat{\mu}_n) - \log f(Y_k - \mu_0)\}1_{\{Y_i > \widehat{\gamma}_n \wedge \gamma_0\}}$$

is asymptotically equivalent to $\sum_{k=1}^{n}\{\log f(Y_k - \widehat{\mu}_n) - \log f(Y_k - \mu_0)\}1_{\{Y_i > \gamma_0\}}$ and it converges to a χ_1^2 variable, from a second order expansion, it is asymptotically independent of $\widehat{u}_{2n} = n(\widehat{\gamma}_n - \gamma_0)$.

The estimator of $\widehat{u}_{2n}$ maximizes the sum of the empirical means on the intervals $]\gamma_0, \gamma]$ and $]\gamma, \gamma_0]$, under P_0 their means are $O(|u_{2n}|)$ and they are bounded in probability, by Theorem 2.6. Under P_0, the process $S_n(\theta_{n,u})$ is tight by the argument of Section 1.4, and its finite dimensional distributions converges weakly to those of the process S, therefore it converges weakly to S. The estimator $\widehat{\gamma}_n$ is such that $n(\widehat{\gamma}_n - \gamma_0)$ converges weakly to the location of the maximum of the process L_X. □

Using the same arguments, the estimator of the parameter γ that maximizes the process $X_n(\widehat{\mu}_{n,\gamma})$ defined with the empirical mean $\widehat{\mu}_{n,\gamma}$ converges weakly to the maximum of a Gaussian process and it is bounded in probability.

2.4 Likelihood ratio test

In the model (2.1), the log-likelihood ratio test for the hypothesis H_0 of a constant mean against the alternative of a change at an unknown threshold γ. According to (2.8), the hypothesis is equivalent to $p_0 = 1$, $\mu_{01} = E_0Y$ and $\mu_{02} = 0$ so the difference of the means is $\mu_{01} - \mu_{02} = E_0Y$ and the density of Y under H_0 is f, $\mu_0 = -E_0Y$. The alternative is equivalent to a difference of means different from E_0Y and a density different from f for Y larger than a finite change-point γ. The test is performed with the log-likelihood ratio statistic

$$\begin{aligned}
T_n &= 2 \sup_{Y_{n:1} < \gamma \le Y_{n:n}} \{\widehat{l}_n(\gamma) - l_{0n}\} \\
&= 2 \sup_{Y_{n:1} < \gamma \le Y_{n:n}} \sum_{k=1}^{n} \log \frac{f(Y_k - \widehat{\mu}_{n,\gamma})}{f(Y_k)} 1_{\{Y_k > \gamma\}}
\end{aligned}$$

where l_{0n} is the logarithm of the density under H_0 reduces to $\sum_{k=1}^{n} \log f(Y_k)$ and $\widehat{l}_n(\gamma)$ is the logarithm of the ratio of the densities at

$\widehat{\mu}_{n,\gamma}$ for an arbitrary γ. If the support of the variable Y is infinite, the statistic T_n is the maximum of the process

$$X_n(\theta) = 2n \int_\gamma^\infty \log \frac{f(y-\mu)}{f(y)} \, d\widehat{F}_n(y), \tag{2.15}$$

under H_0, it has the mean

$$X(\theta) = 2n \int_\gamma^\infty \log \frac{f(y-\mu)}{f(y)} \, dF_0(y).$$

Under the condition of a density f belonging to $C^2(\mathbb{R})$ and (2.8), its maximum under P_0 is achieved as γ is at the upper point γ_0 infinite of the support of Y, for every difference of means μ. The estimator of μ that maximizes X_n converges a.s. under P_0 to $\mu_0 = -E_0 Y$. If γ_0 is finite, $X_n(\theta) = 2n \int_\gamma^{\gamma_0} \{\log f(y-\mu) - \log f(y)\} \, d\widehat{F}_n(y)$.

Under H_0 and under the conditions of a density f in $C^2(\mathbb{R}) \cap L^2(\mathbb{R})$, such that $\int_\mathbb{R} \sup_\mu |\log\{f(y-\mu)f^{-1}(y)\}| \, dF_0(y)$ is finite, on the interval $]\gamma, \gamma_0)$ we have

$$E_0 \sup_\gamma \log \frac{f(Y_k - \widehat{\mu}_{n,\gamma})}{f(Y_k)} 1_{\{Y_k > \gamma\}} \le E_0 \int_\gamma^{\gamma_0} \sup_\gamma \left| \log \frac{f(y - \widehat{\mu}_{n,\gamma})}{f(y)} \right| dF_0(y),$$

if γ_0 is finite, this expression is bounded, it is a $O(|\gamma_0 - \gamma|)$. The inequality (2.13) and Theorem 2.7 are still satisfied and the estimators of the parameters are a.s. consistent, $X'_{n,\mu}(\widehat{\theta}_n) = 0$ and (2.14) is satisfied. If I_0 is positive definite, we have

$$n^{\frac{1}{2}}(\widehat{\mu}_n - \mu_0) = n^{\frac{1}{2}} X'_{n,\mu}(\theta_0) I_0^{-1} + o_p(1)$$

where $n^{\frac{1}{2}} X'_{n,\mu}(\theta_0)$ converges weakly to a centered Gaussian process. More generally, an expansion of $X'_{n,\mu}(\widehat{\theta}_n)$ yields

$$\int_{\widehat{\gamma}_n}^{\gamma_0} \frac{f'}{f}(y - \mu_0) \, d\nu_{0n}(y) = n^{\frac{1}{2}}(\widehat{\mu}_n - \mu_0) X''^{-1}_{n,\mu}(\theta_n)$$

where $\theta_n = (\mu_n, \widehat{\gamma}_n)$ and μ_n is between $\widehat{\mu}_n$ and μ_0.

The variance of $\int_{\widehat{\gamma}_n}^{\gamma_0} (f'f^{-1})(y - \mu_0) \, d\nu_{0n}(y)$ conditionally on $\widehat{\gamma}_n$, is asymptotically equivalent to $\int_{\widehat{\gamma}_n}^{\gamma_0} (f'^2 f^{-1})(y - \mu_0) \, dy - \{\int_{\widehat{\gamma}_n}^{\gamma_0} f'\}^2$, let a_n^2 be its convergence rate to a non-degenerated limit, then the left term of the last equality has the convergence rate a_n and the convergence rate of $X''_{n,\mu}(\mu_0, \widehat{\gamma}_n)$ to a non-zero limit is still a_n^2.

Proposition 2.6. *Under H_0, the statistic T_n converges weakly to a χ_1^2 variable T_0.*

Proof. By the expansion of $X'_{n,\mu}(\mu_0, \widehat{\gamma}_n)$, the variable $n^{\frac{1}{2}} a_n^{-1}(\widehat{\mu}_n - \mu_0)$ converges weakly to a non-degenerated limit under P_0 and a second order expansion of the logarithm of $f(y - \widehat{\mu}_n)$ in the expression of T_n implies

$$T_n = \frac{1}{-X''_{n,\mu}(\mu_0, \widehat{\gamma}_n)} \Big\{ \int_{\widehat{\gamma}_n}^{\gamma_0} (f'f^{-1})(y - \mu_0)\, d\nu_n(y) \Big\}^2 + o_p(1).$$

Its limit follows from the weak convergence of the variables $a_n X'_{n,\mu}(\mu_0, \widehat{\gamma}_n)$ and $a_n^2 X''_{n,\mu}(\mu_0, \widehat{\gamma}_n)$. □

Under the alternative K of a difference of means μ at a change point at γ between $Y_{n:1}$ and $Y_{n:n} < \gamma_0$, $n^{-1} l_{0n}$ converges under P_θ to

$$l_\theta = \int_{-\infty}^{\gamma} \log f(y)\, dF(y) + \int_{\gamma}^{\gamma_0} \log f(y - \mu)\, dF(y - \mu)$$

where the maximum likelihood estimator of μ has the convergence rate $n^{-\frac{1}{2}}$ and the estimator of the change-point has the convergence rate n^{-1}, with a limiting distribution given by Theorem 2.8. Under an alternative with distribution function F_θ, the process $X_n(\theta')$ defined by (2.15) has the mean

$$X_\theta(\theta') = \int_{\gamma'}^{\gamma_0} \log \frac{f(y - \mu')}{f(y)}\, dF_\theta(y),$$

it is maximum at θ and its first derivative with respect to the mean is such that $X_{\theta,\mu'}(\theta) = \int_\gamma^{\gamma_0} f'(y - \mu)\, dy = 0$.

Local alternatives $P_{\theta_{n,u}}$ contiguous to the probability P_0 of the hypothesis H_0 with the rates a_n and b_n tending to infinity as n tends to infinity are defined by parameters $\mu_n = a_n^{-1} v_n$, v_n converging to a non-zero limit v, and $\gamma_n = \gamma_0 - b_n^{-1} u_n$, u_n converging to a non-zero limit u, where γ_0 is the upper point of the support of Y.

Proposition 2.7. *Under a fixed alternative P_θ, the statistic T_n tends to infinity. Under local alternatives $P_{\theta_{n,u}}$ such that $a_n b_n = O(n^{-1})$, the statistic T_n converges weakly to $T_0 + vT_1$ with a non-degenerated variable T_1.*

Proof. Under P_θ with a non-zero parameter μ, $\int_\gamma^{\gamma_0} f'(y - \mu)\, dy = 0$, $\widehat{\mu}_n$ converges to μ and it maximizes $X_n(\mu') = \int_\gamma^{\gamma_0} \log f(y - \mu')\, d\widehat{F}_n(y)$. By an expansion of X'_n in a neighborhood of μ, $\widehat{\mu}_n$ satisfies

$$n^{\frac{1}{2}}(\widehat{\mu}_n - \mu) \int_{\widehat{\gamma}_n}^{\gamma_0} \frac{f'^2 - ff''}{f^2}(y - \mu)\, d\widehat{F}_n(y) = \int_{\widehat{\gamma}_n}^{\gamma_0} \frac{f'}{f}(y - \mu)\, d\nu_n(y) + o_p(1)$$

and under the condition $X''(\theta)$ different from zero, $n^{\frac{1}{2}}(\widehat{\mu}_n - \mu)$ converges weakly to a Gaussian variable. For $\widehat{\mu}_n$ in a neighborhood of μ, the process X_n has a second order expansion

$$\begin{aligned}
nX_n(\widehat{\theta}_n) &= n\int_{\widehat{\gamma}_n}^{\gamma_0} \log \frac{f(y-\mu)}{f(y)}\, d\widehat{F}_n(y) + n\int_{\widehat{\gamma}_n}^{\gamma_0} \log \frac{f(y-\widehat{\mu}_n)}{f(y-\mu)}\, d\widehat{F}_n(y) \\
&= n\int_{\widehat{\gamma}_n}^{\gamma_0} \log \frac{f(y-\mu)}{f(y)}\, d\widehat{F}_n(y) \\
&\quad - \frac{1}{2X''(\mu,\widehat{\gamma}_n)}\Big\{\int_{\widehat{\gamma}_n}^{\gamma_0} (f'f^{-1})(y-\mu)\, dW \circ F(y-\mu)\Big\}^2 + o_p(1),
\end{aligned}$$

the first term tends to infinity under P_θ and the second term converges weakly to a χ_1^2 variable up to a multiplicative scalar.

Under a local alternative H_n, the parameters are μ_n and γ_n. The second term in the expansion of $2nX_n(\widehat{\theta}_n)$ is the same under a fixed alternative and it converges weakly to a χ_1^2 variable, the first term is

$$X_{1n}(\mu_n, \widehat{\gamma}_n) = \int_{\widehat{\gamma}_n}^{\gamma_0} \log \frac{f(y-\mu_n)}{f(y)}\, d\widehat{F}_n(y),$$

it is such that $X_{1n}(0, \widehat{\gamma}_n) = 0$ and

$$\begin{aligned}
n^{\frac{1}{2}} X'_{1n,\mu}(0, \widehat{\gamma}_n) &= n^{\frac{1}{2}} \int_{\widehat{\gamma}_n}^{\gamma_0} f'f^{-1}\, d\widehat{F}_n \\
&= \int_{\widehat{\gamma}_n}^{\gamma_0} f'f^{-1}\, d\nu_n + n^{\frac{1}{2}} \int_{\widehat{\gamma}_n}^{\gamma_0} f'f^{-1} f_{\theta_n}.
\end{aligned}$$

The process X_{1n} converges a.s. under P_{θ_n} to the function

$$X_1(\theta) = \int_{\gamma}^{\gamma_0} \log\{f(y-\mu)f^{-1}(y)\}\, dF_{\theta_n}(y)$$

such that $X''_{1,\mu}(\mu_n, \gamma_n) = -\int_{\gamma_n}^{\gamma_0} (f'^2 f^{-1})(y-\mu_n)\, dy$ tends to zero as n tends to infinity, so the inequality (2.13) is no longer true.

For $\widehat{\gamma}_n$ converging to γ_0, $n^{\frac{1}{2}} X'_{1n,\mu}(0, \widehat{\gamma}_n)$ and $X''_{1n,\mu}(0, \widehat{\gamma}_n)$ converge in probability to zero. By an expansion of $X_{1n}(\mu_n, \widehat{\gamma}_n)$

$$a_n X_{1n}(\mu_n, \widehat{\gamma}_n) = \Big\{v_n X'_{1n,\mu}(0, \widehat{\gamma}_n) + \frac{a_n^{-1} v_n^2}{2} X''_{1n,\mu}(0, \widehat{\gamma}_n)\Big\}\{1 + o_p(1)\},$$

and $nX_{1n}(\mu_n, \widehat{\gamma}_n)$ is asymptotically equivalent to $na_n^{-1} v_n X'_{1n,\mu}(0, \widehat{\gamma}_n)$ as n tends to infinity. We have $na_n^{-1}\int_{\widehat{\gamma}_n}^{\gamma_0} f'f^{-1} f_{\theta_n} = na_n^{-1} b_n^{-1} u_n f'(\widetilde{\gamma}_n)$, where $\widetilde{\gamma}_n$ is between $\widehat{\gamma}_n$ and γ_0, its convergence rate to a non-zero limit depends on the convergence rate to zero of f' in a neighborhood of γ_0 and it has a smaller order than $na_n^{-1}\int_{\widehat{\gamma}_n}^{\gamma_0} f'f^{-1}\,d\nu_n$. The variable $nX_{1n}(\mu_n,\widehat{\gamma}_n)$ is therefore asymptotically equivalent to $na_n^{-1}v_n\int_{\widehat{\gamma}_n}^{\gamma_0} f'f^{-1}\,d\nu_n$ which is a $0(na_n^{-1}b_n^{-1})$. Under the condition $a_nb_n = O(n^{-1})$, the limit of T_n under P_{θ_n} is therefore $T_0 + vT_1$ with a non-degenerated variable T_1. □

Under alternatives in narrower neighbourhoods of H_0 than in Proposition 2.7, the test statistic T_n has the same asymptotic distribution as under H_0 and under alternatives in larger neighbourhoods of H_0, T_n diverges.

2.5 Maximum likelihood with an unknown density

In the previous sections, the density f was known up to parameters depending on a change-point, when the density is unknown it can be estimated by a kernel estimator. Let $\mathcal{I}_Y$ be the finite or infinite support of a variable Y with the density f with respect to the Lebesgue measure and, for $h > 0$, let $\mathcal{I}_h = \{s \in \mathcal{I}_Y; [s-h, s+h] \in \mathcal{I}_Y\}$. Under P_0, the density is supposed to be in $C^2(\mathcal{I}_Y)$ except at γ_0 and it follows equation (2.9) with a non-zero difference of the partial means μ_0.

The common distribution function F of the variables Y such that $Y \le \gamma$, and $Y - \mu$ such that $Y > \gamma$ is supposed to be twice continuously differentiable. It has the empirical estimator

$$\begin{aligned}\widehat{F}_n(y) &= n^{-1}\sum_{k=1}^{n} 1_{\{Y_k \le \gamma \wedge y\}} + n^{-1}\sum_{k=1}^{n} 1_{\{\gamma < Y_k \le y+\mu\}} \\ &= 1_{\{y\le\gamma\}}\widehat{F}_{Y,n}(y) + 1_{\{y>\gamma\}}\widehat{F}_{Y,n}(y+\mu), \qquad (2.16)\end{aligned}$$

it is smoothed by a kernel K with a bandwidth $h = h_n$ satisfying the following Conditions C.

Condition 2.1.

C1 K is a symmetric differentiable density such that $|y|^2K(y)$ converges to zero as $|y|$ tends to infinity or K has a compact support with value zero on its frontier, and $K_h(x) = h^{-1}K(h^{-1}x)$;

C2 The moments $m_{2K} = \int u^2K(u)du$, $k_\alpha = \int K^\alpha(u)du$, for $\alpha \ge 0$, and $\int |K'(u)|^\alpha du$, for $\alpha = 1, 2$, are finite;

C3 As n tends to infinity, h_n converges to zero, nh_n tends to infinity and nh_n^4 converges to zero.

The kernel estimator of the parametric density $f_{Y,\gamma}$ of Y under a probability measure P_γ is defined on $\mathcal{I}_{Y,h}$ by smoothing the empirical distribution function $\widehat{F}_n$ of (2.16) with the function $K_h(y) = h^{-1}K(h^{-1}y)$

$$\widehat{f}_{n,h}(y) = \int K_h(y-s)\, d\widehat{F}_n(s)$$
$$= \frac{1}{n}\sum_{k=1}^{n} 1_{\{Y_k \le \gamma\}} K_h(y - Y_k) + \frac{1}{n}\sum_{k=1}^{n} 1_{\{\gamma < Y_k\}} K_h(y + \mu - Y_k).$$

The kernel estimator of the parametric density $f_{Y,\theta}$ of Y under a probability measure P_θ, with parameter $\theta = (\mu, \gamma)^T$, is deduced as

$$\widehat{f}_{Y,nh,\theta}(y) = \widehat{f}_{n,h}(y) 1_{\{y \le \gamma\}} + \widehat{f}_{n,h}(y - \mu) 1_{\{\gamma < y\}}$$
$$= \frac{1}{n}\sum_{k=1}^{n} 1_{\{Y_k \le \gamma\}} K_h(y - Y_k) + \frac{1}{n}\sum_{k=1}^{n} 1_{\{\gamma < Y_k\}} K_h(y - Y_k),$$

it depends only on the change-point parameter γ.

The expectation of $\widehat{f}_{n,h}(y)$ is

$$f_{n,h}(y) = \int K(u)\{f(y - hu)1_{\{y \le \gamma + hu\}} + f(y - hu)1_{\{y > \gamma + hu\}}\}\, du$$
$$= f(y) + 0(h^2),$$

by a second order expansion of the density f, as h tends to zero. At γ_0, the size of the jump of the mean function $f_{Y,n,h}$ is therefore asymptotically equivalent to $f_Y(\gamma_0^+ - \mu_0) - f_Y(\gamma_0)$ and it can be detected by maximum likelihood replacing the unknown density by its estimator.

The bias of $\widehat{f}_{n,h}(y)$ is

$$b_h(y) = \frac{h^2}{4} m_{2K} f^{(2)}(y) + o(h^2),$$

denoted $h^2 b_f(y) + o(h^2)$, and its variance is

$$Var\{\widehat{f}_{n,h}(y)\} = (nh)^{-1} k_2 f(y) + o((nh)^{-1}),$$

also denoted $(nh)^{-1}\sigma_f^2(y) + o((nh)^{-1})$, these approximations are uniform. The last condition of (2.1) is not the optimal convergence rate of the risk in $L^2(P_0)$, it reduces the bias of the kernel estimator which is necessary to provide the same convergence rate of the change-point estimator as in the model with a known density.

If the kernel has a second order derivative, its first two derivatives are estimated by

$$\widehat{f}_{n,h}^{(k)}(y) = \int K_h^{(k)}(y - s)\, \widehat{F}_n(s).$$

Extending the conditions to the integrability condition to a second derivative such that $\int K^{(2)2}$ is finite, the variance of $\widehat{f}^{(k)}_{n,h}(y)$ is a $O((nh^{k+2})^{-1})$ and its bias is a $O(h^2)$ if f belongs to $C^{k+2}(\mathbb{R})$, for $k = 1, 2$, (Pons, 2011).

Proposition 2.8. *Under Conditions 2.1, the estimator $\widehat{f}_{n,h}$ converges a.s. uniformly to f in $I_{Y,h}$, under P_0. The process $(nh)^{\frac{1}{2}}(\widehat{f}_{n,h} - f)$ converges weakly on $I_{Y,h}$ to a centered Gaussian process with variance function σ_f^2 and with covariance function $E\{W_f(y)W_f(y')\} = 1_{\{y=y'\}}\sigma_f^2(y)$.*

The log-likelihood ratio $X_n(\theta)$ of the sample under P_θ and P_0 is estimated by the process

$$\widehat{X}_{n,h}(\theta) = n^{-1}\sum_{k=1}^{n}\Big\{\log\frac{\widehat{f}_{n,h}(Y_k)}{\widehat{f}_{n,h}(Y_k-\mu_0)}1_{]\gamma_0,\gamma]}(Y_k) + \log\frac{\widehat{f}_{n,h}(Y_k-\mu)}{\widehat{f}_{n,h}(Y_k)}1_{]\gamma,\gamma_0]}(Y_k) + \log\frac{\widehat{f}_{n,h}(Y_k-\mu)}{\widehat{f}_{n,h}(Y_k-\mu_0)}1_{\{Y_k>\gamma\vee\gamma_0\}}\Big\},$$

the parameters are estimated by maximization of

$$\widehat{l}_{n,h}(\theta) = \sum_{k=1}^{n}\Big\{\log\widehat{f}_{n,h}(Y_k)1_{\{Y_k\le\gamma\}} + \log\widehat{f}_{n,h}(Y_k-\mu)1_{\{Y_k>\gamma\}}\Big\}, \qquad (2.17)$$

Under P_0, the process $\widehat{X}_{n,h}$ converges a.s. uniformly on $\Theta\setminus\{\theta_0\}$, to the function X limit of the process X_n in Section 2.3

$$\begin{aligned} X(\theta) = 1_{\{\gamma>\gamma_0\}}&\int_{\gamma_0}^{\gamma}\log\frac{f(y)}{f(y-\mu)}\,dF(y-\mu_0) \\ +1_{\{\gamma<\gamma_0\}}&\int_{\gamma}^{\gamma_0}\log\frac{f(y-\mu)}{f(y)}\,dF(y) \\ +&\int_{\gamma_0}^{\infty}\log\frac{f(y-\mu)}{f(y-\mu_0)}\,dF(y-\mu_0). \end{aligned}$$

Theorem 2.9. *The estimator $\widehat{\theta}_{n,h} = \arg\max_\Theta \widehat{X}_{n,h}(\theta)$ of θ_0 is a.s. consistent.*

This is a direct consequence of Proposition 2.8 and of the consistency of the maximum likelihood estimators of Section 2.3.

Let μ be in a neighborhood of μ_0 and let

$$\widehat{\xi}_{n,h}(y,\mu) = \{\widehat{f}_{n,h}(y-\mu) - \widehat{f}_{n,h}(y-\mu_0)\}\widehat{f}^{-1}_{n,h}(y-\mu_0).$$

Approximations of the mean and the variance under P_0 of $\log \widehat{f}_{n,h}(y-\mu) - \log \widehat{f}_{n,h}(y-\mu_0) = \log\{1+\widehat{\xi}_{n,h}(y,\mu)\}$ are deduced from the expansion of the logarithm

$$\log \widehat{f}_{n,h}(y-\mu) - \log \widehat{f}_{n,h}(y-\mu_0) = \widehat{\xi}_{n,h}(y,\mu) - \frac{1}{2}\widehat{\xi}^2_{n,h}(y,\mu) + o_p(\widehat{\xi}^2_{n,h}(y,\mu)).$$

Denoting $\xi_{n,h} = E_0\widehat{\xi}_{n,h}$, the mean of $\log\{1+\widehat{\xi}_{n,h}(y,\mu)\}$ under P_0 is approximated by

$$\xi_{n,h}(y,\mu) - \frac{1}{2}\{E_0\widehat{\xi}^2_{n,h}(y,\mu)\}\{1+o(1)\}$$

and its variance is $Var_0\widehat{\xi}_{n,h}(y,\mu)\{1+o(1)\}$. Let $\widetilde{\xi}_{n,h} = \{f_{n,h}(y-\mu_0) - f_{n,h}(y-\mu)\}f^{-1}_{n,h}(y-\mu_0)$, then

$$\begin{aligned}
\widehat{\xi}_{n,h}(y,\mu) &= \widetilde{\xi}_{n,h}(y,\mu) + \frac{(\widehat{f}_{n,h}-f_{n,h})(y-\mu) - (\widehat{f}_{n,h}-f_{n,h})(y-\mu_0)}{f_{n,h}(y-\mu_0)} \\
&- \frac{\{\widehat{f}_{n,h}(y-\mu) - \widehat{f}_{n,h}(y-\mu_0)\}\{(\widehat{f}_{n,h}-f_{n,h})(y)\}}{f^2_{n,h}(y-\mu_0)} \\
&+ \frac{\widehat{\xi}_{n,h}(y,\mu)\{(\widehat{f}_{n,h}-f_{n,h})^2(y-\mu_0)\}}{f^2_{n,h}(y-\mu_0)} \\
&- \frac{\Delta_{n,h}(\mu,\mu_0)\{(\widehat{f}_{n,h}-f_{n,h})(y-\mu_0)\}}{f^2_{n,h}(y-\mu_0)}
\end{aligned}$$

where $\Delta_{n,h}(\mu,\mu_0) = (\widehat{f}_{n,h}-f_{n,h})(y-\mu) - (\widehat{f}_{n,h}-f_{n,h})(y-\mu_0)$. Under Condition 2.1, the mean of $\widehat{\xi}_{n,h}(y,\mu)$ satisfies $\xi_{n,h}(y,\mu) = \widetilde{\xi}_{n,h}(y,\mu) + O((nh)^{-\frac{1}{2}})$ and its variance is a $O((nh)^{-1})$. It follows that

$$E_0 \log \frac{\widehat{f}_{n,h}(y-\mu)}{\widehat{f}_{n,h}(y-\mu_0)} = \log \frac{f(y-\mu)}{f(y-\mu_0)} + O(h^2)$$

and its bias has the same order as the bias of the kernel estimator, its variance under P_0 is

$$Var \log \frac{\widehat{f}_{n,h}(y-\mu)}{\widehat{f}_{n,h}(y-\mu_0)} = O((nh)^{-1}),$$

it has the same order as the variance of the kernel estimator.

The space $\mathcal{U}_n = \{u = (u_\gamma, u_\mu)^T : u_\gamma = n(\gamma-\gamma_0), u_\mu = n^{\frac{1}{2}}(\mu-\mu_0); \gamma \neq \gamma_0, \mu \neq 0\}$ is provided with the semi-norm $\|u\| = (|u_\gamma| + \|u_\mu\|^2)^{\frac{1}{2}}$. For $u = (u_\gamma, u_\mu)^T$ in $\mathcal{U}_n$, let $\theta_{n,u} = (\gamma_{n,u}, \mu_{n,u})^T$, with $\gamma_{n,u} = \gamma_0 + n^{-1}u_\gamma$ and $\mu_{n,u} = \mu_0 + n^{-\frac{1}{2}}u_\mu$, the norm in $\mathcal{U}_n$ is $\|u\| = n^{\frac{1}{2}}\rho(\theta,\theta_0)$. For every $\varepsilon > 0$, an

ε neighborhood $V_\varepsilon(\theta_0) = \{\theta : \rho(\theta_{n,u}, \theta_0) \leq \varepsilon\}$ of θ_0 defines a neighborhood of zero in $\mathcal{U}_n$ as

$$\mathcal{U}_{n,\varepsilon} = \{u \in \mathcal{U}_n : \rho(\theta_{n,u}, \theta_0) \leq n^{\frac{1}{2}}\varepsilon\}.$$

Let $\widehat{W}_{n,h} = n^{\frac{1}{2}}(\widehat{X}_{n,h} - X)$.

Proposition 2.9. *Under Conditions 2.1 and* $E_0 \sup_{\theta \in V_\varepsilon(\theta_0)} \log f_\theta(Y)$ *finite, for* $\varepsilon > 0$ *sufficiently small there exists a constant* κ_1 *such that*

$$E_0 \sup_{\rho(\theta,\theta_0)\leq\varepsilon} |\widehat{W}_{n,h}(\gamma)| \leq \kappa_1 \varepsilon^{\frac{1}{2}}. \tag{2.18}$$

Proof. The process $\widehat{W}_{n,h}$ splits as a sum of integrals of expressions of $\widehat{f}_{n,h}$ with respect to $\widehat{F}_{0n}$ on the interval $]\gamma, \gamma_0]$ or $]\gamma_0, \gamma]$ where the density of Y is continuous, and an integral of the logarithm of the ratio of $\widehat{f}_{n,h}(y-\mu)$ and $\widehat{f}_{n,h}(y-\mu_0)$, and $E_0 \sup_{V_\varepsilon(\theta_0)} \widehat{W}_{n,h}(\theta)$ is bounded by

$$\int_{\gamma_0}^{\infty} E_0 \sup_{V_\varepsilon(\theta_0)} n^{\frac{1}{2}} \Big\{\log \frac{\widehat{f}_{n,h}(y-\mu)}{\widehat{f}_{n,h}(y-\mu_0)} - \log \frac{f(y-\mu)}{f(y-\mu_0)}\Big\} \, dF_0(y)$$
$$+ \sup_{V_\varepsilon(\theta_0)} \int_{\gamma_0}^{\gamma} n^{\frac{1}{2}} E_0 \Big\{\log \frac{\widehat{f}_{n,h}(y)}{\widehat{f}_{n,h}(y-\mu_0)} - \log \frac{f(y)}{f(y-\mu_0)}\Big\} \, dF_0(y)$$
$$+ \sup_{V_\varepsilon(\theta_0)} \int_{\gamma}^{\gamma_0} n^{\frac{1}{2}} E_0 \Big\{\log \frac{\widehat{f}_{n,h}(y-\mu)}{\widehat{f}_{n,h}(y)} - \log \frac{f(y-\mu)}{f(y)}\Big\} \, dF_0(y).$$

As the differences of the logarithms have the order h^2 and nh^4 is bounded, the last two terms are $O(|\gamma - \gamma_0|) = O(\varepsilon^{\frac{1}{2}})$ and the first term is bounded by $\varepsilon^{\frac{1}{2}} \int_{\gamma_0}^{\infty} n^{\frac{1}{2}} E_0 |\widehat{f}'_{n,h}(y-\mu_0)\widehat{f}^{-1}_{n,h}(y-\mu_0) - f'(y-\mu_0)f^{-1}(y-\mu_0)| \, dF_0(y)$ where the integral is expanded as $\{\int_{\gamma_0}^{\infty} E_0|\widehat{f}'_{n,h}(y-\mu_0) - f'(y-\mu_0)| \, dy\}\{1 + o(1)\}$ by the a.s. convergence of $\widehat{f}_{n,h}$, the order of the bias of $\widehat{f}'_{n,h}$ implies it is bounded. □

The inequalities (2.11) and (2.18) imply that the estimator $\widehat{\gamma}_{n,h}$ has the convergence rate n^{-1} and the estimator $\widehat{\mu}_{n,h}$ has the convergence rate $n^{-\frac{1}{2}}$, the proof of these convergences is identical to the proof of Theorem 2.7.

Theorem 2.10. *Under Conditions 2.1 and* $E_0 \sup_{\theta \in V_\varepsilon(\theta_0)} \log f_\theta(Y)$ *finite*

$$\overline{\lim}_{n,A\to\infty} P_0(n|\widehat{\gamma}_{n,h} - \gamma_0| > A^2, n^{\frac{1}{2}}|\widehat{\mu}_{n,h} - \mu_0| > A,) = 0.$$

The asymptotic distributions of the estimators are the same as the estimators defined by the maximum likelihood with a known density, given by Theorem 2.8.

Theorem 2.11. *Under the conditions of Theorem 2.10, the variables $\widehat{\gamma}_{n,h}$ and $\widehat{\mu}_{n,h}$ are asymptotically independent under P_0, $n^{\frac{1}{2}}(\widehat{\mu}_{n,h}-\mu_0)$ converges weakly to a centered normal variable with variance I_0^{-1} and $n(\widehat{\gamma}_n-\gamma_0)$ is bounded in probability and it converges weakly to the location of the maximum of the Gaussian process S.*

Proof. The proof is similar to the proof of Theorem 2.8 where the estimated density is asymptotically unbiased under the conditions. A first order expansion of $\widehat{X}_{n,h}$ at θ_{nu} in a neighborhood of θ_0, with the convergence rates given by Theorem 2.10, implies that

$$n^{\frac{1}{2}}(\widehat{\mu}_{n,h}-\mu_0)=n^{\frac{1}{2}}\widehat{X}'_{n,h,\mu}(\theta_0)I_0^{-1}+o_p(1)$$

where $n^{\frac{1}{2}}\widehat{X}'_{n,h,\mu}(\theta_0)$ converges weakly to a centered normal variable with variance I_0. By the consistency of the estimators, $\widehat{X}_{n,h}(\widehat{\theta}_{n,h})$ splits as the sum $\widehat{X}_{1n,h}(\widehat{\mu}_{n,h})+\widehat{X}_{2n,h}(\widehat{\gamma}_{n,h})+o_p(n^{-\frac{1}{2}})$, with the processes

$$\widehat{X}_{1n,h}(\mu)=n^{-1}\sum_{k=1}^{n}\log\frac{\widehat{f}_{n,h}(Y_k-\mu)}{\widehat{f}_{n,h}(Y_k-\mu_0)}1_{\{Y_k>\gamma_0\}},$$

$$\widehat{X}_{2n,h}(\gamma)=n^{-1}\sum_{k=1}^{n}\Big\{\log\frac{\widehat{f}_{n,h}(Y_k)}{\widehat{f}_{n,h}(Y_k-\mu_0)}1_{]\gamma_0,\gamma]}(Y_k)$$
$$+\log\frac{\widehat{f}_{n,h}(Y_k-\mu_0)}{\widehat{f}_{n,h}(Y_k)}1_{]\gamma,\gamma_0]}(Y_k)\Big\}.$$

Let $\gamma_{n,u}=\gamma_0+n^{-1}u$, by Theorem 2.10 the variable $n(\widehat{\gamma}_{n,h}-\gamma_0)$ maximizes $n\widehat{X}_{2n,h}(\gamma_{n,u})$ with respect to u in a set bounded in probability. This process converges weakly under P_0 to the Gaussian process S of Theorem 2.8. The variable $n(\widehat{\gamma}_{n,h}-\gamma_0)$ converges weakly to the location of the maximum of this process, it is bounded in probability by Theorem 2.10. The covariance of the processes $\widehat{X}_{1n,h}$ and $\widehat{X}_{2n,h}$ converge to zero and this entails the asymptotic independence of the estimators. □

A nonparametric likelihood ratio test of the hypothesis H_0 of a constant mean against the alternative of a change in the mean at an unknown threshold γ is performed like in Section 2.4 with the log-likelihood ratio statistic

$$\widehat{T}_n=2n\sup_{\theta\in\Theta}\widehat{X}_{n,h}(\theta)$$

with the estimated densities. Under H_0 and Conditions 2.1, with a density f in C_b^2, the estimator $\widehat{\theta}_n$ converges a.s. to θ_0 and γ_0 is the upper bound of the support of the variable Y, by the same arguments as in Section 2.4 where the unknown density f and its derivatives are replaced by the kernel estimator $\widehat{f}_{n,h}$ and its derivatives.

There exists a sequence $(a_n)_n$ such that $a_n^{-1}n^{\frac{1}{2}}\widehat{X}'_{n,h,\mu}(\mu_0, \widehat{\gamma}_{n,h})$ converges weakly to a centered Gaussian variable and $a_n^{-2}\widehat{X}''_{n,h,\mu}(\mu_0, \widehat{\gamma}_{n,h})$ is asymptotically equivalent to the variance of $a_n^{-1}n^{\frac{1}{2}}\widehat{X}'_{n,h,\mu}(\mu_0, \widehat{\gamma}_{n,h})$, and the integral $\int_{\gamma_0}^{\infty} \widehat{f}''_{n,h}(y-\mu_0)\widehat{f}^{-1}_{n,h}(y-\mu_0)\,dF(y-\mu_0)$ converges a.s. to zero. By the inequality (2.18) and Theorem 2.10, the estimator $\widehat{\mu}_{n,h}$ is such that

$$a_n^{-1}n^{\frac{1}{2}}(\widehat{\mu}_{n,h} - \mu_0) = \widehat{X}''^{-1}_{n,h,\mu}(\mu_n, \widehat{\gamma}_{n,h})n^{\frac{1}{2}}\widehat{X}'_{n,h,\mu}(\theta_0),$$

where μ_n is between $\widehat{\mu}_{n,h}$ and μ_0, then the variable $a_n^{-1}n^{\frac{1}{2}}(\widehat{\mu}_{n,h} - \mu_0)$ converges weakly under H_0 to a centered Gaussian variable with a finite variance $I(\theta)$.

Proposition 2.10. *Under H_0 and Condition 2.1, for a density f in C_b^2, the statistic T_n converges weakly to a χ_1^2 variable T_0.*

Proof. The asymptotic expansion of $\widehat{X}'_{n,h,\mu}(\widehat{\theta}_{n,h})$ is similar to the expansion of $X'_{n,\mu}(\widehat{\theta}_n)$ in Proposition 2.6, by the a.s. consistency of $\widehat{f}_{n,h}$ and its derivatives

$$\int_{\widehat{\gamma}_{n,h}}^{\infty} \frac{\widehat{f}'^2_{n,h} - \widehat{f}_{n,h}\widehat{f}''_{n,h}}{\widehat{f}^2_{n,h}}(y-\mu_0)\,d\widehat{F}_n(y) = \int_{\widehat{\gamma}_{n,h}}^{\infty} \frac{f'^2}{f}(y)(y-\mu_0)\,dy + o_p(a_n^2).$$

Then the variable $n^{\frac{1}{2}}a_n^{-1}(\widehat{\mu}_n - \mu_0)$ converges weakly to a non-degenerated limit under P_0 and a second order expansion of the logarithm of $\widehat{f}_{n,h}(y - \widehat{\mu}_{n,h})$ in the expression of $\widehat{T}_{n,h}$ implies

$$\widehat{T}_{n,h} = -\frac{1}{\widehat{X}''_{n,h,\mu}(\mu_0, \widehat{\gamma}_{n,h})}\Big\{\int_{\widehat{\gamma}_{n,h}}^{\infty} (\widehat{f}'_{n,h}\widehat{f}^{-1}_{n,h})(y-\mu_0)\,d\nu_n(y)\Big\}^2 + o_p(1).$$

The limit of $\widehat{T}_{n,h}$ follows from the weak convergence of the variables $a_n\widehat{X}_{n,h,\mu}(\mu_0, \widehat{\gamma}_{n,h})$ and $a_n^2\widehat{X}''_{n,h,\mu}(\mu_0, \widehat{\gamma}_{n,h})$. □

Under the alternative P_θ of a difference of means μ at an unknown γ, the process $\widehat{X}_{n,h}$ has an expansion similar to the expansion of X_n in Section 2.4 and the limiting distribution of $\widehat{T}_n$ is the same as T_n in Proposition 2.7. Under a fixed alternative P_θ, with distribution function F_θ, the process $\widehat{X}_{n,h}$ converges a.s. to the function X which is maximum at θ. Local alternatives to H_0 are defined as in Proposition 2.7.

Proposition 2.11. *Under fixed alternatives P_θ, the statistic T_n tends to infinity. Under local alternatives $P_{\theta_{n,u}}$ contiguous to H_0, the statistic T_n converges weakly to $T_0 + vT_1$ with a non-degenerated variable T_1.*

Proof. Under P_θ, the process $\widehat{X}_{n,h}$ has a second order expansion

$$\begin{aligned}
\widehat{X}_{n,h}(\widehat{\theta}_{n,h}) &= \int_{\widehat{\gamma}_{n,h}}^{\infty} \log \frac{\widehat{f}_{n,h}(y-\mu)}{\widehat{f}_{n,h}(y)} \, d\widehat{F}_{Y,n}(y) \\
&\quad + \int_{\widehat{\gamma}_{n,h}}^{\infty} \log \frac{\widehat{f}_{n,h}(y-\widehat{\mu}_{n,h})}{\widehat{f}_{n,h}(y-\mu)} \, d\widehat{F}_{Y,n}(y) \\
&= \widehat{X}_{n,h}(\theta) + n \int_{\widehat{\gamma}_{n,h}}^{\gamma} \log \frac{\widehat{f}_{n,h}(y-\mu)}{\widehat{f}_{n,h}(y)} \, d\widehat{F}_{Y,n}(y) \\
&\quad + (\widehat{\mu}_{n,h} - \mu) \int_{\widehat{\gamma}_{n,h}}^{\infty} \frac{\widehat{f}'_{n,h}}{\widehat{f}_{n,h}} (y-\mu) \, d\widehat{F}_{Y,n}(y) \\
&\quad + \frac{n}{2} (\widehat{\mu}_{n,h} - \mu)^2 \widehat{X}''_{n,h,\mu}(\theta_n)
\end{aligned}$$

with $\theta_n = (\mu_n, \widehat{\gamma}_{n,h})$ and μ_n between $\widehat{\mu}_{n,h}$ and μ. We have $\widehat{X}'_{n,h}(\widehat{\theta}_{n,h}) = 0$ and

$$n^{\frac{1}{2}} \{ \widehat{X}'_{n,h,\mu}(\widehat{\theta}_{n,h}) - \widehat{X}'_{n,h,\mu}(\mu, \widehat{\gamma}_{n,h}) \} = (\widehat{\mu}_{n,h} - \mu) \widehat{X}''_{n,h,\mu}(\theta_n),$$

then the variable $n\widehat{X}_{n,h}(\widehat{\theta}_{n,h})$ is approximated by

$$nX(\theta) - \frac{1}{2I(\theta)} \Big\{ \int_\gamma^\infty (f' f^{-1})(y-\mu) \, dW \circ F(y-\mu) \Big\}^2 + o_p(1)$$

therefore $\widehat{T}_n$ diverges.

Under a contiguous alternative with parameters $\mu_n = \mu_0 + a_n^{-1} v_n$, v_n converging to a non-zero limit v and γ_n tending to infinity, the first term of the previous expansion is the process $n\widehat{X}_{1n,h}(\mu_n, \widehat{\gamma}_n) = n \int_{\widehat{\gamma}_n}^\infty \{\log \widehat{f}_{n,h}(y - \mu_n) - \log \widehat{f}_{n,h}(y)\} \, d\widehat{F}_{Y,n}(y)$, it is approximated from the expansion of its derivative according to μ_n by

$$-\frac{1}{2} \widehat{X}''_{n,h,\mu}(\theta) \Big\{ \int_{\widehat{\gamma}_n}^\infty (\widehat{f}'_{n,h} \widehat{f}^{-1}_{n,h})(y - \mu_n) \, d\nu_n(y) \Big\}^2,$$

and it converges weakly under $P_{\theta_{n,u}}$ to a χ_1^2 variable. The second term $n\{\widehat{X}_{n,h}(\widehat{\theta}_n) - \widehat{X}_{n,h}(\mu_n, \widehat{\gamma}_n)\}$ has an asymptotic expansion similar to the asymptotic expansion of $X_{1n}(\mu_n, \widehat{\gamma}_n)$ in Proposition 2.11 and it has the same asymptotic distribution under P_{θ_n}. □

Under alternatives in narrower neighbourhoods of H_0 than in Proposition 2.11, the test statistic T_n has the same asymptotic distribution in Proposition 2.10, and under alternatives in larger neighbourhoods of H_0, T_n diverges.

2.6 Chronological change in the mean

We now consider a change in the mean of a random variable Y related to a variation at a sampling index. In a probability space $(\Omega, \mathcal{A}, P_0)$, a sample of the variable Y splits into two independent subsamples of independent observations of Y such that for $k = 1, \ldots, n$

$$E_0 Y_k = \begin{cases} \mu_1 & \text{if } k \leq k_0, \\ \mu_2 & \text{if } k > k_0, \end{cases}$$

where the change-point occurs at the index

$$k_0 = \min\{k > 1 : EY_k \neq E_0 Y_{k-1}\}$$

such that $k_0 < n$, then $n^{-1}k_0$ is a real in $]0,1[$. Under a probability P, the variables $Y_k - EY_k$ are identically distributed with a continuous density f, for $k = 1, \ldots, n$. We denote

$$\pi_{0n} = n^{-1}k_0$$

and k_0 is the integer part of nt_0, for a real t_0 in $]0,1[$. For a real t in $]0,1[$, k denotes the integer part of nt in $\{2, \ldots, n-1\}$, and for an integer k in $\{2, \ldots, n-1\}$, t is the limit of $\pi_{nt} = n^{-1}k$ as n tends to infinity.

The density of an observation Y_k under P_0 is $g_0(y) = f(y - \mu_{01}) 1_{\{k \leq k_0\}} + f(y - \mu_{02}) 1_{\{k > k_0\}}$ and the empirical distribution of the sample is

$$\widehat{G}_{0n}(y) = \pi_{0n} \widehat{F}_{1,k_0}(y) + (1 - \pi_{nt}) \widehat{F}_{2,n-k_0}(y)$$

where

$$\widehat{F}_{1,k_0}(y) = k_0^{-1} \sum_{i=1}^{k_0} 1_{\{Y_i \leq y\}}, \quad \widehat{F}_{2,n-k_0}(y) = (n - k_0)^{-1} \sum_{i=k_0+1}^{n} 1_{\{Y_i \leq y\}}.$$

Under the hypothesis of a change-point at an arbitrary $k = [nt]$ with t in $]0,1[$, the empirical distribution is

$$\widehat{G}_{nt}(y) = n^{-1}[nt] \widehat{F}_{1,[nt]}(y) + n^{-1}(n - [nt]) \widehat{F}_{2,n-[nt]}(y),$$

the means under P are the parameters $\mu_{1t} = E(Y_i)$, for $i \leq [nt]$, and $\mu_{2t} = E(Y_i)$, for $i > [nt]$, they are estimated by the empirical means

$$\widehat{\mu}_{1n,t} = [nt]^{-1} \sum_{i=1}^{[nt]} Y_i,$$

$$\widehat{\mu}_{2n,t} = (n - [nt])^{-1} \sum_{i=[nt]+1}^{n} Y_i.$$

Under P_0, the bias of the estimators at t is

$$E_0(\widehat{\mu}_{1n,t} - \widehat{\mu}_{1n,t_0}) = \Big(1 - \frac{[nt_0]}{[nt]}\Big)(\mu_{01} - \mu_{02})1_{\{t_0<t\}},$$

$$E_0(\widehat{\mu}_{2n,t} - \widehat{\mu}_{2n,t_0}) = \frac{[nt_0] - [nt]}{n - [nt]}(\mu_{01} - \mu_{02})1_{\{t<t_0\}}.$$

The variance of the sample with a change of mean at t is a function of θ_t, the vector with components μ_{1t}, μ_{2t} and t. The variance of the sample is

$$\sigma^2_{n,\theta_t} = n^{-1}\Big\{\sum_{i=1}^{[nt]}(Y_i - \mu_{1t})^2 + \sum_{i=[nt]+1}^{n}(Y_i - \mu_{2t})^2\Big\},$$

it is also denoted $\sigma^2_{n,\theta_t} = \pi_{nt}\sigma^2_{1n,\mu_{1t}} + (1-\pi_{nt})\sigma^2_{2n,\mu_{2t}}$ with

$$\sigma^2_{1n,\mu_{1t}} = [nt]^{-1}\sum_{i=1}^{[nt]}(Y_i - \widehat{\mu}_{1n,t})^2,$$

$$\sigma^2_{2n,\mu_{2t}} = (n - [nt])^{-1}\sum_{i=[nt]+1}^{n}(Y_i - \widehat{\mu}_{2n,t})^2.$$

Under P_0 and if $E_0(Y^2)$ is finite, σ^2_{n,θ_t} has the mean $\sigma^2_{\theta_t} = t\sigma^2_{1t} + (1-t)\sigma^2_{2t}$ where $\sigma^2_{1t} = E_0\{(Y_i - \mu_{1t})^2\}$ for $i \le k_0$, and $\sigma^2_{2t} = E_0\{(Y_i - \mu_{2t})^2\}$ for $i > k_0$. The quadratic error of estimation with a change-point at $[nt]$ is

$$\widehat{\sigma}^2_{n,t} = n^{-1}\Big\{\sum_{i=1}^{[nt]}(Y_i - \widehat{\mu}_{1n,t})^2 + \sum_{i=[nt]+1}^{n}(Y_i - \widehat{\mu}_{2n,t})^2\Big\},$$

it is denoted as $\widehat{\sigma}^2_{n,t} = t\widehat{\sigma}^2_{1n,t} + (1-t)\widehat{\sigma}^2_{2n,t}$, and the unknown change-point parameter t is estimated by minimization of the empirical variance

$$\widehat{t}_n = \arg\inf_t \widehat{\sigma}^2_{n,t}.$$

Then the empirical estimators of the means are $\widehat{\mu}_{in} = \widehat{\mu}_{in,\widehat{t}_n}$, for $i = 1, 2$, and the variance is estimated by $\widehat{\sigma}^2_n = \widehat{\sigma}^2_{n,\widehat{t}_n}$.

For all t and t_0 in $]0,1[$, the process

$$X_n(t) = \widehat{\sigma}^2_{n,t} - \widehat{\sigma}^2_{n,t_0} \tag{2.19}$$

develops as

$$X_n(t) = (-1)^{1_{\{t<t_0\}}}n^{-1}\Big\{\sum_{i=[nt_0]+1}^{[nt]}(Y_i - \widehat{\mu}_{1n,t_0})^2 - \sum_{i=[nt_0]+1}^{[nt]}(Y_i - \widehat{\mu}_{2n,t_0})^2\Big\}$$
$$-\pi_{nt}(\widehat{\mu}_{1n,t} - \widehat{\mu}_{1n,t_0})^2 - (1-\pi_{nt})(\widehat{\mu}_{2n,t} - \widehat{\mu}_{2n,t_0})^2, \tag{2.20}$$

where the sums are from $[nt_0]$ to $[nt]$ and from $[nt_0]+1$ to $[nt]$ if $t_0 < t$ or from $[nt]$ to $[nt_0]$ and from $[nt]+1$ to $[nt_0]$ if $t_0 > t$, and $\pi_0 = \pi_{t_0}$. From (2.20), we have

$$X_n(t) = (-1)^{1_{\{t<t_0\}}}(\widehat{\mu}_{1n,t_0} - \widehat{\mu}_{2n,t_0})\Big\{(\pi_{nt} - \pi_{n0})\widehat{\mu}_n - 2n^{-1}\sum_{i=[nt_0]+1}^{[nt]} Y_i\Big\}$$
$$-\pi_{nt}(\widehat{\mu}_{1n,t} - \widehat{\mu}_{1n,t_0})^2 - (1-\pi_{nt})(\widehat{\mu}_{2n,t} - \widehat{\mu}_{2n,t_0})^2, \tag{2.21}$$

where

$$n^{-1}\sum_{i=[nt_0]+1}^{[nt]} Y_i = \pi_{nt}\widehat{\mu}_{1n,t} - \pi_{n0}\widehat{\mu}_{1n,t_0},$$
$$n^{-1}\sum_{i=[nt]+1}^{[nt_0]} Y_i = \widehat{\mu}_{2n,t} - \widehat{\mu}_{2n,t_0} - (\pi_{nt}\widehat{\mu}_{2n,t} - \pi_{n0}\widehat{\mu}_{2n,t_0}).$$

Under P_0, the mean of $\widehat{\mu}_{1n,t}$ is $\mu_{1t} = \mu_{01} - [nt]^{-1}([nt] - [nt_0])(\mu_{01} - \mu_{02})1_{\{t>t_0\}}$, it is asymptotically equivalent to $\mu_{01} - t^{-1}(t-t_0)(\mu_{01} - \mu_{02})1_{\{t>t_0\}}$ and $\pi_{nt} - \pi_{n0}$ is asymptotically equivalent to $t - t_0$, therefore $n^{-1}\sum_{i=[nt_0]+1}^{[nt]} Y_i$ is asymptotically equivalent to

$$\mu_{01}(t-t_0) - t^{-1}(t-t_0)(\mu_{01} - \mu_{02})1_{\{t>t_0\}} = 0(|t-t_0|),$$

in the same way, $n^{-1}\sum_{i=[nt]+1}^{[nt_0]} Y_i = 0(|t-t_0|)$, as n tends to infinity.

Under P_0, $X_n(t)$ converges a.s. uniformly on every interval $]a,b[$, with $a > 0$ and $b < 1$, to the function

$$X(t) = \sigma_t^2 - \sigma_0^2. \tag{2.22}$$

The function X being minimum at t_0, the estimators $\widehat{t}_n$ and therefore $\widehat{\theta}_n$ are a.s. consistent. The mean of $X_n(t)$ is $E_0 X_n(t) = X(t) + o(1)$. For ε sufficiently small to ensure that a neighborhood $V_\varepsilon(t_0)$ belongs to an interval $]a,b[$, with $a > 0$ and $b < 1$, the expansion of X_n implies the existence of a constant κ_0 such that

$$\inf_{|t-t_0|\le\varepsilon} X(t)| \ge \kappa_0|t-t_0|. \tag{2.23}$$

Let

$$W_n(t) = n^{\frac{1}{2}}\{X_n(t) - X(t)\},$$

if $E_0 Y^4$ is finite, the process $W_n(t)$ converges weakly to a centered Gaussian process with a finite variance function $v_t = O(|t - t_0|)$ then for every $\varepsilon > 0$, there exists a constant κ_1 such that for n sufficiently large

$$E_0 \sup_{|t-t_0|\le\varepsilon} |W_n(t)| \le \{E_0 \sup_{|t-t_0|\le\varepsilon} |W_n(t)|^2\}^{\frac{1}{2}}$$
$$\le \kappa_1 |t - t_0|^{\frac{1}{2}}. \quad (2.24)$$

By the same arguments as in Section 2.1, the inequalities (2.23) and (2.24) imply

Theorem 2.12. *If $E_0 Y^4$ is finite*

$$\overline{\lim}_{n,A\to\infty} P_0(n|\widehat{t}_n - t_0| > A^2, n^{\frac{1}{2}}|\widehat{\mu}_n - \mu| > A) = 0.$$

Let $t_{n,u} = t_0 + n^{-1}u$, u in $\mathcal{U}_n^A$. The process W_n is defined by integrals with respect to the empirical processes $\nu_{1,[nt]}$ and $\nu_{2,n-[nt]}$, given in (1.4), for the partial sums of the variables Y_i according to $i \le k_0$ and $i > k_0$, it converges weakly to a Gaussian process depending on integrals of the composed Brownian motion $\nu_{1,t}$ and $\nu_{2,t}$ of Proposition 1.2.

By Theorem 2.12, the process $W_n(t_{n,u})$ converges weakly in $\mathcal{U}_n^A$ to a Gaussian process with variance $v(u) = O(|u|)$, as n and A tend to infinity. The asymptotic distribution of the variable $n(\widehat{t}_n - t_0)$ is deduced from the limiting distribution of the process X_n in the bounded set $\mathcal{U}_n^A$, as n and A tend to infinity.

Theorem 2.13. *If $E_0 Y^4$ is finite, the variable $n(\widehat{t}_n - t_0)$ converges weakly to the location u_0 of the minimum of an uncentered Gaussian process, it is bounded in probability.*

Proof. From (2.21), the mean of the process $X_n(t)$ is asymptotically a $0(|t - t_0|)$. At $t_{n,u} = t_0 + n^{-1}u$, the mean of $nX_n(t)$ and its variance converge to a finite limit $O|u|)$. By the central limit theorem for the empirical process, the process $nX_n(t_{n,u})$ converges weakly to an uncentered Gaussian process and it is minimum with respect to u is $\widehat{u}_n = n(\widehat{t}_n - t_0)$ which converges weakly to the location of the minimum of the limiting process, $\widehat{u}_n$ is bounded in probability by Theorem 2.12. □

Theorem 2.13 entails the weak convergence of the random variable $n^{\frac{1}{2}}(\widehat{\mu}_n - \mu_0)$ to a centered Gaussian variable with independent components and with marginal variances σ_j^2, for $j = 1, 2$.

A mean square error test of the hypothesis H_0 of a constant mean against the alternative of a change in the mean at an unknown threshold index of the sample is performed like in Section 2.2 with the difference of the process X_n at the estimators of the parameters under the alternative and the hypothesis

$$T_n = \inf_{t \in]0,1]} n(\widehat{\sigma}^2_{n,t} - \widehat{\sigma}^2_n) = n(\widehat{\sigma}^2_{n,\widehat{t}_n} - \widehat{\sigma}^2_n)$$

where $\widehat{\sigma}^2_n$ is the empirical variance of the sample under H_0 and $t_0 = 1$ under H_0. Let u_0 be the location of the minimum of the Gaussian process limit of the variable $n(1 - \widehat{t}_n)$ given by Theorem 2.13.

Proposition 2.12. *If $E_0 Y^4$ is finite, the statistic T_n converges weakly under H_0 to $T_0 = -\mu_0^2([u_0] + 1)$.*

Proof. Under H_0, the process $X_n(t) = \widehat{\sigma}^2_{n,t} - \widehat{\sigma}^2_n$ converges a.s. uniformly under P_0 to the function

$$\begin{aligned} X(t) = tE_0\{(Y_i - \mu_{1t})^2\} \\ +(1-t)E_0\{(Y_i - \mu_{2t})^2\} - E_0\{(Y - \mu_0)^2\}, \end{aligned}$$

this limit is minimum as the parameters are $t_0 = 1$, $\mu_{01} = \mu_0$ and $\mu_{02} = 0$, according to $k_0 = n$. Then the estimator $\widehat{\theta}_n$ converges a.s. to $\theta_0 = (\mu_0, 0, 1)^T$. Under H_0, for every t in $]0,1[$, $X_n(t)$ reduces to

$$X_n(t) = \widehat{\mu}^2_n - \widehat{\mu}^2_{1n,t} + (1 - \pi_{nt})(\widehat{\mu}^2_{2n,t} - \widehat{\mu}^2_{1n,t}) - 2\widehat{\mu}^2_{2n,t}.$$

At $\widehat{t}_n$, let $\widehat{\pi}_n = \pi_{n\widehat{t}_n}$. Under H_0, $\widehat{F}_{1,[n\widehat{t}_n]}$ converge in probability to F_0 and $\widehat{F}_{2,n-[n\widehat{t}_n]}$ converges in probability to zero. The variable

$$n(1 - \widehat{\pi}_n) = n - [n\widehat{t}_n] = [n(1 - \widehat{t}_n)] + 1$$

converges weakly to $[u_0] + 1$ where u_0 is the limit of $n(1 - \widehat{t}_n)$ defined by Theorem 2.13. As $(n - [nt])\widehat{\mu}_{2n}$ is asymptotically equivalent to a finite sum of variables, $\widehat{\mu}_{2n}$ and $n^{\frac{1}{2}}\widehat{\mu}_{2n}$ converge to zero in probability. The empirical mean of the sample is $\widehat{\mu}_n = \widehat{\pi}_n\widehat{\mu}_{1n} + (1 - \widehat{\pi}_n)\widehat{\mu}_{2n}$, it converges a.s. to μ_0 and the differences of the empirical means at t are

$$\widehat{\mu}_n - \widehat{\mu}_{1n,t} = (1 - \pi_{nt})(\widehat{\mu}_{2n,t} - \widehat{\mu}_{1n,t}), \quad \widehat{\mu}_n - \widehat{\mu}_{2n,t} = \pi_{nt}(\widehat{\mu}_{1n,t} - \widehat{\mu}_{2n,t})$$

which yields

$$\begin{aligned} nX_n(\widehat{t}_n) &= n(1 - \widehat{\pi}_n)(\widehat{\mu}_{2n} - \widehat{\mu}_{1n})(\widehat{\mu}_{1n} + \widehat{\mu}_n) \\ &\quad -n(1 - \widehat{\pi}_n)(\widehat{\mu}^2_{1n} - \widehat{\mu}^2_{2n}) - 2n\widehat{\mu}^2_{2n} \\ &= -n(1 - \widehat{\pi}_n)\mu_0^2 + o_p(1), \end{aligned}$$

the limit of T_n follows. □

Under the alternative K of a finite change point at t between $Y_{n:1}$ and $Y_{n:n} < t_0$, $\bar{Y}_n$ converges under P_t to $\mu_t = t\mu_{1t} + (1-t)\mu_{2t}$, the limit of $n(\widehat{t}_n - t)$ under P_t is given by Theorem 2.13 and the two-dimensional variable $n^{\frac{1}{2}}(\widehat{\mu}_n - \mu)$ converges weakly to centered Gaussian variable with variance (σ_1^2, σ_2^2).

Proposition 2.13. *Under fixed alternatives, the statistic T_n tends to infinity as n tends to infinity. Under local alternatives P_{θ_n} contiguous to H_0, with $t_n = 1 - n^{-1}v_{0n}$ and v_{0n} converging to v_0, the statistic T_n converges weakly to the variable $T_0 + [v_0]\mu_0^2$ as n tends to infinity.*

Proof. Under P_θ, with $\theta = (\mu_1, \mu_2, t)^T$ and t in $]0,1[$, the process $X_n(t', t) = \widehat{\sigma}^2_{n,t'} - \widehat{\sigma}^2_{n,t}$ has the expansion

$$X_n(t', t) = \pi_{nt}\widehat{\mu}^2_{1n,t} - \pi_{nt'}\widehat{\mu}^2_{1n,t'} + (1-\pi_{nt})\widehat{\mu}^2_{2n,t} - (1-\pi_{nt'})\widehat{\mu}^2_{1n,t'}$$

and it converges a.s. uniformly to the function $X = E_\theta X_n$ minimum as $t' = t$, so $X_n(t_0, t)$ is strictly positive, the statistic T_n is not centered and it tends to infinity as n tends to infinity.
Local alternatives P_{θ_n} have parameters $\theta_n = (\mu_{1n}, \mu_{2n}, t_n)^T$ converging to θ_0 with $t_n = 1 - n^{-1}v_{0n}$ and means $\mu_{1n} = \mu_0 + n^{-\frac{1}{2}}v_{1n}$ and $\mu_{2n} = n^{-\frac{1}{2}}v_{2n}$ such that the sequence $(v_n)_n$ converges to a vector v with non-zero components, and μ_2 differs from μ_0. The mean of Y under P_{θ_n} is $\mu_n = t_n\mu_{1n} + (1-t_n)\mu_{2n} = \mu_0 + n^{-\frac{1}{2}}v_{1n} + o(n^{-\frac{1}{2}})$ and $\pi_n = n^{-1}[nt_n] = 1 - n^{-1}[v_{0n}]$. At the estimated value, the test statistic is asymptotically equivalent to $T_0 + [v_{0n}]\mu_0^2$. □

2.7 Maximum likelihood for a chronological change

The log-likelihood of the sample with a change of mean at t in $]0,1[$ is

$$l_n(\theta) = \sum_{i=1}^{[nt]} \log f(Y_i - \mu_1) + \sum_{i=[nt]+1}^{n} \log f(Y_i - \mu_2), \tag{2.25}$$

at $\theta = (\mu, t)$. The maximum likelihood estimators of the means $\widehat{\mu}_{1n,t}$ and $\widehat{\mu}_{2n,t}$, at an arbitrary t, are solutions of the estimating equations

$$\sum_{i=1}^{[nt]} \frac{f'}{f}(Y_i - \mu_1) = 0, \qquad \sum_{i=[nt]+1}^{n} \frac{f'}{f}(Y_i - \mu_2) = 0$$

and

$$\widehat{t}_n = \arg\max_t l_n(\widehat{\mu}_{nt}, t).$$

Tho process

$$X_n(\theta) = n^{-1}\{l_n(\theta) - l_n(\theta_0)\}$$

is the sum

$$X_n(\theta) = n^{-1} \sum_{i=1}^{[nt]\wedge[nt_0]} \log \frac{f(Y_i - \mu_1)}{f(Y_i - \mu_{01})} + n^{-1} \sum_{i=[nt]\vee[nt_0]+1}^{n} \log \frac{f(Y_i - \mu_2)}{f(Y_i - \mu_{02})}$$

$$+1_{\{t<t_0\}} n^{-1} \sum_{i=[nt]+1}^{[nt_0]} \log \frac{f(Y_i - \mu_2)}{f(Y_i - \mu_{01})}$$

$$+1_{\{t>t_0\}} n^{-1} \sum_{i=[nt_0]+1}^{[nt]} \log \frac{f(Y_i - \mu_1)}{f(Y_i - \mu_{02})},$$

it converges a.s. uniformly under P_0 to

$$X(\theta) = (t \wedge t_0) E_0 \Big\{ \log \frac{f(Y_i - \mu_1)}{f(Y_i - \mu_{01})} 1_{\{i \le k \wedge k_0\}} \Big\}$$

$$+(1 - t \vee t_0) E_0 \Big\{ \log \frac{f(Y_i - \mu_2)}{f(Y_i - \mu_{02})} 1_{\{i > k \vee k_0\}} \Big\}$$

$$+1_{\{k<k_0\}} (t_0 - t) E_0 \Big\{ \log \frac{f(Y_i - \mu_2)}{f(Y_i - \mu_{01})} 1_{\{k < i \le k_0\}} \Big\}$$

$$+1_{\{k>k_0\}} (t - t_0) E_0 \Big\{ \log \frac{f(Y_i - \mu_1)}{f(Y_i - \mu_{02})} 1_{\{k_0 < i \le k\}} \Big\},$$

where $k = [nt]$ and $t = \lim_{n\to\infty} n^{-1}[nt]$.

Theorem 2.14. *If f is C_b^2 in an interval including the means μ_{01} and μ_{02}, the maximum likelihood estimator $\widehat{\gamma}_n$ of γ_0 is a.s. consistent.*

Proof. The function X is such that $X'_\mu(\theta_0) = 0$ and $-X''_\mu(\theta_0)$ is a definite positive matrix, the function X is concave in an interval including μ_{01} and μ_{02}. By the a.s. uniform convergence of X_n in an interval including the means and their estimators, $\widehat{\theta}_n$ is a.s. consistent under P_0. □

Let $U_n(\theta) = n^{-\frac{1}{2}} l'_n(\theta)$ and $I_n(\theta) = -n^{-1} l''_{n,\mu}(\theta)$, $I_n(\theta_0)$ converges a.s. under P_0 to a positive definite matrix I_0 and $U_n(\theta_0)$ converges weakly to a centered Gaussian variable with variance I_0, according to the classical theory of the maximum likelihood estimation

$$n^{\frac{1}{2}}(\widehat{\mu}_{nt_0} - \mu_0) = I_0^{-1} U_n(\theta_0) + o_p(1)$$

and it converges weakly to a centered Gaussian variable with variance I_0^{-1}. Under P_0 and at an arbitrary t, $\widehat{\mu}_{n,t}$ converges a.s. uniformly on $]0,1[$ to its mean $\mu_{0,t}$ and $n^{\frac{1}{2}}(\widehat{\mu}_{nt} - \mu_{0,t}) = I_{0,t}^{-1}U_n(\mu_t, t) + o_p(1)$, where $I_{0,t}$ is the limit of $-n^{-1}l''_{n,\mu}(\mu_t, t)$ under P_0, and it converges weakly to a centered Gaussian variable with variance $I_{0,t}^{-1}$.

By a second order expansion, since $X'(\theta_0) = 0$ and X is a concave function, the function X is negative in a neighborhood of θ_0 and for ε sufficiently small there exists a constant $\kappa_0 > 0$ such that

$$\sup_{\rho(\theta,\theta_0)<\varepsilon} X(\theta) \geq -\kappa_0\varepsilon. \tag{2.26}$$

Under the condition $E_0 \sup_{\mu\in V_\varepsilon(\mu_0)} \log^2 f(Y-\mu)$ finite and by monotonicity with respect to t of μ_t and the sums defining the process X_n, the process

$$W_n = n^{\frac{1}{2}}(X_n - X)$$

has a bounded variance then for every $\varepsilon > 0$ there exists a constant $\kappa_1 > 0$ such that and for n sufficiently large

$$E_0 \sup_{\rho(\theta,\theta_0)\leq\varepsilon} W_n^2(\theta) \leq \kappa_1^2\varepsilon,$$

$$E_0 \sup_{\rho(\theta,\theta_0)\leq\varepsilon} |W_n(\theta)| \leq \kappa_1\varepsilon^{\frac{1}{2}}, \tag{2.27}$$

by a first order expansion of $\log f(y-\mu) - \log f(y-\mu_0)$ with respect to μ and by the Cauchy–Schwarz inequality. Arguing like in Theorem 2.2, the inequalities (2.26) and (2.27) provide the convergence rate of $\widehat{t}_n$.

Theorem 2.15. $\overline{\lim}_{n,A\to\infty} P_0(n|\widehat{t}_n - t_0| > A) = 0.$

Under P_0, the estimator of the mean has an expansion conditionally on $\widehat{t}_n$ similar to $\widehat{\mu}_{n,t_0}$

$$n^{\frac{1}{2}}(\widehat{\mu}_{n,\widehat{t}_n} - \mu_0) = I_{\widehat{t}_n}^{-1}U_n(\widehat{t}_n) + o_p(1).$$

Let $Z_i(\mu)$ denote the vector with components $(f^{-1}f')(Y_i - \mu_1)$ and $(f^{-1}f')(Y_i - \mu_2)$, let $t_{n,u} = t_0 + n^{-1}u$ and $\mu_{n,v} = \mu_0 + n^{-\frac{1}{2}}v$, and let $\theta_{n,u,v}$ be the vector with components $\mu_{n,v}$ and $t_{n,u}$. The process $U_n(\theta_{n,u,v})$ develops as

$$U_n(\theta_{n,u,v}) = U_n(\theta_0) - n^{-\frac{1}{2}}\Big\{1_{\{t_0<t_{n,u}\}} \sum_{i=[nt_0]+1}^{[nt_{n,u}]} Z_i(\mu_{n,v})$$

$$+1_{\{t_{n,u}<t_0\}} \sum_{i=[nt_{n,u}]+1}^{[nt_0]} Z_i(\mu_{n,v})\Big\},$$

the variance $Var_0\{U_n(\theta_{n,u,v})-U_n(\theta_0)\} = o(n^{-1}|u|)$ converges to zero hence $U_n(\theta_{n,u,v}) = U_n(\theta_0)+o_p(1)$. It follows that on $\mathcal{U}_n^A$, $U_n(\widehat{\theta}_n) = U_n(\theta_0)+o_p(1)$ and

$$n^{\frac{1}{2}}(\widehat{\mu}_n - \mu_0) = I_0^{-1}U_n(t_0) + o_p(1), \tag{2.28}$$

it converges weakly to a centered Gaussian variable with variance I_0^{-1}.

The process W_n is such that

$$\begin{aligned}
W_n(\theta_{n,u,v}) &= W_{1n}(\theta) + (1_{\{t_0<t_{n,u}\}} - 1_{\{t_0>t_{n,u}\}}) \\
&\quad \times n^{-\frac{1}{2}} \sum_{i=[nt_0]+1}^{[nt_{n,u}]} \Big\{\log \frac{f(Y_i-\mu_{01})}{f(Y_i-\mu_{02})} - E_0 \log \frac{f(Y_i-\mu_{01})}{f(Y_i-\mu_{02})}\Big\} \\
&\quad +(1_{\{t_0<t_{n,u}\}} - 1_{\{t_0>t_{n,u}\}})n^{-\frac{1}{2}} \sum_{i=[nt_0]+1}^{[nt_{n,u}]} \Big\{\log \frac{f(Y_i-\mu_1)f(Y_i-\mu_{02})}{f(Y_i-\mu_{01})f(Y_i-\mu_2)} \\
&\quad -E_0 \log \frac{f(Y_i-\mu_1)f(Y_i-\mu_{02})}{f(Y_i-\mu_{01})f(Y_i-\mu_2)}\Big\}
\end{aligned}$$

where W_{1n} does not depend on t and the last sum has a smaller order than the previous one. As n tends to infinity, $[nt_{n,u}] - [nt_0]$ is an interval of length $[u] \pm 1$ and the variance of $W_n(\theta_{n,u,v}) - W_n(\theta_0))$ is a $O(n^{-1})$.

Theorem 2.16. *Under P_0, the variable $n(\widehat{t}_n - t_0)$, with t_0 in $]0,1[$, is asymptotically independent of $\widehat{\eta}_n$ and it converges weakly to the location U_0 of the maximum of an uncentered Gaussian process.*

Proof. The process X_n satisfies $X_n(\theta) = X_{1n}(\theta) + X_{2n}(\theta)$ with

$$X_{1n}(\theta) = n^{-1}\sum_{i=1}^{[nt_0]} \log \frac{f(Y_i-\mu_1)}{f(Y_i-\mu_{01})} + n^{-1} \sum_{i=[nt_0]+1}^{n} \log \frac{f(Y_i-\mu_2)}{f(Y_i-\mu_{02})},$$

$$\begin{aligned}
X_{2n}(\theta) &= (1_{\{t_0<t\}} - 1_{\{t_0>t\}})\, n^{-1} \sum_{i=[nt_0]+1}^{[nt]} \log \frac{f(Y_i-\mu_{01})}{f(Y_i-\mu_{02})} \\
&\quad +(1_{\{t_0<t\}} - 1_{\{t_0>t\}})n^{-1} \sum_{i=[nt_0]+1}^{[nt]} \log \frac{f(Y_i-\mu_1)f(Y_i-\mu_{02})}{f(Y_i-\mu_{01})f(Y_i-\mu_2)},
\end{aligned}$$

at $\theta_{n,u}$, the last sum of the process X_{2n} has a smaller order than the previous one, with μ_j in a neighbourhood of μ_{j0}, for $j = 1, 2$. Under P_0 and as n tends to infinity, $[nt_{n,u}] - [nt_0]$ is an interval of length $[u] \pm 1$, the mean and the variance of the process $Z_n(u,v) = nX_{2n}(\theta_{n,u,v})$ are $O(|u|)$. By Proposition 1.4, it converges weakly on $\mathcal{U}_n^A$ to an uncentered Gaussian process Z_A.

The process $nX_{1n}(\theta_{n,u,v})$ is asymptotically independent of the process Z_n which determines the estimator $\widehat{t}_n$ and it is asymptotically free of the mean parameter. As the estimators of the means maximize $X_{1n}(\theta_{n,u,v})$, they are asymptotically independent of $\widehat{t}_n$. The maximum of the process Z_n is achieved at $\widehat{u}_n = n(\widehat{t}_n - t_0)$ and it converges weakly on $\mathcal{U}_n^A$ to the maximum of the process Z_A, theorem 2.15 ends the proof. □

The log-likelihood ratio test of the hypothesis H_0 of a constant mean μ_0 against the alternative of a change at an unknown index k_0, with distinct means μ_1 and μ_2, is performed with the statistic

$$T_n = 2 \sup_{t\in]0,1[} \{l_n(\widehat{\theta}_{n,t}) - \widehat{l}_{0n}\} = 2\{l_n(\widehat{\theta}_n) - \widehat{\theta}_{0n}\},$$

where $\widehat{l}_{0n} = \sup_{\mu\in\mathbb{R}} \sum_{i=1}^n \log f(Y_i - \mu)$ is the maximum of the likelihood under the hypothesis H_0, it has the approximation $\widehat{l}_{0n} = I_{0n}^{-1} U_{0n}^2 + o_p(1)$ where $U_{0n} = n^{-\frac{1}{2}} \sum_{i=1}^n (f^{-1}f')(Y_i - \mu_0)$ is centered and converges weakly under P_0 to a centered Gaussian variable with variance the limit of l_{0n}, $\widehat{l}_{0n}$ converges weakly to a χ_1^2 variable.

The process X_n is the sum

$$X_n(\theta) = n^{-1} \sum_{i=1}^{[nt]} \log \frac{f(Y_i - \mu_1)}{f(Y_i - \mu_0)} + n^{-1} \sum_{i=[nt]+1}^{n} \log \frac{f(Y_i - \mu_2)}{f(Y_i - \mu_0)}$$

it converges a.s. uniformly under P_0 to

$$X(\theta) = tE_0\Big\{\log \frac{f(Y_i - \mu_1)}{f(Y_i - \mu_0)} 1_{\{i\le k\}}\Big\} + (1-t)E_0\Big\{\log \frac{f(Y_i - \mu_2)}{f(Y_i - \mu_0)} 1_{\{i>k\}}\Big\}$$

where $k = [nt]$ and $t_0 = 1$ so $X(\theta_0) = 0$. Let $X'_\mu(\theta)$ and $X''_\mu(\theta)$ be the first derivatives of $X(\theta)$ with respect to mean parameter μ. As t tends to one, the maximum likelihood estimator of μ_1 converges a.s. to μ_0 and the expansion (2.28) of the estimator of the mean vector μ is no longer valid with a singular matrix I_0, as π_{0n} converges a.s. to one.

Proposition 2.14. *The statistic T_n converges weakly under H_0 to a χ_1^2 variable T_0.*

Proof The process X_n splits according to the components of the mean parameter as $X_n = X_{1n} + X_{2n}$ with

$$X_{1n}(\mu_1, t) = \sum_{i=1}^{[nt]} \log \frac{f(Y_i - \mu_1)}{f(Y_i - \mu_0)},$$

$$X_{2n}(\mu_2, t) = \sum_{i=[nt]+1}^{n} \log \frac{f(Y_i - \mu_2)}{f(Y_i - \mu_0)}.$$

For every t, the process X_{1n} provides an a.s. consistent estimator $\widehat{\mu}_{1n,t}$ of μ_0 from the sub-sample $(Y_1, \ldots, Y_{[nt]})$, the process X_{2n} provides an estimator $\widehat{\mu}_{2n,t}$ of the parameter μ_2 from the sub-sample $(Y_{[nt]+1}, \ldots, Y_n)$.

At t_n converging to one, the variable $X_{2n}(\mu_{2,t_n}, t_n)$ is asymptotically equivalent to a sum of $n(1 - t_n)$ terms. The first derivative

$$\dot{X}_{2n,\mu_2}(\mu_2, t) = - \sum_{i=[nt]+1}^{n} \frac{f'(Y_i - \mu_2)}{f(Y_i - \mu_2)}$$

is zero at $(\widehat{\mu}_{2n}, \widehat{t}_n)$. The variable $\dot{X}_{2n,\mu_2}(\mu_0, \widehat{t}_n)$ is asymptotically equivalent to a centered Gaussian variable with a strictly positive variance I_{02} and $-I_{02}$ is the limit in probability of the second derivative $\ddot{X}_{2n,\mu_2}(\widehat{\mu}_{2n}, \widehat{t}_n)$, under H_0. The variable $\widehat{\mu}_{2n}$ satisfies

$$\widehat{\mu}_{2n} - \mu_0 = I_{02}^{-1} \dot{X}_{2n,\mu_2}(\mu_0, \widehat{t}_n) + o_p(1).$$

The variable $\widehat{l}_{2n} = X_{2n}(\widehat{\mu}_{2n}, \widehat{t}_n)$ has the asymptotic expansion

$$\widehat{l}_{2n} = \dot{X}^T_{2n,\mu_2}(\mu_0, \widehat{t}_n) \ddot{X}^{-1}_{2n,\mu_2}(\widehat{\mu}_{2n}, \widehat{t}_n) \dot{X}^T_{2n,\mu_2}(\mu_0, \widehat{t}_n) + o_p(1)$$

and it converges in weakly to a χ_1^2 variable under H_0.

The estimator $\widehat{\mu}_{1n}$ converges a.s. to μ_0 and the first two derivatives of $W_{1n} = n^{-\frac{1}{2}} X_{1n}$ with respect to μ_1 are

$$\dot{W}_{1n,\mu_1}(\mu_1, t) = -n^{-\frac{1}{2}} \sum_{i=1}^{[nt]} \frac{f'(Y_i - \mu_1)}{f(Y_i - \mu_1)},$$

and $\ddot{W}_{1n,\mu_1}$ such that $-n^{-1} \ddot{X}_{1n,\mu_1}(\widehat{\mu}_{1n}, \widehat{t}_n)$ converges in probability to a strictly positive limit $I_{01} = \int f'^2(y - \mu_0) f^{-1}(y - \mu_0)\, dy$. A first order expansion in a neighbourhood of μ_0 implies

$$n^{\frac{1}{2}}(\widehat{\mu}_{1n} - \widehat{\mu}_{0n}) = I_{01}^{-1} \dot{W}_{1n,\mu_1}(\widehat{\mu}_{0n}, \widehat{t}_n) + o_p(1).$$

Let

$$W_{0n}(\mu_1, \mu_0) = n^{-\frac{1}{2}} \sum_{i=1}^{n} \log \frac{f(Y_i - \mu_1)}{f(Y_i - \mu_0)},$$

its first derivative with respect to μ_0, $\dot{W}_{0n,\mu_0}(\mu_0)$ is such that $\dot{W}_{0n,\mu_0}(\widehat{\mu}_{0n}) = 0$ and $n^{-\frac{1}{2}}\ddot{W}_{0n,\mu_0}(\mu_0)$ converges in probability to $-I_{01}$, hence $\dot{W}_{1n,\mu_1}(\widehat{\mu}_{0n}, \widehat{t}_n)$ is a $o_p(1)$ from the convergence rate of $\widehat{t}_n$.

It follows that $n^{\frac{1}{2}}(\widehat{\mu}_{1n} - \widehat{\mu}_{0n})$ converges in probability to zero. By a second order asymptotic expansion

$$X_{1n}(\widehat{\mu}_{1n}, \widehat{\mu}_{0n}) = \dot{W}^T_{1n,\mu_1}(\widehat{\mu}_{0n}, \widehat{t}_n) I_{01}^{-1} \dot{W}_{1n,\mu_1}(\widehat{\mu}_{0n}, \widehat{t}_n) + o_p(1),$$

it converges therefore in probability to zero. □

Proposition 2.15. *Under fixed alternative, the statistic T_n tends to infinity as n tends to infinity and under local alternatives P_{θ_n}, T_n converges weakly to $T_0 + T$ where T is a non-degenerated variable.*

Proof. Under a fixed alternative P_θ with t in $]0,1[$ and distinct means, one of the processes W_{jn} is uncentered and T_n diverges. Under local alternatives $P_n = P_{\theta_n}$, the parameters are t_n in $]0,1[$ converging to one and distinct means μ_{1n} and μ_{1n}, such that

$$\mu_{1n} = \mu_0 + n^{-\frac{1}{2}} v_{1n},\ \mu_{2n} = \mu_0 + n^{-\frac{1}{2}} v_{2n},\ t_n = 1 - n^{-1} u_n,$$

where u_n, respectively v_{1n} and v_{2n}, converge to non-null limits u, respectively v_1 and v_2, as n tends to infinity. The test statistic is the sum $T_n = T_{1n} + T_{2n}$ with

$$T_{1n} = 2 \sum_{i=1}^{[n\widehat{t}_n]} \log \frac{f(Y_i - \widehat{\mu}_{1n})}{f(Y_i - \widehat{\mu}_{0n})},$$

$$T_{2n} = 2 \sum_{i=[n\widehat{t}_n]+1}^{n} \log \frac{f(Y_i - \widehat{\mu}_{2n})}{f(Y_i - \widehat{\mu}_{0n})}.$$

The variable $n^{\frac{1}{2}}(\widehat{\mu}_{1n} - \mu_{1n})$ and $n^{\frac{1}{2}}(\widehat{\mu}_{0n} - \mu_0)$ have the same asymptotic distribution under P_n and $n^{\frac{1}{2}}(\widehat{\mu}_{1n} - \widehat{\mu}_{0n})$ converges in probability under P_n to v_1. By a second order expansion of T_{1n}, it converges in probability under P_n to $v_1^T I_{01}^{-1} v_1$.

The variable T_{2n} is expanded like X_{2n} at $\widehat{t}_n$, $\widehat{\mu}_{2n}$, $\widehat{\mu}_{0n}$. Under P_n, $n(1-\widehat{t}_n) = u_n + n(t_n - \widehat{t}_n)$ converges to $u + u_0$, with u_0 defined by Theorem 2.16, and $\widehat{\mu}_{2n}$ has the same convergence rate as $\widehat{t}_n$ hence $n\widehat{\mu}_{2n}$ converges in probability to zero under P_n. The variable $\dot{X}_{2n}(\mu_{2n}, \widehat{t}_n)$ is the sum

$$\sum_{i=[n\widehat{t}_n]+1}^{[nt_n]} \frac{f'(Y_i - \mu_{2n})}{f(Y_i - \mu_2)} + \sum_{i=[nt_n]+1}^{n} \frac{f'(Y_i - \mu_{2n})}{f(Y_i - \mu_{2n})}.$$

The second term is asymptotically equivalent to a finite sum of u centered variables with finite variance and the first sum has an expansion similar to the expansion of $\widehat{l}_{2n}$ under H_0, T_{2n} converges weakly to the sum of T_0 and a non-degenerated variable under P_n. □

2.8 Nonparametric maximum likelihood

When the density f is unknown, the log-likelihood (2.25) of the sample with a change of mean at t in $]0,1[$ is replaced by

$$\widehat{l}_{nh}(\theta) = \sum_{i=1}^{[nt]} \log \widehat{f}_{nh}(Y_i - \mu_{1t}) + \sum_{i=[nt]+1}^{n} \log \widehat{f}_{nh}(Y_i - \mu_{2t}), \tag{2.29}$$

with a kernel estimator $\widehat{f}_{nh}$ of the density. The estimator $\widehat{\mu}_{nh,t}$ of the means vector is solution of the equations

$$\sum_{i=1}^{[nt]} \frac{\widehat{f}'_{nh}}{\widehat{f}_{nh}}(Y_i - \mu_{1t}) = 0, \qquad \sum_{i=[nt]+1}^{n} \frac{\widehat{f}'_{nh}}{\widehat{f}_{nh}}(Y_i - \mu_{2t}) = 0$$

where the bias of $\widehat{f}'_{n,h}$ is a $O(h^2)$ and its variance is a $O((nh^3)^{-1})$. The location of the change is estimated by

$$\widehat{t}_{nh} = \arg\max_{t\in]0,1[} \widehat{l}_{nh}(\widehat{\mu}_{nh,t}, t)$$

and $\widehat{\mu}_{nh} = \widehat{\mu}_{nh,\widehat{t}_{nh}}$.

Under P_0, by the consistency of the kernel estimator of the density, the process $\widehat{X}_{nh}(\theta) = n^{-1}\{\widehat{l}_{nh}(\theta) - \widehat{l}_{nh}(\theta_0)\}$ converges a.s. to the concave function X of the previous section, the estimators $\widehat{\mu}_{nh,t}$ and $\widehat{t}_{nh}$ are therefore a.s. consistent. The process $\widehat{W}_{nh} = n^{\frac{1}{2}}(\widehat{X}_{nh} - X)$ has the mean $E_0\widehat{W}_{nh}(\theta) = O(n^{\frac{1}{2}}h^2)$ tending to zero by Conditions 2.1.

Under the conditions $E_0 \sup_{\mu\in V_\varepsilon(\mu_0)} \log^2 f(Y-\mu)$ finite and 2.1, for every $\varepsilon > 0$, a first order expansion of the logarithms, the monotonicity of the sums with respect to t and the Cauchy–Schwarz inequality imply the existence of a constant $\kappa_1 > 0$ such that for n large enough

$$E_0 \sup_{|t-t_0|\le\varepsilon} |\widehat{W}_{nh}(\theta)| \le \kappa_1 \varepsilon^{\frac{1}{2}}. \tag{2.30}$$

Like in Theorem 2.15, the inequalities (2.26) and (2.30) provide the convergence rate of the estimator $\widehat{t}_n$.

Theorem 2.17. $\overline{\lim}_{n,A\to\infty} P_0(n|\widehat{t}_n - t_0| > A) = 0$.

The derivatives of the estimated log-likelihood with respect to the mean parameter define the process

$$\widehat{U}_{nh}(\theta) = n^{-\frac{1}{2}}\widehat{l}'_{nh}(\theta),$$

the process $\widehat{I}_{nh}(\theta) = -n^{-1}\widehat{l}''_{nh,\mu}(\theta)$ such that $\widehat{I}_{nh}(\theta_0)$ converges a.s. under P_0 to I_0 and the process

$$\widehat{U}_{nh}(\theta) = n^{-\frac{1}{2}}\sum_{i=1}^{[nt]}\frac{\widehat{f}'_{nh}}{\widehat{f}_{nh}}(Y_i - \mu_{1t}) + n^{-\frac{1}{2}}\sum_{i=[nt]+1}^{n}\frac{\widehat{f}'_{nh}}{\widehat{f}_{nh}}(Y_i - \mu_{2t}).$$

Under the conditions that nh tends to infinity and nh^4 tends to zero, $\widehat{U}_{nh}(\theta) - U_n(\theta)$ converges in probability to zero under P_0, uniformly on bounded parameter intervals and the estimator of the means maximizing $\widehat{l}_{nh}(\mu, t_0)$ is such that $n^{\frac{1}{2}}(\widehat{\mu}_n - \mu_0)$ converges weakly to a centered Gaussian process with variance I_0. Under P_0, the estimator of μ_0 has the expansion

$$n^{\frac{1}{2}}(\widehat{\mu}_n - \mu_0) = \widehat{I}_{nh}^{-1}(\mu_0, \widehat{t}_n)\widehat{U}_{nh}(\mu_0, \widehat{t}_n) + o_p(1)$$

where

$$\widehat{U}_{nh}(\mu_0, \widehat{t}_n) = U_n(\theta_0) + o_p(1)$$

by Theorem 2.17, and $\widehat{I}_{nh}(\mu_0, \widehat{t}_n) = I(\theta_0) + o_p(1)$ then $n^{\frac{1}{2}}(\widehat{\mu}_n - \mu_0)$ converges weakly to a centered Gaussian variable with variance I_0^{-1}.

Theorem 2.18. *For t_0 in $]0,1[$, the variable $n(\widehat{t}_n - t_0)$ is asymptotically independent of $\widehat{\eta}_n$ and it converges weakly to the location of the maximum of a Gaussian process, as n and A tend to infinity.*

Proof. Let $t_{n,u} = t_0 + n^{-1}u$, let $\mu_{n,v} = \mu_0 + n^{-\frac{1}{2}}v$ and let $\theta_{n,u,v}$ be the parameter with components $t_{n,u}$ and $\mu_{n,v}$. The process $n\widehat{X}_{nh}(\theta_{n,u,v})$ has an expansion similar to $X_n(\theta_{n,u,v})$ in Theorem 2.16 and the condition nh^4 converging to zero implies n$\{\widehat{X}_{nh}(\theta_{n,u,v}) - X_n(\theta_{n,u,v})\}$ converges weakly to a Gaussian process in $\mathcal{U}_n^A$, Theorem 2.17 ends the proof. □

A test of the hypothesis H_0 against the alternative of a change at an unknown index k_0, with an unknown density f is performed with the statistic

$$\widehat{T}_n = 2\{\widehat{l}_{nh}(\widehat{\theta}_n) - \widehat{l}_{0nh}\},$$

where $\widehat{l}_{0nh} = \sup_{\mu\in\mathbb{R}}\sum_{i=1}^{n}\log\widehat{f}_{nh}(Y_i - \mu)$ is the maximum of the estimated likelihood under the hypothesis H_0. Under the previous conditions, it has the approximation $\widehat{l}_{0nh} = I_0^{-1} U_{0n}^2 + o_p(1)$ and it converges weakly to a χ_1^2 variable. The process

$$\begin{aligned}
\widehat{X}_{nh}(\theta) &= n^{-1}\{\widehat{l}_{nh}(\theta) - \widehat{l}_{nh}(\theta_0)\} \\
&= n^{-1} \sum_{i=1}^{[nt]} \log \frac{\widehat{f}_{nh}(Y_i - \mu_1)}{\widehat{f}_{nh}(Y_i - \mu_0)} + n^{-1} \sum_{i=[nt]+1}^{n} \log \frac{\widehat{f}_{nh}(Y_i - \mu_2)}{\widehat{f}_{nh}(Y_i - \mu_0)}
\end{aligned}$$

converges a.s. uniformly under P_0 to the function $X(\theta)$ limit of the process $X_n(\theta)$ for the test statistic T_n of the previous section. The asymptotic behaviour of $\widehat{T}_n$ is the same as T_n (Propositions 2.14 and 2.15).

Proposition 2.16. *The statistic $\widehat{T}_n$ converges weakly under H_0 to a χ_1^2 variable T_0. Under fixed alternatives, T_n tends to infinity as n tends to infinity and under local alternatives P_{θ_n}, $\widehat{T}_n$ converges weakly to the sum of T_0 and a non-centered variable.*

Chapter 3

Change-points for parametric densities

Abstract. This chapter studies the maximum likelihood estimation and the likelihood ratio test for the parametric density of a real variable in a model with a change of parameter according to an unknown threshold of the variable and in a model with a chronological change at an unknown sampling index. First, the parametric model is supposed to be known then it depends on an unknown density function and the likelihood is estimated using a kernel estimator of the density. In all cases, the convergence rates of the estimator of the change-point is n^{-1} under suitable conditions on the convergence rate for bandwidth of the density estimator. The weak convergence of the estimators and the test statistics under the hypothesis and alternatives are proved.

3.1 Maximum likelihood estimation

Let Y be a real variable with a parametric density

$$f_\theta(y) = f(y)1_{\{y\leq\gamma\}} + f_\eta(y)1_{\{y>\gamma\}} \tag{3.1}$$

where the density before a change at γ is supposed to be known and the parameter θ of the model with a change has the components γ, the location of the change of density, and η a parameter which modifies the density after the change-point. The parameter η belongs to an open subset $\mathcal{H}$ of $\mathbb{R}^d$ such that f differs from f_η for every η in $\mathcal{H}$ and the function $\eta \mapsto f_\eta(\cdot)$ belongs to $C^2(\mathcal{H})$ uniformly in $\mathbb{R}$, with a second order derivative uniformly bounded in $\mathcal{H}\times\mathbb{R}$. Under the probability measure P_0 the observations, the density is $f_0 = f_{\theta_0}$ defined by (3.1) with the parameters γ_0 and η_0, and the variance of Y is σ_0^2.

The parameter θ of $\Theta = \mathcal{H} \times \mathbb{R}$ is estimated from a sample $Y_1, \ldots, Y_n$ with density f_θ by maximization of the log-likelihood process

$$l_n(\theta) = \sum_{k=1}^n \Big\{ \log f(Y_k) 1_{\{Y_k \le \gamma\}} + \log f_\eta(Y_k) 1_{\{Y_k > \gamma\}} \Big\}. \tag{3.2}$$

The model will later be generalized by replacing the known density f of the first phase by a parametric density with an unknown parameter distinct from the parameter of the density in the second phase, the estimators will then be defined by the same method.

The logarithm of the likelihood ratio of the sample under P_θ and P_0 defines the process

$$\begin{aligned}
X_n(\theta) &= n^{-1}\{l_n(\theta) - l_n(\theta_0)\} \\
&= n^{-1} \sum_{k=1}^n \Big\{ \log \frac{f(Y_k)}{f_{\eta_0}(Y_k)} 1_{]\gamma_0,\gamma]}(Y_k) + \log \frac{f_\eta(Y_k)}{f(Y_k)} 1_{]\gamma,\gamma_0]}(Y_k) \\
&\qquad + \log \frac{f_\eta(Y_k)}{f_{\eta_0}(Y_k)} 1_{\{Y_k > \gamma \vee \gamma_0\}} \Big\} \\
&= n^{-1} \sum_{k=1}^n \Big\{ 1_{(\gamma_0,\gamma)}(Y_k) \log \frac{f(Y_k)}{f_\eta(Y_k)} + 1_{\{Y_k > \gamma_0\}} \log \frac{f_\eta(Y_k)}{f_{\eta_0}(Y_k)} \Big\}.
\end{aligned}$$

For a function $\eta \mapsto f_\eta$ belonging to $C^2(\mathcal{H})$ with a second order derivative bounded on $\mathcal{H}$ uniformly on $\mathbb{R}$, under P_0, the process X_n converges a.s. uniformly on Θ to the function

$$\begin{aligned}
X(\theta) &= 1_{\{\gamma > \gamma_0\}} \Big\{ \int_{\gamma_0}^{\gamma} \log \frac{f}{f_{\eta_0}} \, dF_{\eta_0} + \int_{\gamma}^{\infty} \log \frac{f_\eta}{f_{\eta_0}} \, dF_{\eta_0} \Big\} \\
&\quad + 1_{\{\gamma < \gamma_0\}} \Big\{ \int_{\gamma}^{\gamma_0} \log \frac{f_\eta}{f} \, dF + \int_{\gamma_0}^{\infty} \log \frac{f_\eta}{f_{\eta_0}} \, dF_{\eta_0} \Big\} \\
&= \int_{\gamma_0}^{\gamma} \log \frac{f}{f_\eta} \, dF_0 + \int_{\gamma_0}^{\infty} \log \frac{f_\eta}{f_{\eta_0}} \, dF_{\eta_0},
\end{aligned}$$

and $X(\theta_0) = 0$. We assume furthermore that the matrix

$$I_0 = \int \dot{f}_\eta^2 f_\eta^{-1} \, dF_0$$

is finite.

Theorem 3.1. *Under P_0 and the integrability and differentiability properties of the densities f_η, the maximum likelihood estimator $\widehat{\theta}_n$ of θ_0 is a.s. consistent.*

Proof. The function X is concave with respect to f and it is zero at $f_0 = f_{\theta_0}$ where it reaches its maximum. By the uniform convergence of X_n to X, we have

$$0 \leq X_n(\widehat{\theta}_n) \leq \sup_{\Theta} |X_n(\theta) - X(\theta)| + X(\widehat{\theta}_n)$$

therefore $\lim_{n\to\infty} X(\widehat{\theta}_n) \geq 0$, moreover X is maximum at f_0 with $0 = X(\theta_0)$ which implies $0 \geq X(\widehat{\theta}_n)$ and it follows that $X(\widehat{\theta}_n)$ converges a.s. to zero. If $\widehat{\theta}_n$ did not converge a.s. to θ_0, we could extract a sub-sequence converging to θ_1 distinct from θ_0 and such that $X(\theta_1) = 0$ but this is impossible for X to achieve maximum at f_0. □

The function X is differentiable on $\{\gamma < \gamma_0\}$ and on $\{\gamma_0 < \gamma\}$, its derivative with respect to γ is

$$X'_\gamma(\theta) = \{1_{\{\gamma<\gamma_0\}} f(\gamma) + 1_{\{\gamma_0<\gamma\}} f_{\eta_0}(\gamma)\} \log \frac{f(\gamma)}{f_\eta(\gamma)}$$

and $X'_\gamma(\theta)$ is zero if and only if $\gamma_0 = \gamma$, for f different from every f_η, η in $\mathcal{H}$. If $\gamma < \gamma_0$, $X'_\gamma(\theta_0) > 0$ and if $\gamma_0 < \gamma$, $X'_\gamma(\theta_0) < 0$ therefore the function X is concave at γ_0 where X reaches its maximum with respect to γ.

The derivatives of $X(\theta)$ with respect to η are expressed with the derivative f'_η and f''_η of f_η with respect to η

$$X'_\eta(\theta) = \int_\gamma^\infty \frac{f'_\eta}{f_\eta}\, dF_0,$$

$$X''_\eta(\theta) = -\int_\gamma^\infty \frac{f'^2_\eta - f_\eta f''_\eta}{f^2_\eta}\, dF_0,$$

$$X''_{\eta,\gamma}(\theta) = -\frac{f'_\eta}{f_\eta}(\gamma)\,\{1_{\{\gamma<\gamma_0\}} f(\gamma) + 1_{\{\gamma_0<\gamma\}} f_{\eta_0}(\gamma)\}$$

where $X'_\eta(\theta_0) = 0$ and $I(\theta) = -X''_\eta(\theta)$ is a definite positive matrix of $\mathbb{R}^d \times \mathbb{R}^d$. The function X is therefore concave at θ_0. By concavity of X, the second order left-derivative $X''_\gamma(\theta_0)$ is negative.

Let $\varepsilon > 0$ and let $V_\varepsilon(\theta_0)$ be a neighborhood of θ_0 for the semi-norm $\rho(\theta, \theta_0) = \{(\eta - \eta_0)^2 + |\gamma - \gamma_0|\}^{\frac{1}{2}}$.

Lemma 3.1. *For ε small enough, there exists a constant $\kappa_0 > 0$ such that for every θ in Θ*

$$X(\theta) \geq -\kappa_0 \rho^2(\theta, \theta_0). \tag{3.3}$$

Proof. A second order expansion of X with respect to η in a neighborhood of θ_0 is written as

$$X(\theta) = X(\eta_0,\gamma) + (\eta-\eta_0)^T X'_\eta(\eta_0,\gamma) f - \frac{1}{2}(\eta-\eta_0)^T I(\eta_0,\gamma)(\eta-\eta_0) + o(\|\eta-\eta_0\|^2)$$
$$\leq \int_{\gamma_0}^{\gamma} \log\frac{f}{f_{\eta_0}}\, dF_0 - \frac{1}{2}(\eta-\eta_0)^T I(\eta_0,\gamma)(\eta-\eta_0) + o(\|\eta-\eta_0\|^2)$$

where $\int_{\mathbb{R}} f'_{\eta_0} = 0$ hence

$$X'_\eta(\eta_0,\gamma) = \int_{\gamma}^{\infty} f'_{\eta_0} < 0$$

by concavity, and

$$X(\eta_0,\gamma) = 1_{\{\gamma>\gamma_0\}} \int_{\gamma_0}^{\gamma} \log\frac{f}{f_{\eta_0}}\, dF_{\eta_0} + 1_{\{\gamma<\gamma_0\}} \int_{\gamma}^{\gamma_0} \log\frac{f_\eta}{f}\, dF$$

is always negative, with the maximum zero at γ_0. It follows that there exists a constant $\kappa_0 > 0$ such that for every θ in $V_\varepsilon(\theta_0)$ the inequality (3.3) is true. The constant κ_0 can be chosen as $\sup_{y\gamma_0} f_{\eta_0}(y)(\log f - \log f_{\eta_0})(y) + \sup_{\theta\in V_\varepsilon(\theta_0)} I(\eta_0,\gamma)$. □

Let $\nu_{0n} = n^{\frac{1}{2}}(\widehat{F}_{Y,n} - F_0)$ is the empirical process of the sample under P_0. The process $W_n = n^{\frac{1}{2}}(X_n - X)$ is developed as

$$W_n(\theta) = 1_{\{\gamma>\gamma_0\}} \int_{\gamma_0}^{\gamma} \log\frac{f}{f_{\eta_0}}\, d\nu_{0n} + 1_{\{\gamma<\gamma_0\}} \int_{\gamma}^{\gamma_0} \log\frac{f_\eta}{f}\, d\nu_{0n} + \int_{\gamma\vee\gamma_0}^{\infty} \log\frac{f_\eta}{f_{\eta_0}}\, d\nu_{0n}.$$

Lemma 3.2. *Under the condition $E_0 \sup_{\theta\in V_\varepsilon(\theta_0)} \log^2 f_\theta(Y)$ finite, there exists a constant $\kappa_1 > 0$ such that for n sufficiently large*

$$E_0 \sup_{\theta\in V_\varepsilon(\theta_0)} W_n(\theta) \leq \kappa_1 \varepsilon^{\frac{1}{2}}. \tag{3.4}$$

Proof. Under the condition, there exists a constant $\kappa_1 > 0$ such that for n sufficiently large $E_0 \sup_{\theta\in V_\varepsilon(\theta_0)} W_n^2(\theta) \leq \kappa_1^2\varepsilon$, by a first order expansion of $f_\eta f_{\eta_0}^{-1}$ in $V_\varepsilon(\theta_0)$, and the inequality (3.4) follows from the Cauchy–Schwarz inequality. □

The matrix $X''(\theta_0)$ of the second order derivatives of X with respect to the components of θ is singular, with $X''_{\eta,\gamma}(\theta_0) = 0$, then we consider the maximum likelihood estimator $\widehat{\eta}_{n,\gamma}$ of the parameter η_γ, at fixed γ, it satisfies

$$X'_{n,\eta}(\widehat{\eta}_{n,\gamma},\gamma) = 0.$$

Theorem 3.1 and a first order expansion of $X'_n(\theta)$ with respect to η in a neighbourhood of η_0 imply that for every $\varepsilon > 0$ and for n sufficiently large $\widehat{\theta}_n$ belongs to $V_\varepsilon(\theta_0)$ and

$$\begin{aligned} -n^{\frac{1}{2}} X'_{n,\eta}(\eta_0, \widehat{\gamma}_n) &= n^{\frac{1}{2}} (\widehat{\eta}_n - \eta_0)^T X''_{n,\eta}(\eta_0, \widehat{\gamma}_n) + o_p(n^{\frac{1}{2}} \|\widehat{\eta}_n - \eta_0\|), \\ n^{\frac{1}{2}} X'_{n,\eta}(\eta_0, \widehat{\gamma}_n) &= n^{\frac{1}{2}} (\widehat{\eta}_n - \eta_0)^T I_0 + o_p(n^{\frac{1}{2}} \|\widehat{\eta}_{n,\gamma} - \eta_0\|), \qquad (3.5) \end{aligned}$$

where the variable

$$\begin{aligned} X'_{n,\eta}(\eta_0, \widehat{\gamma}_n) &= n^{-1} \sum_{k=1}^n \Big\{ \frac{f'_\eta(Y_k)}{f_\eta(Y_k)} 1_{]\widehat{\gamma}_n, \gamma_0]}(Y_k) + \frac{f'_\eta(Y_k)}{f_\eta(Y_k)} 1_{\{Y_k > \widehat{\gamma}_n \vee \gamma_0\}} \Big\} \\ &= n^{-1} \sum_{k=1}^n \frac{f'_\eta(Y_k)}{f_\eta(Y_k)} 1_{\{Y_k > \widehat{\gamma}_n\}} \end{aligned}$$

converges a.s. under P_0 to $X'_\eta(\theta_0) = \int_{\gamma_0}^\infty f'_{\eta_0}(y)\, dy = 0$ and $n^{\frac{1}{2}} X'_{n,\eta}(\theta_0)$ converges weakly to a centered Gaussian variable.

The convergence rate of $\widehat{\gamma}_n$ is proved like Theorem 2.2, the next theorem extends it to the parameter θ, using the inequalities (3.3) and (3.4). Let

$$\mathcal{U}_n = \{u_n = (n^{\frac{1}{2}} (\eta - \eta_0)^T, n(\gamma - \gamma_0))^T, \eta \in \mathcal{H}, \gamma \in \mathbb{R}\}$$

and for u in $\mathcal{U}_n$, let $\theta_{n,u}$ be the vector with components $\eta_0 + n^{-\frac{1}{2}} u_1$ and $\gamma_0 + n^{-1} u_2$, with $u = (u_1^T, u_2)^T$, and let $\|u\| = n^{\frac{1}{2}} \rho(\theta_{n,u}, \theta_0)$, reversely $u_{n,\theta}$ denotes the vector of $\mathcal{U}_n$ such that $\theta = \theta_{n,u_{n,\theta}}$. For $\varepsilon > 0$ let

$$\mathcal{U}_{n,\varepsilon} = \{u \in \mathcal{U}_n : \|u\| \le n^{\frac{1}{2}} \varepsilon\},$$

there is equivalence between u belongs to $\mathcal{U}_{n,\varepsilon}$ and $\theta_{n,u}$ belongs to $V_\varepsilon(\theta_0)$.

Theorem 3.2. *Under the conditions of Lemmas 3.1 and 3.2, for $\varepsilon > 0$ sufficiently small*

$$\begin{aligned} \overline{\lim}_{n,A\to\infty} P_0(\sup\nolimits_{\theta \in V_\varepsilon(\theta_0), \|u_{n,\theta}\| > A} X_n(\gamma) \ge 0) &= 0, \\ \overline{\lim}_{n,A\to\infty} P_0\{n^{\frac{1}{2}} \rho(\widehat{\theta}_n, \theta_0) > A) &= 0. \end{aligned}$$

Proof. Let $\widehat{u}_n = (n^{\frac{1}{2}} (\widehat{\eta}_n - \eta_0)^T, n(\widehat{\gamma}_n - \gamma_0))^T$. For every $\eta > 0$ the consistency of the estimators implies that for $\varepsilon > 0$ sufficiently small

$$P_0\{\widehat{u}_n \in \mathcal{U}_{n,\varepsilon}\} = P_0\{\widehat{\theta}_n \in V_\varepsilon(\theta_0)\} > 1 - \eta$$

therefore

$$P_0(\|\widehat{u}_n\| > A) \le P_0\Big(\sup_{u \in \mathcal{U}_{n,\varepsilon}, \|u\| > A} l_n(\theta_{n,u}) \ge 0 \Big) + \eta.$$

Let g be an increasing function such that $\sum_{g(j)>A} g(j+1)g^{-2}(j)$ tends to zero as A tends to infinity, and let

$$H_{n,j} = \{u \in \mathcal{U}_{n,\varepsilon} : g(j) < \|u\| \le g(j+1)\},\ j \in \mathbb{N}.$$

For every u belonging to $H_{n,j}$, $n^{-\frac{1}{2}}g(j) \le \rho(\theta_{n,u}, \theta_0) \le n^{-\frac{1}{2}}g(j+1)$ and the inequality (3.3) implies $X(\theta_{n,u}) \le -\kappa_0 n^{-1} g^2(j)$, with $X(\theta_0) = 0$. For every positive $\varepsilon \le n^{-\frac{1}{2}}g(j+1)$, the sets $H_{n,j}$ split the probability as a sum

$$\begin{aligned}
P_0\Big(\sup_{u\in\mathcal{U}_{n,\varepsilon},\|u\|>A} l_n(\theta_{n,u}) \ge 0\Big) &\le \sum_{g(j)>A} P_0\Big(\sup_{u\in H_{n,j}} l_n(\theta_{n,u}) \ge 0\Big)\\
&\le \sum_{g(j)>A} P_0\Big(\sup_{u\in H_{n,j}} |W_n(\theta_{n,u})| \ge n^{-\frac{1}{2}} g^2(j)\kappa_0\Big)\\
&\le \frac{n^{\frac{1}{2}}}{\kappa_0} \sum_{g(j)>A} g^{-2}(j) E_0 \sup_{u\in H_{n,j}} |W_n(\theta_{n,u})| \le \frac{\kappa_1}{\kappa_0} \sum_{g(j)>A} \frac{g(j+1)}{g^2(j)}
\end{aligned}$$

by the inequality (3.4). This bound tends to zero as A tends to infinity. □

Proposition 3.1. *The variable $n^{\frac{1}{2}}(\widehat{\eta}_n - \eta_0)$ converges weakly under P_0 to a centered Gaussian variable with variance I_0^{-1}.*

Proof. From Theorem 3.2 and the expansion (3.5) of the process $X'_{n,\eta}$, we have

$$\begin{aligned}
n^{\frac{1}{2}}\{X'_{n,\eta}(\eta_0, \widehat{\gamma}_n) - X'_{n,\eta}(\theta_0)\} &= n^{\frac{1}{2}} \int_{\widehat{\gamma}_n}^{\gamma_0} \frac{f'_{\eta_0}}{f_{\eta_0}}\, d\widehat{F}_n\\
&= n^{\frac{1}{2}} \int_{\widehat{\gamma}_n}^{\gamma_0} \frac{f'_{\eta_0}}{f_{\eta_0}}\, d\{F_0 + o_p(1)\} = o_p(1),
\end{aligned}$$

$$n^{\frac{1}{2}}(\widehat{\eta}_n - \eta_0) = n^{\frac{1}{2}} X'_{n,\eta}(\theta_0)^T I_0^{-1} + o_p(1) \tag{3.6}$$

and the variance of $n^{\frac{1}{2}} X'_{n,\eta}(\theta_0)$ is I_0, with $\int_{\gamma_0}^{\infty} f'_{\eta_0} = \int_{\gamma_0}^{\infty} f''_{\eta_0} = 0$. □

For every $A > 0$, let $\mathcal{U}_n^A = \{u \in \mathcal{U}_n : \|u\| \le A\}$. We consider the process defined on $\mathcal{U}_n^A$ as

$$\widetilde{W}_n(u) = W_n(\widehat{\eta}_{n,\gamma_{n,u}}, \gamma_{n,u}),$$

with $\gamma_{n,u}$ in a neighborhood of γ_0.

We assume that the integrals $\int \log^2 f_\eta(y)\, dF_0(y)$ and $\int (f'^2_\eta f_\eta^{-1})$ are finite for η in a neighborhood of η_0.

Proposition 3.2. *On $\mathcal{U}_n^A$, for $A > 0$, the processes $\widetilde{W}_n$ has the expansion*

$$\widetilde{W}_n(u) = \int_{\gamma_0}^{\gamma_{n,u}} \log \frac{f}{f_{\eta_0}}\, d\nu_{0n} + (\widehat{\mu}_{n,\gamma_0} - \mu_0)^T \int_{\gamma_0}^{\infty} \frac{f'_{\eta_0}}{f_{\eta_0}}\, d\nu_{0n} o_p(1).$$

Proof. By an expansion of $f_{\widehat{\eta}_{n,\gamma_{n,u}}}$ for u in $\mathcal{U}_n^A$, the process $\widetilde{W}_n(u)$ is written as

$$\widetilde{W}_n(u) = \int_{\gamma_0}^{\infty} \log \frac{f_{\widehat{\eta}_{n,\gamma_{n,u}}}}{f_{\eta_0}} d\nu_{0n} + 1_{\{\gamma_0 < \gamma_{n,u}\}} \int_{\gamma_0}^{\gamma_{n,u}} \log \frac{f}{f_{\eta_0}} d\nu_{0n}$$

$$+1_{\{\gamma_{n,u} < \gamma_0\}} \int_{\gamma_{n,u}}^{\gamma_0} \log \frac{f_{\widehat{\eta}_{n,\gamma_{n,u}}}}{f} d\nu_{0n} + o_p(1),$$

$$= \int_{\gamma_0}^{\gamma_{n,u}} \log \frac{f}{f_{\eta_0}} d\nu_{0n} + (\widehat{\mu}_{n,\gamma_0} - \mu_0)^T \Big\{ \int_{\gamma_0}^{\infty} \frac{f'_{\eta_0}}{f_{\eta_0}} d\nu_{0n} + o_p(1) \Big\}$$

$$+o_p(1),$$

where f'_{η_0} is the value at η_0 of the derivative, its convergence is a consequence of the convergence of the empirical process ν_{0n} to $W \circ F_0$ where W is a Gaussian process and (3.5). The weak convergence of $n^{\frac{1}{2}}(\widehat{\eta}_{n,\gamma_0} - \eta_0)$ to a centered Gaussian variable and the order of $\gamma_{n,u} - \gamma_0$ imply that the last term is a $o_p(n^{-\frac{1}{2}})$ and the asymptotic variance of the first term in the expansion of $\widetilde{W}_n$ is a $O(n^{-1})$. □

The restriction of the process nX_n on the intervals $]\gamma_{n,u}, \gamma_0]$ and $]\gamma_0, \gamma_{n,u}]$ is the sum

$$S_n(u) = \sum_{k=1}^{n} \log \frac{f_{\eta_0}(Y_k)}{f(Y_k)} 1_{]\gamma_{n,u},\gamma_0]}(Y_k) + \sum_{k=1}^{n} \log \frac{f(Y_k)}{f_{\eta_0}(Y_k)} 1_{]\gamma_0,\gamma_{n,u}]}(Y_k),$$

for $\gamma_{n,u} = \gamma_0 + n^{-1}u$, $|u| \leq A$. By Proposition 1.1, the process S_n converges weakly in $D([-A, A])$ to an uncentered Gaussian process S.

Theorem 3.3. *Under P_0, the variable $n(\widehat{\gamma}_n - \gamma_0)$ is asymptotically independent of $\widehat{\eta}_n$, it is bounded in probability and it converges weakly to the location u_0 of the maximum of the Gaussian process S.*

Proof. The process nX_n is the sum

$$nX_n(\theta) = n \int_{\gamma_0}^{\gamma} \log \frac{f}{f_\eta} d\widehat{F}_n + n \int_{\gamma_0}^{\infty} \log \frac{f_{\eta_0}}{f_\eta} d\widehat{F}_n,$$

the estimator $\widehat{u}_n = n(\widehat{\gamma}_n - \gamma_0)$ maximizes the first term of this expression at $\theta_{n,u}$ and the $n^{\frac{1}{2}}(\widehat{\eta}_n - \eta_0)$ maximizes its second term. By an expansion of X'_n for θ in a neighborhood of θ_0 and the equality $X'_n(\theta_0) = 0$, under P_0 the variable $n^{\frac{1}{2}}(\widehat{\eta}_n - \eta_0)$ converges weakly to a centered Gaussian variable and the second term of the expression of $2nX_n(\theta_{n,u})$ converges to a χ_d^2 variable. The estimators are therefore asymptotically independent and bounded in probability, by Theorem 3.2. By the weak convergence of the process S_n, the estimator $\widehat{u}_n$ converges weakly to the location of the maximum of its limit. □

Roplacing the model (3.1) by a model where the real variable Y has a parametric density with an unknown parameter before and after the threshold γ, we now consider the model

$$f_\theta(y) = f_{\eta_1}(y)1_{\{y\leq\gamma\}} + f_{\eta_2}(y)1_{\{y>\gamma\}} \tag{3.7}$$

and we denote η the vector of $\mathbb{R}^{2d}$ with components those of η_1 and η_2. The log-likelihood process

$$l_n(\theta) = \sum_{k=1}^{n}\Big\{\log f_{\eta_1}(Y_k)1_{\{Y_k\leq\gamma\}} + \log f_{\eta_2}(Y_k)1_{\{Y_k>\gamma\}}\Big\} \tag{3.8}$$

is maximum at the estimator $\widehat{\theta}_n$ of the parameter θ. The process

$$X_n(\theta) = n^{-1}\{l_n(\theta) - l_n(\theta_0)\}$$

has a behaviour similar to the process defined with the log-likelihood process (3.2), its limit is the function

$$\begin{aligned} X(\theta) = \int_{-\infty}^{\gamma\wedge\gamma_0} \log\frac{f_{\eta_1}}{f_{\eta_{01}}}\,dF_0 + 1_{\{\gamma<\gamma_0\}}\int_{\gamma}^{\gamma_0} \log\frac{f_{\eta_2}}{f_{\eta_{01}}}\,dF_0 \\ +1_{\{\gamma_0<\gamma\}}\int_{\gamma_0}^{\gamma} \log\frac{f_{\eta_1}}{f_{\eta_{02}}}\,dF_0 + \int_{\gamma\vee\gamma_0}^{\infty} \log\frac{f_{\eta_2}}{f_{\eta_{02}}}\,dF_0, \end{aligned}$$

where the terms between γ and γ_0 is negative, the inequality (3.3) is still satisfied and the asymptotic behaviour of $\widehat{\theta}_n$ is the same as above.

3.2 Likelihood ratio test

We consider the log-likelihood ratio test for the hypothesis H_0 of a density f_0 without change of parameter against the alternative of a change at an unknown threshold γ according to the model (3.1) and with f distinct from the densities f_η for every η of $\mathcal{H}$. The statistic for the log-likelihood ratio test of H_0 is

$$\begin{aligned} T_n &= 2 \sup_{Y_{n:1}<\gamma\leq Y_{n:n}} \{\widehat{l}_n(\gamma) - l_{0n}\} \\ &= 2 \sup_{Y_{n:1}<\gamma\leq Y_{n:n}} \sum_{k=1}^{n} \log\frac{f_{\widehat{\eta}_{n,\gamma},\gamma}(Y_k)}{f(Y_k)} 1_{\{Y_k>\gamma\}}, \end{aligned}$$

it is defined by the logarithm of the density $l_{0n} = \sum_{k=1}^{n}\log f(Y_k)$ under H_0 and the process $\widehat{l}_n(\gamma) = l_n(\widehat{\eta}_{n,\gamma},\gamma)$. Under H_0, $T_n = 2n\sup_\gamma \widehat{X}_n(\gamma)$. If the upper bound of the density f is infinite, the process $\widehat{X}_n(\gamma) = X_n(\widehat{\eta}_{n,\gamma}, \gamma)$ reduces to

$$\widehat{X}_n(\gamma) = \int_\gamma^\infty \log \frac{f_{\widehat{\eta}_{n,\gamma}}}{f}\, d\widehat{F}_{Y,n}, \tag{3.9}$$

the function $X(\theta) = \int_\gamma^\infty \log(f^{-1} f_\eta)\, dF_Y$ is negative under H_0 and it is maximum with the value zero as γ is the upper limit τ of the support of the variable Y, then the parameter value η_0 of the density under H_0 is such that $f_0 = f_{\eta_0}$, or it does not exist. If the density has a finite support with upper bound γ_0 under H_0, the function X is $X(\theta) = \int_\gamma^{\gamma_0} \log(f^{-1} f_\eta)\, dF_Y$ and $\widehat{X}_n(\gamma) = \int_\gamma^{\gamma_0} \log\{f^{-1} f_{\widehat{\eta}_{n,\gamma}}\}\, d\widehat{F}_{Y,n}$.

Under H_0 and under the conditions of a regular density f uniformly in $C^2(\mathcal{H}')$ in a set $\mathcal{H}'$ including $\mathcal{H}$, and such that $\int_{\mathbb{R}} f_\eta'^2(y) f_\eta^{-1}(y)\, dy$ is finite on $\mathcal{H}'$, the estimator $\widehat{\theta}_n$ maximizing X_n in the model (3.1) is such that $\widehat{\gamma}_n$ tends to τ, $X'_{n,\eta}(\widehat{\theta}_n) = 0$. The derivatives of X_n with respect to η converge a.s. under H_0 to X'_η and respectively X''_η, and there exists a parameter value η^* in $\mathcal{H}'$ but not in $\mathcal{H}$ where X achieves its maximum. With $\gamma^* = \tau$, the derivatives satisfy $X'_\eta(\theta^*) = 0$ and $X''_\eta(\theta^*) = 0$ under H_0. The maximum of X_n with respect to η, at a fixed γ is achieved at $\widehat{\theta}_{n,\gamma}$ converging a.s. to a limit $\theta_{0,\gamma}$ under H_0.

For every sequence $(\theta_n)_n$ converging to θ_0, the derivatives at θ_n of the variable X limit of the process X_n converge to zero if γ_0 is infinite so the inequality (3.3) and Theorem 3.2 are not satisfied. Let c_n^{-1} be the convergence rate of $n^{\frac{1}{2}} X'_{n,\eta}(\eta^*, \widehat{\gamma}_n)$ to a non-degenerated limit, then the convergence rate of $X''_{n,\eta}(\eta^*, \widehat{\gamma}_n)$ to a non-zero limit is c_n^{-2} and $I^* = n c_n^2 E_0 X'^{\otimes 2}_{n,\eta}(\theta^*)$. If γ_0 is finite, Theorem 3.2 is satisfied and $c_n = 1$. The behaviour of the test statistic T_n is similar to the statistic of Section 2.4 for the hypothesis of a change in the mean of the variable Y.

Proposition 3.3. *The statistic T_n converges weakly under H_0 to a χ^2_d variable T_0.*

Proof. By a first order expansion of $X'_{n,\eta}(\widehat{\theta}_n)$ with respect to the parameter η, the variable

$$n^{\frac{1}{2}} c_n^{-1} (\widehat{\eta}_n - \eta^*) = I^{*-1} c_n n^{\frac{1}{2}} X'_{n,\eta}(\eta^*, \widehat{\gamma}_n) + o_p(1)$$

converges weakly to a non-degenerated limit under P_0 and a second order expansion of the logarithm of $f_{\widehat{\eta}_n}$ in the expression of T_n implies

$$nX_n(\widehat{\theta}_n) = n^{\frac{1}{2}}c_n^{-1}(\widehat{\eta}_n - \eta^*)^T c_n n^{\frac{1}{2}} X'_{n,\eta}(\eta^*, \widehat{\gamma}_n) + \frac{nc_n^{-2}}{2}(\widehat{\eta}_n - \eta^*)^T I^*(\widehat{\eta}_n - \eta^*) + o_p(1)$$

and by plug-in the expression of $\widehat{\eta}_n - \eta^*$ in this expression, we obtain

$$T_n = X'^T_{n,\eta}(\eta^*, \widehat{\gamma}_n) I_n^{*-1} X'_{n,\eta}(\eta^*, \widehat{\gamma}_n) + o_p(1).$$

Its limit follows from the weak convergence of the variables $c_n n^{\frac{1}{2}} X'_{n,\eta}(\theta^*)$ and $c_n^2 X''_{n,\eta}(\widehat{\theta}_n)$. □

Under the alternative K of a change point at γ between $Y_{n:1}$ and $Y_{n:n}$, $n^{-1}l_{0n}$ converges under P_θ to $l(\theta) = \int_{-\infty}^{\gamma} \log f\, dF + \int_{\gamma}^{\tau} \log f\, dF_\eta$. Under the alternative F_θ, the process $X_n(\theta')$ defined by (3.9) has the mean

$$X_\theta(\theta') = \int_{\gamma'}^{\tau} \log \frac{f_{\eta'}}{f}\, dF_\theta,$$

it is maximum at θ. Under P_θ such that the integral $X''(\theta) = \int_{\gamma}^{\tau} (f'^2_\eta) f_\eta^{-1}\, dy$ is not zero, the maximum likelihood estimator of η has the convergence rate $n^{-\frac{1}{2}}$ and the estimator of the change-point has the convergence rate n^{-1}, with a limiting distribution given by Theorem 3.3.

Local alternatives P_{θ_n} contiguous to the probability P_0 of the hypothesis H_0 with the rates a_n and b_n tending to infinity as n tends to infinity are defined by parameters $\eta_n = \eta^* + a_n^{-1} v_n$, v_n converging to a non-zero limit v, and $\gamma_n = \tau - b_n^{-1} u_n$, u_n converging to a non-zero limit u

Proposition 3.4. *Under a fixed alternative P_θ, the statistic T_n tends to infinity. Under local alternatives $P_{\theta_{n,u}}$ contiguous to H_0, with rates such that $a_n b_n^{\frac{1}{2}} = O(n^{-1})$, the statistic T_n converges weakly to $T_0 + vT_1$ where T_1 is a non-degenerated variable.*

Proof. Under P_θ, $\int_\gamma^\tau f'_\eta(y)\, dy = 0$, and $\widehat{\eta}_n$ converges to η. By an expansion of X'_n in a neighborhood of η, $\widehat{\eta}_n$ satisfies

$$\int_{\widehat{\gamma}_n}^{\tau} \frac{f'_\eta}{f_\eta}\, d\nu_{n,\theta} = \Big\{ n^{\frac{1}{2}}(\widehat{\eta}_n - \eta)^T \int_{\widehat{\gamma}_n}^{\tau} \frac{f'^2_\eta - f_\eta f''_\eta}{f_\eta^2}\, d\widehat{F}_{Y,n} \Big\}\{1 + o_p(1)\},$$

under the condition of a second derivative $X''_\eta(\theta)$ different from zero, $n^{\frac{1}{2}}(\widehat{\eta}_n - \eta)$ converges weakly to a Gaussian variable. A second order expansion of the process X_n entails

$$\begin{aligned} nX_n(\widehat{\theta}_n) &= n \int_{\widehat{\gamma}_n}^{\tau} \log \frac{f_\eta}{f} \, d\widehat{F}_{Y,n} + n \int_{\widehat{\gamma}_n}^{\tau} \log \frac{f_{\widehat{\eta}_n}}{f_\eta} \, d\widehat{F}_{Y,n} \\ &= \frac{1}{2} \Big\{ \int_{\widehat{\gamma}_n}^{\tau} (f'_\eta f_\eta^{-1}) \, dW \circ f_\eta \Big\}^T I(\theta) \Big\{ \int_{\widehat{\gamma}_n}^{\tau} (f'_\eta f_\eta^{-1}) \, dW \circ f_\eta \Big\} \\ &\quad + n \int_{\widehat{\gamma}_n}^{\tau} \log \frac{f_\eta}{f} \, dF_\eta + o_p(1) \end{aligned}$$

the first term of the last equality converges weakly to a χ^2_d variable and the last term tends to infinity under P_θ.

Let P_{θ_n} be alternatives with densities f_{η_n} in $C^2(\mathcal{H})$ converging to the density $f = f_{\eta^*}$ of the hypothesis H_0, with parameters γ_n converging to τ and η_n converging to η^*. The variable

$$Z_n(\eta_n, \widehat{\gamma}_n) = \int_{\widehat{\gamma}_n}^{\tau} \log \frac{f_{\eta_n}}{f} \, d\widehat{F}_{Y,n} = \int_{\widehat{\gamma}_n}^{\tau} \Big(\log \frac{f_{\eta_n}}{f_{\eta^*}} + \log \frac{f_{\eta^*}}{f} \Big) \, d\widehat{F}_{Y,n} \quad (3.10)$$

of the previous expression of $X_n(\widehat{\theta}_n)$ is denoted $Z_{1n}(\eta_n, \widehat{\gamma}_n) + Z_{2n}(\widehat{\gamma}_n)$ where Z_{1n} is such that $Z_{1n}(\eta^*, \widehat{\gamma}_n) = 0$, $n^{\frac{1}{2}} Z'_{1n,\eta}(\eta^*, \widehat{\gamma}_n)$ and $Z''_{1n,\eta}(\eta^*, \widehat{\gamma}_n)$ converge in probability to zero as n tends to infinity. By an expansion of the logarithm of f_{η_n}, $Z_{1n}(\eta_n, \widehat{\gamma}_n)$ has the expansion

$$a_n Z_{1n}(\eta_n, \widehat{\gamma}_n) = \Big\{ v_n Z'_{1n,\eta}(\eta^*, \widehat{\gamma}_n) + \frac{a_n^{-1} v_n^2}{2} Z''_{1n,\eta}(\eta^*, \widehat{\gamma}_n) \Big\} \{1 + o_p(1)\},$$

and $nZ_{1n}(\eta_n, \widehat{\gamma}_n)$ is asymptotically equivalent to $na_n^{-1} v_n Z'_{1n,\eta}(\eta^*, \widehat{\gamma}_n)$ as n tends to infinity. We have

$$na_n \int_{\widehat{\gamma}_n}^{\tau} f'_{\eta^*} f_{\eta^*}^{-1} f_{\theta_n} = \{ na_n^{-1} b_n^{-1} u_n f'_{\eta^*}(\widetilde{\gamma}_n) \} \{1 + o_p(1)\}$$

where $\widetilde{\gamma}_n$ is between $\widehat{\gamma}_n$ and τ. The convergence rate of the variable $na_n^{-1} v_n \int_{\widehat{\gamma}_n}^{\tau} f' f^{-1} \, d\nu_n$ to a non-zero limit depends on the rate of its variance which has the order $n^2 a_n^{-2} b_n^{-1}$, it is a $O(1)$ according to the choice of the sequences $(a_n)_n$ and $(b_n)_n$. It follows that $na_n \int_{\widehat{\gamma}_n}^{\tau} f'_{\eta^*} f_{\eta^*}^{-1} f_{\theta_n} = o_p(1)$ and the variable $nZ_{1n}(\eta_n, \widehat{\gamma}_n)$ is asymptotically equivalent to $na_n^{-1} v_n \int_{\widehat{\gamma}_n}^{\tau} f' f^{-1} \, d\nu_n$, it converges weakly to vT_1 with a non-degenerated variable T_1. By the same arguments, the variable $nZ_{2n}(\widehat{\gamma}_n)$ is asymptotically equivalent under P_{θ_n} to

$$n \int_{\widehat{\gamma}_n}^{\gamma_0} \log \frac{f_{\eta^*}}{f_{\eta_n}} \, d\nu_n = 0(nb_n^{-1}) = o_p(1).$$

A second order expansion of $nX_n(\widehat{\theta}_n)$ ends the proof as previously. □

Replacing the model (3.1) by the model (3.7) with two unknown density parameters η_1 and η_2, the log-likelihood ratio test of the hypothesis H_0 of a single parametric density against the alternative of a change under (3.7) is performed by the test statistic T_n defined from the log-likelihood process (3.8) as

$$\begin{aligned} T_n &= 2\{l_n(\widehat{\theta}_n) - l_{0n}(\widehat{\eta}_{0n})\} \\ &= 2\Big\{\sum_{k=1}^n \log \frac{f_{\widehat{\eta}_{1n}}(Y_k)}{f_{\widehat{\eta}_{0n}}(Y_k)} 1_{\{Y_k \le \widehat{\gamma}_n\}} + \sum_{k=1}^n \log \frac{f_{\widehat{\eta}_{2n}}(Y_k)}{f_{\widehat{\eta}_{0n}}(Y_k)} 1_{\{Y_k > \widehat{\gamma}_n\}}\Big\}, \end{aligned}$$

where $\widehat{\eta}_{0n}$ is the maximum likelihood estimator of η_0 under H_0 for the log-likelihood process $l_{0n}(\eta_0) = \sum_{k=1}^n \log f_{\eta_0}(Y_k)$. Under H_0, the parameters of the alternative are $\eta_{01} = \eta_0$, $\eta_{02} = \eta^*$ in $\mathcal{H}'$ and $\gamma_0 = \tau$. The test statistic is also written as

$$T_n = 2\Big\{\sum_{k=1}^n \log \frac{f_{\widehat{\eta}_{1n}}(Y_k)}{f_{\widehat{\eta}_{0n}}(Y_k)} + \sum_{k=1}^n \log \frac{f_{\widehat{\eta}_{2n}}(Y_k)}{f_{\widehat{\eta}_{1n}}(Y_k)} 1_{\{Y_k > \widehat{\gamma}_n\}}\Big\},$$

which is denoted $T_n = T_{1n} + T_{2n}$ and $n^{-1}T_n$ converges a.s. to zero, the estimators $\widehat{\eta}_{1n}$ and $\widehat{\gamma}_n$ are then a.s. consistent.

Proposition 3.5. *The statistic T_n converges weakly under H_0 to variable T_0 with distribution χ^2_d.*

Proof. The variable T_{2n} is similar to the test statistic for the hypothesis H_0 in the model (3.1) and it converges weakly under H_0 to a χ^2_d variable. The variable T_{1n} is the value at $\widehat{\eta}_{0n}$ and $\widehat{\eta}_{1n}$ of the log-likelihood ratio

$$l_{0n}(\eta_0, \eta_1) = \sum_{k=1}^n \log \frac{f_{\eta_1}(Y_k)}{f_{\eta_0}(Y_k)}, \tag{3.11}$$

Let

$$l_{1n}(\eta_0, \eta_1) = \sum_{k=1}^n 1_{\{Y_k \le \widehat{\gamma}_n\}} \log \frac{f_{\eta_1}(Y_k)}{f_{\eta_0}(Y_k)},$$

at η_{0n} and η_{1n} converging to η_0 with the rate $n^{-\frac{1}{2}}$, the variable l_{1n} has the approximation $l_{1n}(\eta_{0n}, \eta_{1n}) = l_{0n}(\eta_{0n}, \eta_{1n}) + o_p(1)$. Let $U_{jn}(\eta_1)$ be the first derivative of $n^{-\frac{1}{2}} l_{jn}(\eta_0, \eta_1)$ with respect to η_1, their difference converges to zero in probability therefore $U_{0n}(\widehat{\eta}_{0n}) = 0$ and $U_{0n}(\widehat{\eta}_{1n}) = o_p(1)$.

At η_0, the second derivative of $n^{-1}l_{0n}$ with respect to η_1 converges in probability under P_0 to $-I_{01}$ and $I_{01} = \int_{\mathbb{R}} f'^2_{\eta_0} f^{-1}_{\eta_0}$ is the asymptotic variance of $U_{0n}(\eta_{1n})$, for η_{1n} converging to η_0 with the rate $n^{-\frac{1}{2}}$, it is a positive definite matrix. Then $n^{\frac{1}{2}}(\widehat{\eta}_{1n} - \widehat{\eta}_{0n}) = o_p(1)$.

By a second order expansion of the process l_{0n}, the variable

$$T_{1n} = U_{0n}^T(\eta_n) I_{01}^{-1} U_{0n}(\eta_n) + o_p(1)$$

converges in probability to zero under H_0. □

Proposition 3.6. *Under fixed alternatives, the statistic T_n tends to infinity as n tends to infinity. Under local alternatives P_{θ_n} contiguous to H_0 such that $a_n b_n^{\frac{1}{2}} = O(n^{-1})$, the statistic T_n converges weakly to $T_0 + T_1$ with a non-degenerated variable T_1.*

Proof. Under a fixed alternative P_θ with a finite change-point γ between $Y_{n:1}$ and $Y_{n:n}$ and with distinct parameters η_1 and η_2, η_0 differs from η_1 or η_2 and T_n diverges.

Let P_{θ_n} be a local alternative with $\gamma_n = \tau - b_n^{-1} u_n$ where u_n converges to u different from zero, $\eta_{jn} = \eta_0 + n^{-\frac{1}{2}} v_{jn}$, where v_{jn} converges to a non-zero limit v_1, for $j = 0, 1$, and $\eta_{2n} = \eta_{02} + a_{2n}^{-1} v_{2n}$ where η_{02} is an arbitrary limit and v_{2n} converges to a limit v_2. Under P_{θ_n}, the variables $n^{\frac{1}{2}}(\widehat{\eta}_{jn} - \eta_{jn})$ converges weakly to centered Gaussian variables with the same variance, for $j = 1, 2$, and $n^{\frac{1}{2}}(\widehat{\eta}_{1n} - \widehat{\eta}_{0n})$ converges in probability to $v_1 - v_0$. By the expansion $n^{\frac{1}{2}}(\widehat{\eta}_{1n} - \widehat{\eta}_{0n})^T I_{01} = U_{1n}(\widehat{\eta}_{1n}) + o_p(1)$, the variable $U_{1n}(\widehat{\eta}_{1n})$ converges in probability to $(v_1 - v_0)^T I_{01}$.

The process

$$T_{1n}(\theta_n) = 2 \sum_{k=1}^{n} \log \frac{f_{\eta_{1n}}(Y_k)}{f_{\eta_0}(Y_k)} 1_{\{Y_k \leq \gamma_{n,u}\}}$$

has the approximation

$$T_{1n}(\theta_n) = 2 l_{1n}(\eta_{1n}, \eta_{0n}) + o_p(1)$$

with l_{1n} defined by (3.11). Like under H_0, a second order expansion of $l_{1n}(\eta_{1n}, \eta_{0n})$ implies $T_{1n} = U_{1n}^T(\widehat{\eta}_{1n}) I_{01}^{-1} U_{1n}(\widehat{\eta}_{1n}) + o_p(1)$ and it converges in probability to $(v_1 - v_0)^T I_{01}(v_1 - v_0)$ which is strictly positive.

The process

$$T_{2n}(\theta_n) = 2 \sum_{k=1}^{n} \log \frac{f_{\eta_{2n}}(Y_k)}{f_{\eta_0}(Y_k)} 1_{\{Y_k > \gamma_{n,u}\}}$$

is similar to the process $Z_n(\theta_n)$ defined by (3.10) in the expansion of the test statistic of Proposition 3.4. Under P_{θ_n} and under the conditions, it converges weakly to the sum of a χ_d^2 variable and a non-degenerated variable. □

3.3 Maximum likelihood with an unknown density

The model (3.1) for the density can be viewed as a regular transform of the variable after the change-point

$$f_\theta(y) = f(y)1_{\{y\le\gamma\}} + f\circ\varphi_\eta(y)1_{\{y>\gamma\}} \tag{3.12}$$

with a known function φ defined from $\mathcal{H}\times\mathbb{R}$ to $\mathbb{R}$. The common distribution function F of the real variables Y such that $Y\le\gamma$ and $\varphi_\eta(Y)$, for Y such that $Y>\gamma$, has the empirical estimator

$$\begin{aligned}\widehat{F}_n(y) &= n^{-1}\sum_{k=1}^n 1_{\{Y_k\le\gamma\wedge y\}} + n^{-1}\sum_{k=1}^n 1_{\{\gamma<Y_k\le\varphi_\eta^{-1}(y)\}} \qquad (3.13)\\ &= 1_{\{y\le\gamma\}}\widehat{F}_{Y,n}(y) + 1_{\{y>\gamma\}}\{\widehat{F}_{Y,n}\circ\varphi_\eta^{-1}(y) - \widehat{F}_{Y,n}\circ\varphi_\eta^{-1}(\gamma)\},\end{aligned}$$

depending on the inverse of the function φ_η. The density f of F is estimated by the kernel estimator $f_{n,h}$ smoothing $\widehat{F}_n$ under the Conditions 2.1

$$\begin{aligned}\widehat{f}_{n,h}(y) &= \int K_h(y-s)\,\widehat{F}_n(s) = n^{-1}\sum_{k=1}^n K_h(Y_k-y)1_{\{Y_k\le\gamma\wedge y\}}\\ &\quad + n^{-1}\sum_{k=1}^n K_h(\varphi_\eta(Y_k)-y)1_{\{\gamma<Y_k\le\varphi_\eta^{-1}(y)\}}, \qquad (3.14)\end{aligned}$$

The parameters are estimated by maximization of the estimated likelihood

$$\widehat{l}_{n,h}(\theta) = \sum_{k=1}^n\Big\{\log\widehat{f}_{n,h}(Y_k)1_{\{Y_k\le\gamma\}} + \log\widehat{f}_{n,h}\circ\varphi_\eta(Y_k)1_{\{Y_k>\gamma\}}\Big\}, \tag{3.15}$$

and the log-likelihood ratio $X_n(\theta)$ of the sample under P_θ and P_0 is estimated by the process

$$\begin{aligned}\widehat{X}_{n,h}(\theta) &= n^{-1}\sum_{k=1}^n\Big\{\log\frac{\widehat{f}_{n,h}(Y_k)}{\widehat{f}_{n,h}\circ\varphi_{\eta_0}(Y_k)}1_{]\gamma_0,\gamma]}(Y_k)\\ &\quad + \log\frac{\widehat{f}_{n,h}\circ\varphi_\eta(Y_k)}{\widehat{f}_{n,h}(Y_k)}1_{]\gamma,\gamma_0]}(Y_k) + \log\frac{\widehat{f}_{n,h}\circ\varphi_\eta(Y_k)}{\widehat{f}_{n,h}\circ\varphi_{\eta_0}(Y_k)}1_{\{Y_k>\gamma\vee\gamma_0\}}\Big\}.\end{aligned}$$

Under P_0, the process $\widehat{X}_{n,h}$ converges a.s. uniformly on $\Theta\setminus\{\theta_0\}$, to the function X limit of the process X_n in Section 3.1

$$\begin{aligned}X(\theta) &= 1_{\{\gamma>\gamma_0\}}\int_{\gamma_0}^{\gamma}\log\frac{f(y)}{f_\eta(y)}\,df_{\eta_0}(y) + 1_{\{\gamma<\gamma_0\}}\int_{\gamma}^{\gamma_0}\log\frac{f_\eta(y)}{f(y)}\,dF(y)\\ &\quad + \int_{\gamma_0}^{\infty}\log\frac{f_\eta(y)}{f_{\eta_0}(y)}\,df_{\eta_0}(y).\end{aligned}$$

We assume the following regularity conditions for the functions f and φ : f belongs to $C^2(\mathbb{R})$, $\theta \mapsto \varphi_\theta(y)$ belongs to $C^2(\mathcal{H})$ for every y, and their second order derivatives are uniformly bounded.

Theorem 3.4. *Under the regularity conditions for the functions f and φ and Conditions 2.1, the estimator $\widehat{\theta}_{n,h} = \arg\max_\Theta \widehat{X}_{n,h}(\theta)$ of θ_0 is a.s. consistent.*

This is a direct consequence of Proposition 2.8 for the consistency of kernel estimator (3.14), and Theorem 3.1 the maximum likelihood estimators of Section 3.1.

Under the above conditions, the bias of $\widehat{f}_{n,h}(y)$ is written as $h^2 b_{f\circ\varphi}(y) + o(h^2)$ and its variance is

$$Var\{\widehat{f}_{n,h}(y)\} = (2nh)^{-1} k_2 \, f \circ \varphi_\eta(y) + o((nh)^{-1}),$$

it is denoted $(nh)^{-1}\sigma^2_{f\circ\varphi_\eta}(y) + o((nh)^{-1})$, with uniform approximations. Moreover, the process $U_{n,h} = (nh)^{\frac{1}{2}}\{\widehat{f}_{n,h} - f\}I\{\mathcal{I}_{X,h}\}$ converges weakly to $W_{f\circ\varphi_\eta}$, where W_f is a centered Gaussian process on $\mathcal{I}_X$ with variance $\sigma^2_{f\circ\varphi_\eta}$ and covariance zero.

Like in Section 3.1, we have $\log \widehat{f}_{n,h} \circ \varphi_\eta(y) - \log \widehat{f}_{n,h}(y) = \log\{1 + \widehat{\xi}_{n,h}(y,\eta)\}$, where

$$\widehat{\xi}_{n,h}(y,\eta) = \{\widehat{f}_{n,h} \circ \varphi_\eta(y) - \widehat{f}_{n,h}(y)\}\widehat{f}^{-1}_{n,h}(y),$$

under Conditions 2.1, its mean $\xi_{n,h}(y,\eta)$ has the expansion

$$\xi_{n,h}(y,\eta) = \{f_{n,h}(y) - f_{n,h} \circ \varphi_\eta(y)\} f^{-1}_{n,h}(y) + O((nh)^{-\frac{1}{2}})$$

and its variance is a $O((nh)^{-1})$. It follows

$$E_0 \log \frac{\widehat{f}_{n,h} \circ \varphi_\eta(y)}{\widehat{f}_{n,h}(y)} = \log \frac{f_\eta(y)}{f(y)} + O(h^2)$$

its bias has the same order as the bias of the kernel estimator and its variance is

$$Var \log \frac{\widehat{f}_{n,h} \circ \varphi_\eta(y)}{\widehat{f}_{n,h}(y)} = O((nh)^{-1}),$$

it has the same order as the variance of the kernel estimator. Let

$$\mathcal{U}_n = \{u = (u_\eta, u_\gamma)^T : u_\eta = n^{\frac{1}{2}}(\eta - \eta_0), u_\gamma = n(\gamma - \gamma_0); f_\eta \neq f, \gamma \neq \gamma_0\}$$

it is endowed with the semi-norm $\|u\| = (|u_\gamma| + \|u_\eta\|^2)^{\frac{1}{2}}$. For $u = (u_\eta^T, u_\gamma)^T$ in $\mathcal{U}_n$, $\theta_{n,u}$ is the vector with components $\eta_{n,u}$ and $\gamma_{n,u}$, with $\eta_{n,u} = \eta_0 +$

$n^{-\frac{1}{2}}u_\eta$ and $\gamma_{n,u} = \gamma_0 + n^{-1}u_\gamma$. For every $\varepsilon > 0$, an ε-neighborhood $V_\varepsilon(\theta_0)$ of θ_0 defines a neighborhood $\mathcal{U}_{n,\varepsilon}$ of zero in $\mathcal{U}_n$. The process $\widehat{X}_{n,h}$ splits like X_n as a sum of integrals on intervals of length $|\gamma - \gamma_0|$ and an integral of the logarithm of the ratio of $\widehat{f}_{n,h} \circ \varphi_\eta(y)$ and $\widehat{f}_{n,h} \circ \varphi_{\eta_0}(y)$. Let

$$\widehat{W}_{n,h}(\theta) = n^{\frac{1}{2}}\{\widehat{X}_{n,h}(\theta) - X(\theta)\}.$$

For every $\varepsilon > 0$, under Conditions 2.1 and $E_0 \sup_{\theta \in V_\varepsilon(\theta_0)} \log^2 f_\theta(Y)$ finite and by the approximation of the moments of $\widehat{f}_{n,h}$ and $\widehat{\xi}_{n,h}(y, \eta)$, there exists a constant $\kappa_1 > 0$ such that for n large enough

$$E_0 \sup_{\theta \in V_\varepsilon(\theta_0)} \widehat{W}_{n,h}(\theta) \le \kappa_1 \rho(\theta, \theta_0), \tag{3.16}$$

so the nonparametric estimation of the density does not modify this bound, the parameters of the model belonging to a finite dimensional real space. The convergence rates of the maximum likelihood estimator $\widehat{\gamma}_{n,h}$ and $\widehat{\eta}_{n,h}$ are deduced from the inequalities (3.3) and (3.16). The proof is the same as for Theorem 3.2.

Theorem 3.5. *Under the conditions 2.1 and the regularity conditions of f and φ, the estimator $\widehat{\theta}_{n,h}$ is such that*

$$\overline{\lim}_{n,A\to\infty} P_0(n|\widehat{\gamma}_{n,h} - \gamma_0| > A^2, n^{\frac{1}{2}}\|\widehat{\eta}_{n,h} - \eta_0\| > A,) = 0.$$

The process $\widehat{W}_{n,h} = n^{\frac{1}{2}}(\widehat{X}_{n,h} - X)$ has an expansion $\widehat{W}_{n,h}(\theta) = W_{1n,h}(\gamma) + W_{2n,h}(\eta) + o_p(1)$ with

$$\begin{aligned}
\widehat{W}_{1n,h}(\gamma) &= \int_{\gamma_0}^{\gamma} \log \frac{\widehat{f}_{n,h}}{\widehat{f}_{n,h} \circ \varphi_{\eta_0}} \, d\nu_{n0} \\
&\quad + n^{\frac{1}{2}} \int_{\gamma_0}^{\gamma} \Big\{ \log \frac{\widehat{f}_{n,h}}{\widehat{f}_{n,h} \circ \varphi_{\eta_0}} - \log \frac{f}{f_{\eta_0}} \Big\} \, dF_0, \\
\widehat{W}_{2n,h}(\eta) &= \int_{\gamma_0}^{\infty} \log \frac{\widehat{f}_{n,h} \circ \varphi_\eta}{\widehat{f}_{n,h} \circ \varphi_{\eta_0}} \, d\nu_{n0} \\
&\quad + n^{\frac{1}{2}} \int_{\gamma_0}^{\infty} \Big\{ \log \frac{\widehat{f}_{n,h} \circ \varphi_\eta}{\widehat{f}_{n,h} \circ \varphi_{\eta_0}} - \log \frac{f_\eta}{f_{\eta_0}} \Big\} \, dF_0,
\end{aligned}$$

where the mean of the first terms is zero and at $\gamma_{nu} = \gamma_0 + n^{-1}u$, $|u| < A^2$, $nVar\widehat{W}_{1n,h}(\gamma) = 0(|u|)$. The limiting distribution of $\widehat{\gamma}_{nh}$ is deduced from the behaviour of the process $\widehat{W}_{1n,h}$, it is similar to the limiting distribution of the maximum likelihood estimator, with a modified variance.

The restriction of the process nX_n on the intervals $]\gamma_{n,u}, \gamma_0]$ and $]\gamma_0, \gamma_{n,u}]$ is also expressed on $\mathcal{U}_n^A$ with the partial sums

$$S_n(u) = \sum_{k=1}^n \log \frac{f \circ \varphi_{\eta_0}(Y_k)}{f(Y_k)} 1_{]\gamma_{n,u},\gamma_0]}(Y_k) + \sum_{k=1}^n \log \frac{f(Y_k)}{f \circ \varphi_{\eta_0}(Y_k)} 1_{]\gamma_0,\gamma_{n,u}]}(Y_k),$$

it is a difference of two processes on $\mathbb{R}^+$, and respectively $\mathbb{R}^-$.

Theorem 3.6. *The variable $n(\widehat{\gamma}_{n,h} - \gamma_0)$ is asymptotically independent of $\widehat{\eta}_{n,h}$ and converges weakly to the location of the maximum of the uncentered Gaussian process S.*

Proof. For θ in a neighborhood of θ_0, the process $n^{\frac{1}{2}}\widehat{X}_{nh}$ is expanded as

$$n^{\frac{1}{2}}\widehat{X}_{nh}(\theta) = n^{\frac{1}{2}} \int_{\gamma_0}^{\gamma} \log \frac{\widehat{f}_{n,h}}{\widehat{f}_{n,h} \circ \varphi_{\eta_0}} \, d\widehat{F}_n$$
$$+ n^{\frac{1}{2}} \int_{\gamma_0}^{\infty} \log \frac{\widehat{f}_{n,h} \circ \varphi_\eta}{\widehat{f}_{n,h} \circ \varphi_{\eta_0}} \, d\widehat{F}_n + o_p(1),$$

where the first term does not depend on η and the second one does not depend on γ. By the arguments of Theorem 2.8, the first term of the expression of $n\widehat{X}_{nh}(\theta)$ converges weakly on $\mathcal{U}_n^A$ to L_X and it is maximum at $\widehat{\gamma}_{n,h}$, then $n(\widehat{\gamma}_{n,h} - \gamma_0)$ converges weakly to the location of the maximum of the process $S(u)$. $\square$

At $\eta_{nv} = \eta_0 + n^{-\frac{1}{2}} v$ with $\|v\| < A$, a first order expansion of the process $\widehat{W}_{2n,h}(\eta)$ yields

$$n^{\frac{1}{2}}\widehat{W}_{2n,h}(\eta) = v^T \int_{\gamma_0}^{\infty} \varphi'_{\eta_0} \frac{\widehat{f}'_{n,h}}{\widehat{f}_{n,h}} \circ \varphi_{\eta_0} \, d\nu_{n0}$$
$$+ n^{\frac{1}{2}} v^T \int_{\gamma_0}^{\infty} \varphi'_{\eta_0} \frac{\widehat{f}'_{n,h}}{\widehat{f}_{n,h}} \circ \varphi_{\eta_0} \, dF_0 + O(h^2 v^2)$$
$$= v^T \int_{\gamma_0}^{\infty} \varphi'_{\eta_0} \frac{\widehat{f}'_{n,h}}{\widehat{f}_{n,h}} \circ \varphi_{\eta_0} \, d\nu_{n0} + 0(v(nh^4)^{\frac{1}{2}}).$$

Under the condition of convergence to zero of nh^4, $n^{\frac{1}{2}} E_0 \widehat{W}_{2n,h}(\eta)$ converges to zero and $nVar\widehat{W}_{2n,h}(\eta) = 0(\|v\|^2)$. The process $n^{\frac{1}{2}} W_{2n,h}$ has therefore a second order expansion

$$n^{\frac{1}{2}}\widehat{W}_{2n,h}(\eta) = v^T \int_{\gamma_0}^{\infty} \varphi'_{\eta_0} \frac{\widehat{f}'_{n,h}}{\widehat{f}_{n,h}} \circ \varphi_{\eta_0} \, d\nu_{0n}$$
$$+ \frac{v^T}{2} \Big[\int_{\gamma_0}^{\infty} \Big\{ \varphi'_{\eta_0} \frac{\widehat{f}'_{n,h}}{\widehat{f}_{n,h}} \circ \varphi_{\eta_0} \Big\} \, d\nu_{0n} \Big] v + o_p(1)$$

and the variable $n^{\frac{1}{2}}(\widehat{\eta}_{nh} - \eta_0)$ converges weakly to a centered Gaussian variable with variance I_0^{-1}.

In the model (3.7) where the density of the variable Y is defined by two density parameters according to the threshold γ, and by an unknown density function f, the estimated log-likelihood process is

$$\widehat{l}_{nh}(\theta) = \sum_{k=1}^{n} \Big\{ \log \widehat{f}_{nh} \circ \varphi_{\eta_1}(Y_k) 1_{\{Y_k \leq \gamma\}} + \log \widehat{f}_{nh} \circ \varphi_{\eta_2}(Y_k) 1_{\{Y_k > \gamma\}} \Big\}.$$

The estimator $\widehat{\theta}_{nh}$ maximizing $\widehat{l}_{nh}$ have the same asymptotic behaviour as the maximum likelihood estimators under the Conditions 2.1 and the conditions of functions f ad φ in C^2.

A nonparametric likelihood ratio test of the hypothesis H_0 of a common density for all observations against the alternative of a change of density at an unknown threshold γ is performed with the statistic

$$\widehat{T}_{n,h} = 2n \sup_{\theta \in \Theta} \widehat{X}_{n,h}(\theta).$$

Under H_0 and under the conditions for functions f and φ, and Condition 2.1, the estimator $\widehat{\eta}_{n,h}$ satisfies (3.5) where the unknown density f and its derivatives are replaced by the kernel estimator $\widehat{f}_{n,h}$ and its derivatives

$$n^{\frac{1}{2}}(\widehat{\eta}_{n,h} - \eta_0)\widehat{X}''_{n,\eta}(\widehat{\theta}_{n,h}) = \widehat{X}'_{n,\eta}(\eta_0) + o_p(1).$$

where f_{η_0} is the density under H_0, $X'_\eta(\eta_0) = 0$ and $-X''_\eta(\eta_0)$ is positive definite. By consistency of the density estimator, the process $\widehat{W}_{n,h}$ satisfies the expansion (3.16) and $\widehat{T}_{n,h}$ converges weakly to T_0 defined in Proposition 3.3. Under fixed or local alternatives of a change of density at an unknown γ, the expansion of the process $\widehat{X}_{n,h}$ is similar to the expansion of X_n and the limiting distribution of $\widehat{T}_{n,h}$ is the same as T_n in Proposition 3.4. These properties extend to the likelihood ratio test of an i.i.d. sample under model (3.7).

Let Y be a real variable with a parametric density (3.1) on a finite support I. The properties of the estimators are the same as in Section 3.1 if f is known or Section 3.3 if f is unknown. For a test of the hypothesis H_0, the inequalities (3.3) for the function X and (3.4) for the process W_n are still satisfied under H_0 and under local alternatives, the convergence rate of the breakdown point is therefore n^{-1}. The convergence rate of the estimators of the parameter η is $n^{-\frac{1}{2}}$ and the expansions of the test statistic are written accordingly. Condition $na_nb_n^{\frac{1}{2}} = O(1)$ is fulfilled and the asymptotic distributions of T_n follow by the same arguments. The results extend to the model with an unknown density.

3.4 Maximum likelihood for a chronological change

In a model with a chronological change of parameter for the density of a random variable Y, the sample splits in two independent subsamples of independent and identically distributed observations

$$f_\theta(Y_i) = f_{\eta_1}(Y_i)1_{\{i\le k\}} + f_{\eta_2}(Y_i)1_{\{i>k\}} \tag{3.17}$$

where $n^{-1}k$ converges to a real t in $]0,1[$ as n tends to infinity and it is estimated by $\widehat{t}_n = n^{-1}k$, and k is the integer part of nt. The log-likelihood of the sample with a change of parameter at t in $]0,1[$ is

$$l_n(\theta) = \sum_{i=1}^{[nt]} \log f_{\eta_1}(Y_i) + \sum_{i=[nt]+1}^{n} \log f_{\eta_2}(Y_i), \tag{3.18}$$

at the vector parameter $\theta = (\eta_1^T, \eta_2^T, t)^T$ in $\mathcal{H}^2\times]0,1[$ and the parameter value under P_0 is $\theta_0 = (\eta_{01}^T, \eta_{02}^T, t_0)^T$. The maximum likelihood estimators $\widehat{\eta}_{1n,t}$ and $\widehat{\eta}_{2n,t}$ of η_1 and respectively η_2, at an arbitrary t, are solutions of the estimating equations

$$\sum_{i=1}^{[nt]} \frac{f'}{f} \circ \varphi_{\eta_1}(Y_i)\varphi'_{\eta_1}(Y_i) = 0, \qquad \sum_{i=[nt]+1}^{n} \frac{f'_\eta}{f} \circ \varphi_{\eta_2}(Y_i)\varphi'_{\eta_2}(Y_i) = 0$$

and the point of change is estimated by

$$\widehat{t}_n = \arg\max_t l_n(\widehat{\eta}_{nt}, t),$$

then $\widehat{\eta}_{1n} = \widehat{\eta}_{1n,\widehat{t}_n}$ and $\widehat{\eta}_{2n} = \widehat{\eta}_{2n,\widehat{t}_n}$.

The process $X_n(\theta) = n^{-1}\{l_n(\theta) - l_n(\theta_0)\}$ is written as

$$\begin{aligned}
X_n(\theta) = n^{-1} \sum_{i=1}^{[nt]\wedge[nt_0]} \log \frac{f_{\eta_1}(Y_i)}{f_{\eta_{01}}(Y_i)} &+ n^{-1} \sum_{i=[nt]\vee[nt_0]+1}^{n} \log \frac{f_{\eta_2}(Y_i)}{f_{\eta_{02}}(Y_i)} \\
&+1_{\{t<t_0\}} n^{-1} \sum_{i=[nt]+1}^{[nt_0]} \log \frac{f_{\eta_2}(Y_i)}{f_{\eta_{01}}(Y_i)} \\
&+1_{\{t>t_0\}} n^{-1} \sum_{i=[nt_0]+1}^{[nt]} \log \frac{f_{\eta_1}(Y_i)}{f_{\eta_{02}}(Y_i)}.
\end{aligned}$$

Under P_0 and the condition that the logarithms are integrable, the process X_n converges a.s. uniformly as t tends to infinity to the function

$$X(\theta) = (t \wedge t_0)E_0\Big\{\log \frac{f_{\eta_1}(Y_i)}{f_{\eta_{01}}(Y_i)} 1_{\{i \le k \wedge k_0\}}\Big\}$$
$$+(1 - t \vee t_0)E_0\Big\{\log \frac{f_{\eta_2}(Y_i)}{f_{\eta_{02}}(Y_i)} 1_{\{i > k \vee k_0\}}\Big\}$$
$$+1_{\{k<k_0\}}(t_0 - t)E_0\Big\{\log \frac{f_{\eta_2}(Y_i)}{f_{\eta_{01}}(Y_i)} 1_{\{k<i\le k_0\}}\Big\}$$
$$+1_{\{k>k_0\}}(t - t_0)E_0\Big\{\log \frac{f_{\eta_1}(Y_i)}{f_{\eta_{02}}(Y_i)} 1_{\{k_0<i\le k\}}\Big\},$$

where $k = [nt]$ and $t = \lim_{n\to\infty} n^{-1}k$. For a function $\eta \mapsto f_\eta$ belonging to $C^2(\mathcal{H})$ with a second order derivative bounded on $\mathcal{H}$, uniformly on $\mathbb{R}$, and the function X satisfies $X'_\eta(\theta_0) = 0$ and

$$I_0 = -X''_\eta(\theta_0)$$

is a definite positive matrix, then the function $\eta \mapsto X(\eta, t)$ is concave in a neighborhood of η_0 such that η_{01} and η_{02} are distinct. By the uniform convergence of X_n to X, $\widehat{\eta}_n$ is a.s. consistent. Moreover, X is maximum at θ_0 where it is zero, then $\widehat{t}_n$ converges a.s. to t_0 as t tends to infinity.

Let $U_n(\theta) = n^{-\frac{1}{2}} l'_{n,\eta}(\theta)$ and $I_n(\theta) = -n^{-1} l''_{n,\eta}(\theta)$, $I_n(\theta_0)$ converges a.s. under P_0 to the positive definite matrix I_0 and $U_n(\theta_0)$ converges weakly to a centered Gaussian variable with variance I_0 as t tends to infinity. At t_0, according to the classical theory of the maximum likelihood estimation, a first order expansion of $X'_n(\widehat{\eta}_{nt_0}, t_0)$ entails

$$n^{\frac{1}{2}}(\widehat{\eta}_{nt_0} - \eta_0) = I_0^{-1}U_n(\theta_0) + o_p(1)$$

and it converges weakly to a centered Gaussian variable with variance I_0^{-1}. Under P_0 and at an arbitrary t, $\widehat{\eta}_{n,t}$ converges a.s. uniformly on $]0,1[$ to its mean $\eta_{0,t}$ and $n^{\frac{1}{2}}(\widehat{\eta}_{nt} - \eta_{0,t}) = I_{0,t}^{-1}U_n(\eta_t, t) + o_p(1)$, where $I_{0,t}$ is the limit of $-X''_{n,\eta}(\eta_t, t)$ under P_0, and $n^{\frac{1}{2}}(\widehat{\eta}_{nt} - \eta_{0,t})$ converges weakly to a centered Gaussian variable with variance $I_{0,t}^{-1}$. By a second order expansion, the function X is negative and there exists a constant $\kappa_0 > 0$ such that for every t in a neighborhood of t_0

$$X(\theta_t) \ge -\kappa_0|t - t_0|, \tag{3.19}$$

since $X'_\eta(\theta_0) = 0$.

For every $\varepsilon > 0$, under the condition $E_0 \sup_{\eta \in V_\varepsilon(\eta_0)} \log^2 f_\eta$ finite and by monotonicity of the sums with respect to t, there exists a constant $\kappa_1 > 0$ such that for n large enough, the process

$$W_n = n^{\frac{1}{2}}(X_n - X)$$

satisfies

$$E_0 \sup_{\rho(\theta,\theta_0) \leq \varepsilon} W_n(\theta) \leq \kappa_1 \varepsilon^{\frac{1}{2}}, \tag{3.20}$$

using the Cauchy–Schwarz inequality. Like in Theorem 2.2, the convergence rate of the estimator $\widehat{t}_n$ is deduced from the inequalities (3.19) and (3.20).

Theorem 3.7. $\overline{\lim}_{n,A\to\infty} P_0(n|\widehat{t}_n - t_0| > A) = 0$.

The weak convergence of the estimator $\widehat{\eta}_n$ relies on the behaviour of the processes of the partial sums

$$X_{1,n}(\theta) = \sum_{i=1}^{[nt]\wedge[nt_0]} \xi_i(\eta_1) + \sum_{i=[nt]\vee[nt_0]+1}^{n} \xi_i(\eta_2)$$

of independent and identically distributed centered variables $\xi_1(\eta_1)$, and respectively $\xi_2(\eta_2)$, defined as log-likelihood ratios. At $\theta_{n,u}$, $|u| \leq A$, the have finite means and variances. By Donsker's theorem, the process $X_{1,n}(\theta_{n,u})$ converges weakly to the difference of two weighted Brownian motions. By Proposition 1.4, the process

$$X_{2n}(\theta) = 1_{\{t<t_0\}} \sum_{i=[nt]+1}^{[nt_0]} \log \frac{f_{\eta_2}(Y_i)}{f_{\eta_{01}}(Y_i)}$$

$$+1_{\{t>t_0\}} n^{-1} \sum_{i=[nt_0]+1}^{[nt]} \log \frac{f_{\eta_1}(Y_i)}{f_{\eta_{02}}(Y_i)}$$

converges weakly in $D([-A, A])$ to the difference of two uncentered Gaussian processes with finite variances.

Theorem 3.8. *The variable $n^{\frac{1}{2}}(\widehat{\eta}_n - \eta_0)$ is asymptotically independent of $\widehat{\eta}_n$ and converges weakly to an uncentered Gaussian variable with variance I_0^{-1} and, for t_0 in $]0,1[$, $n(\widehat{t}_n - t_0)$ converges weakly to the location u_0 of the maximum of an uncentered Gaussian process, it is bounded in probability.*

Proof. Under P_0, the estimator of the density parameter has an expansion conditionally on $\widehat{t}_n$ similar to $\widehat{\eta}_{n,t_0}$

$$n^{\frac{1}{2}}(\widehat{\eta}_n - \eta_0) = I_0^{-1} U_n(\eta_0, \widehat{t}_n) + o_p(1)$$

where U_n is the vector with components

$$U_{1n}(\theta) = n^{-\frac{1}{2}} \sum_{i=1}^{[nt]} \frac{f'_{\eta_1}}{f_{\eta_1}}(Y_i),$$

$$U_{2n}(\theta) = n^{-\frac{1}{2}} \sum_{i=[nt]+1}^{n} \frac{f'_{\eta_2}}{f_{\eta_2}}(Y_i).$$

By Donsker's theorem, if the variables Y_i have the density f_{η_k}, the process $U_{kn}(\theta)$ is centered and it converges weakly to a vector of weighted Brownian bridges, for $k = 1, 2$. Under a density $f_{\eta_{0k'}}$ with $k' \neq k$, the process $U_{kn}(\theta) - U_{kn}(\eta, t_0)$ is a weighted sum of a number $[n\widehat{t}_{n,u}] - [nt_0]$ of variables, by Theorem 3.7 this number is bounded in probability therefore its mean converges a.s. to zero and its variance is finite, then $U_{kn}(\theta) - U_{kn}(\eta, t_0)$ converges weakly to a vector of Gaussian processes, for $k = 1, 2$. Furthermore, the variable $U_{kn}(\theta_0)$ converges weakly to a Gaussian variable and $U_{kn}(\widehat{\theta}_n) = 0$, it follows that $U_{kn}(\widehat{\eta}_n, t_0)$ converges weakly to a Gaussian variable. By the expansion

$$U_{kn}(\widehat{\eta}_n, t_0) - U_{kn}(\theta_0) = n^{\frac{1}{2}}(\widehat{\eta}_n - \eta_0)^T I_{0k} + o_p(1),$$

where $-I_{0k}$ is the limit of the variance of $U_{kn}(\theta_0)$ as n tends to infinity, the variable $n^{\frac{1}{2}}(\widehat{\eta}_n - \eta_0)$ converges weakly to a centered Gaussian variable with variance I_0^{-1}.

Let $t_{n,u} = t_0 + n^{-1}u$ and $\eta_{n,v} = \eta_0 + n^{-\frac{1}{2}}v$, and let $\theta_{n,u,v}$ be the vector with components $\eta_{n,v}$ and $t_{n,u}$. The process $Z_n(u,v) = nX_n(\theta_{n,u,v})$ is written as

$$Z_n(u,v) = (1_{\{t_0 < t_{n,u}\}} - 1_{\{t_0 > t_{n,u}\}}) \sum_{i=[nt_0]+1}^{[nt_{n,u}]} \log \frac{f_{\eta_{01}}(Y_i)}{f_{\eta_{02}}(Y_i)}$$

$$+(1_{\{t_0 < t_{n,u}\}} - 1_{\{t_0 > t_{n,u}\}}) \sum_{i=[nt_0]+1}^{[nt_{n,u}]} \log \frac{f_{\eta_{1n,v}}(Y_i) f_{\eta_{02}}(Y_i)}{f_{\eta_{01}}(Y_i) f_{\eta_{2n,v}}(Y_i)}.$$

As n tends to infinity, $[nt_{n,u}] - [nt_0]$ is an interval of length $[u_n]$, it is bounded in probability and, for $k = 1, 2$, the logarithms of the ratios of densities $\log f_{\widehat{\eta}_{n,k}}(Y_i) f_{\eta_{0k}}^{-1}(Y_i)$ converges in probability to zero under P_0, by consistency of the estimator $\widehat{\eta}_n$.

Under P_0, the first term of the process $Z_n(u)$ is asymptotically free of the parameter η, it converges weakly on $\mathcal{U}_n^A$ to an uncentered Gaussian process with a finite and non-degenerated variance. The maximum on $\mathcal{U}_n^A$ of the process Z_n is achieved at $\widehat{u}_n = n(\widehat{t}_n - t_0)$ which converges weakly on $\mathcal{U}_n^A$ to the location of the maximum of its limit, as n and A tend to infinity. Theorem 3.7 ends the proof. □

3.5 Likelihood ratio test for a chronological change

The log-likelihood ratio test of the hypothesis H_0 of density without change of parameter against the alternative of a change at an unknown index k_0, with distinct parameters η_1 and η_2, is performed with the statistic

$$T_n = 2\{l_n(\widehat{\theta}_n) - \widehat{l}_{0n}\},$$

where $\widehat{l}_{0n} = \sup_{\eta\in\mathbb{R}} \sum_{i=1}^n \log f_\eta(Y_i) = \sum_{i=1}^n \log f_{\widehat{\eta}_{0n}}(Y_i)$ is the maximum of the likelihood under the hypothesis H_0 and the estimator $\widehat{\eta}_{0n}$ of the parameter η_0 is a.s. consistent under H_0, $\widehat{l}_{0n}$ has the approximation

$$l_{0n} = U_{0n}^T I_{0n}^{-1} U_{0n} + o_p(1)$$

where $U_{0n} = n^{-\frac{1}{2}} \sum_{i=1}^n (f_{\eta_0}^{-1} f'_{\eta_0})(Y_i)$ is centered and it converges weakly under P_0 to a centered Gaussian variable with variance the limit of I_{0n}, the weak convergence of l_{0n} to a χ_d^2 variable follows.

Under H_0, $t_0 = 1$ and the process $X_n(\theta) = n^{-1}\{l_n(\theta) - l_n(\theta_0)\}$ is the sum

$$X_n(\theta) = n^{-1} \sum_{i=1}^{[nt]} \log \frac{f_{\eta_1}(Y_i)}{f_{\eta_0}(Y_i)} + n^{-1} \sum_{i=[nt]+1}^{n} \log \frac{f_{\eta_2}(Y_i)}{f_{\eta_0}(Y_i)},$$

it converges a.s. uniformly under P_0 to its expectation

$$X(\theta) = tE_0\Big\{\log \frac{f_{\eta_1}(Y_i)}{f_{\eta_0}(Y_i)} 1_{\{i\le[nt]\}}\Big\} + (1-t)E_0 \log\Big\{\frac{f_{\eta_2}(Y_i)}{f_{\eta_0}(Y_i)} 1_{\{i>[nt]\}}\Big\},$$

let $X'_\eta(\theta)$ and $X''_\eta(\theta)$ be its derivatives with respect to the components of η. At t tends to one, η_2 is arbitrary and the second sum in the expression of $X_n(\theta)$ tends to zero and the matrix $I_0 = X''_\eta(\theta_0)$ is singular in $\mathbb{R}^{2d} \times \mathbb{R}^{2d}$. Under H_0, the maximum likelihood estimator of η_1 converges a.s. to η_0 and the estimator $\widehat{\eta}_{2n}$ converges in probability to zero.

For $\varepsilon > 0$, let $V_\varepsilon(\eta_{0k})$ be an ε-neighborhood of η_{0k}, we assume that the integrals I_0 and

$$E_0 \sup_{\eta_k\in V_\varepsilon(\eta_{0k})} \{\log f_{\eta_k} - \log f_{\eta_{0k}}\}^2(Y),\ k=1,2,$$

are finite. Let u_0 be the limit of $\widehat{u}_n = n(t_0 - \widehat{t}_n)$, it is the maximum of a Gaussian process and it is bounded in probability.

Proposition 3.7. *The statistic T_n converges weakly under H_0 to a χ^2_d variable T_0.*

Proof. The process $Z_n = nX_n$ is a sum $Z_n = Z_{1n} + Z_{2n}$ where

$$Z_{1n}(\eta_1, t) = \sum_{i=1}^{[nt]} \log \frac{f_{\eta_1}(Y_i)}{f_{\eta_0}(Y_i)},$$

$$Z_{2n}(\eta_2, t) = \sum_{i=[nt]+1}^{n} \log \frac{f_{\eta_2}(Y_i)}{f_{\eta_0}(Y_i)}.$$

For every t, the process Z_{2n} is negative under P_0 and it is maximum at t_0 where it is zero, for every η_2 distinct from η_0, it provides the estimator $\widehat{\eta}_{2n}$ from the sub-sample $(Y_{[nt]+1}, \ldots, Y_n)$, $\widehat{\eta}_{2n}$ converges a.s. to an arbitrary limit η_{02} distinct from η_0 under P_0. The first term is maximum at $\widehat{\eta}_{1n}$ which converges a.s. to η_0 under P_0, and $\widehat{t}_n$ is a.s. consistent. The process Z_n is also written as

$$Z_n(\theta) = \sum_{i=1}^{n} \log \frac{f_{\eta_1}(Y_i)}{f_{\eta_0}(Y_i)} + \sum_{i=[nt]+1}^{n} \log \frac{f_{\eta_2}(Y_i)}{f_{\eta_1}(Y_i)}$$

and a first order expansion of the derivative of the first term at $\widehat{\eta}_{0n}$ and $\widehat{\eta}_{1n}$ under H_0 implies

$$n^{\frac{1}{2}}(\widehat{\eta}_{0n} - \widehat{\eta}_{1n}) = I_{01}^{-1} U_{1n}(\widehat{\eta}_{1n}) + o_p(1)$$

where U_{0n} is the first derivative of $\sum_{i=1}^{n}\{\log f_{\eta_1}(Y_i) - \log f_{\eta_0}(Y_i)\}$ with respect to η_1 and $-I_{01}$ is the limit of its second derivative, as n tends to infinity. The inequalities (3.19) and (3.20) and therefore Theorem 3.7 are still satisfied under H_0 therefore $\widehat{t}_n$ converges to one with the rate n^{-1}. We have $U_{0n}(\widehat{\eta}_{0n}) = 0$ and from the convergence rate of $\widehat{t}_n$ and the weak convergence of $n^{\frac{1}{2}}(\widehat{\eta}_{1n} - \eta_0)$, $U_{0n}(\widehat{\eta}_{1n})$ converges to zero in probability. By a second order expansion of the first term

$$T_{1n} = 2 \sum_{i=1}^{n} \log \frac{f_{\widehat{\eta}_{1n}}(Y_i)}{f_{\widehat{\eta}_{0n}}(Y_i)}$$

is asymptotically equivalent to $U_{0n}^T(\widehat{\eta}_{1n}) I_{01}^{-1} U_{0n}(\widehat{\eta}_{1n})$ and it converges in probability to zero under H_0.

The variable

$$T_{2n} = 2 \sum_{i=[n\widehat{t}_n]+1}^{n} \log \frac{f_{\widehat{\eta}_{2n}}(Y_i)}{f_{\widehat{\eta}_{1n}}(Y_i)}$$

is the sum of $\widehat{u}_n = n(t_0 - \widehat{t}_n)$ variables, logarithms of the ratios of the densities $\log\{f_{\widehat{\eta}_{2n}}(Y_i) f_{\eta_{02}}^{-1}(Y_i)\} - \log\{f_{\widehat{\eta}_{1n}}(Y_i) f_{\eta_0}^{-1}(Y_i)\} + \log\{f_{\eta_{02}}(Y_i) f_{\eta_0}^{-1}(Y_i)\}$ and $\widehat{u}_n$ converges weakly to a variable u_0 defined in Theorem 3.8. By the consistency of the estimators and second order asymptotic expansions, the first two sums of logarithms are $o_p(1)$ under H_0 and

$$T_{2n} = 2 \sum_{i=[n\widehat{t}_n]+1}^{n} \log \frac{f_{\eta_{02}}(Y_i)}{f_{\eta_0}(Y_i)} + o_p(1),$$

where η_{02} is the limit of the estimator $\widehat{\eta}_{2n}$. Let

$$W_{2n}(\eta_2, \eta_0) = \sum_{i=[n\widehat{t}_n]+1}^{n} \log \frac{f_{\eta_{02}}(Y_i)}{f_{\eta_0}(Y_i)},$$

the first derivative $\dot{W}_{2n}(\mu_{2,t_n}, t_n)$ with respect to μ_2 is asymptotically equivalent to a centered Gaussian variable with finite variance I_{02} and $-I_{02}$ is the limit in probability of the second derivative $\ddot{W}_{2n,\mu_2}(\mu_{2,t_n}, t_n)$ with respect to μ_2, under H_0. Using the arguments of Proposition 2.14

$$\widehat{\mu}_{2n} = I_{02}^{-1} \dot{W}_{2n,\mu_2}(0, \widehat{t}_n) + o_p(1),$$
$$T_{2n} = \dot{W}_{2n,\mu_2}^{T}(0, \widehat{t}_n) I_{02}^{-1} \dot{W}_{2n,\mu_2}^{T}(0, \widehat{t}_n) + o_p(1),$$

it converges weakly to a χ_1^2 variable under H_0. □

Under a probability P_θ, the process $X_n(\theta')$ converges weakly to its expectation

$$X_\theta(\theta') = t' \int \Big\{ \log \frac{f_{\eta_1'}(y)}{f_{\eta_1}(y)} + \log \frac{f_{\eta_1}(y)}{f_{\eta_0}(y)} \Big\} \, dF_\theta(y)$$
$$+ (1 - t') \int \Big\{ \log \frac{f_{\eta_2'}(y)}{f_{\eta_2}(y)} + \log \frac{f_{\eta_2}(y)}{f_{\eta_0}(y)} \Big\} \, dF_\theta(y),$$

the function X_θ is maximum as $\theta' = \theta$ and, by the same arguments as in Section 3.4, the maximum likelihood estimator $\widehat{\theta}_n$ converges a.s. under P_θ to θ. The convergence rates of its components $\widehat{\eta}_n$ and $\widehat{t}_n$ are still given by Theorem 3.7 and their asymptotic distributions are similar to those of Theorem 3.8 under P_θ.

Proposition 3.8. *Under fixed alternatives, the statistic T_n tends to infinity as n tends to infinity and under local alternatives P_{θ_n} contiguous to H_0, the statistic T_n converges weakly to $T_0 + S$ where S is a non-degenerated variable.*

Proof Under a fixed alternative P_θ with t in $]0,1[$ and distinct parameters, the ratio of the densities $f_{\widehat{\eta}_{1n}}$ and $f_{\widehat{\eta}_{0n}}$ is asymptotically equivalent to $f_{\eta_1} f_0^{-1}$ by the consistency of the estimators and several cases must be distinguished. If $\eta_1 = \eta_0$, a second order expansion of $Z_{1n}(\widehat{\eta}_{1n}, \widehat{t}_n)$ implies its weak convergence to a χ_d^2 variable, like in Proposition 3.7. The process Z_{2n} is not centered and by Theorem 3.7 and Donsker's theorem for the weak convergence of the empirical process, $Z_{2n}(\widehat{\eta}_{2n}, \widehat{t}_n)$ diverges. If η_0 is distinct from η_1 and η_2, the processes Z_{1n} and Z_{2n} are not centered and they diverge. If $\eta_0 = \eta_2$ is different from η_1, the variable $Z_{1n}(\widehat{\eta}_{1n}, \widehat{t}_n)$ still diverges and $Z_{2n}(\widehat{\eta}_{2n}, \widehat{t}_n)$ converges to a χ_d^2 variable, like in Proposition 3.7. In all cases, the divergence of one variable implies that the statistic T_n diverges.

Under a local alternative P_{θ_n} with t_n in $]0,1[$ converging to one and distinct parameters, let $\eta_{1n} = \eta_0 + n^{-\frac{1}{2}} v_{1n}$, $\eta_{2n} = \eta_{02} + n^{-\frac{1}{2}} v_{2n}$ and let $t_n = 1 - n^{-1} u_n$, where η_{02} is different from η_0, u_n, v_{1n} and v_{2n} converge to finite limits as n tends to infinity. The test statistic is written as

$$T_n = T_{1n} + T_{2n} = 2 \sum_{i=1}^{n} \log \frac{f_{\widehat{\eta}_{1n}}}{f_{\widehat{\eta}_{0n}}}(Y_i) + 2 \sum_{i=[n\widehat{t}_n]+1}^{n} \log \frac{f_{\widehat{\eta}_{2n}}}{f_{\widehat{\eta}_{1n}}}(Y_i).$$

By the consistency of the estimators under the alternative and the convergence of η_{1n} to η_0, the variable T_{1n} has the same expansion under K_n as under H_0 and it converges in probability to zero.

As t_n converges to 1, T_{2n} is the sum of a finite number $n - [n\widehat{t}_n] = [\widehat{u}_n]$ variables logarithm of the ratio of the densities

$$\{f_{\widehat{\eta}_{2n}}(Y_i) f_{\eta_{2n}}^{-1}(Y_i)\}\{f_{\widehat{\eta}_{1n}}(Y_i) f_{\eta_{1n}}^{-1}(Y_i)\}^{-1}\{f_{\eta_{2n}}(Y_i) f_{\eta_{1n}}^{-1}(Y_i)\}.$$

By second order expansions, the sums of the logarithm of the first two ratios are $o_p(1)$ and $\frac{1}{2} T_{2n}$ is asymptotically equivalent to the sum of the uncentered variables

$$\begin{aligned}\log\{f_{\eta_{2n}}(Y_i) f_{\eta_{1n}}^{-1}(Y_i)\} &= \log\{f_{\eta_{02}}(Y_i) f_{\eta_0}^{-1}(Y_i)\} + \log\{f_{\eta_{2n}}(Y_i) f_{\eta_{02}}^{-1}(Y_i)\} \\ &\quad - \log\{f_{\eta_{1n}}(Y_i) f_{\eta_1}^{-1}(Y_i)\}.\end{aligned}$$

Let $U_{kn} = n^{-\frac{1}{2}} \sum_{i=[n\widehat{t}_n]+1}^{n} \dot{f}_{\eta_{0k}}(Y_i) f_{\eta_{0k}}^{-1}(Y_i)$, it converges weakly to a Gaussian variable with variance I_k, $k = 1, 2$, under P_{θ_n} and

$$\begin{aligned}T_{2n} = 2 \sum_{i=[n\widehat{t}_n]+1}^{n} \log \frac{f_{\eta_{02}}(Y_i)}{f_{\eta_{01}}(Y_i)} - 2 v_{1n}^T U_{1n} + n^{-1} v_{1n}^T I_{1n} v_{1n} \\ + 2 v_{2n}^T U_{2n} - n^{-1} v_{2n}^T I_{2n} v_{2n} + o_p(1),\end{aligned}$$

it converges weakly to the variable $S = -2v_1^T U_1 + v_{1n}^T I_1 v_1 + 2v_2^T U_2 - v_2^T I_2 v_2$ where U_{kn} converges weakly to a Gaussian variable U_k and the matrix I_{kn} converges in probability to I_k, for $k = 1, 2$. □

3.6 Nonparametric maximum likelihood

When the density f is unknown in the model (3.17), the log-likelihood (3.18) of the sample with a change of parameter at an unknown index $k = [nt]$, t in $]0,1[$, is estimated by

$$\widehat{l}_{nh}(\theta) = \sum_{i=1}^{[nt]} \log \widehat{f}_{nh} \circ \varphi_{\eta_1}(Y_i) + \sum_{i=[nt]+1}^{n} \log \widehat{f}_{nh} \circ \varphi_{\eta_2}(Y_i), \qquad (3.21)$$

with a kernel estimator $\widehat{f}_{nh}$ of the density defined by (3.14). The estimators $\widehat{\eta}_{1nh,t}$ and $\widehat{\eta}_{2nh,t}$ of the components of the parameter $\eta = (\eta_1^T, \eta_2^T)^T$ at a fixed value t are solutions of the equations

$$\sum_{i=1}^{[nt]} \frac{\widehat{f}'_{nh}}{\widehat{f}_{nh}} \circ \varphi_{\eta_1}(Y_i) = 0, \qquad \sum_{i=[nt]+1}^{n} \frac{\widehat{f}'_{nh}}{\widehat{f}_{nh}} \circ \varphi_{\eta_2}(Y_i) = 0$$

and the location of the change is estimated by

$$\widehat{t}_{nh} = \arg \max_{t \in]0,1[} \widehat{l}_{nh}(\widehat{\eta}_{nh,t}, t).$$

Under P_0, by the consistency of the kernel estimator of the density, the process

$$\begin{aligned}\widehat{X}_{nh}(\theta) &= n^{-1}\{\widehat{l}_{nh}(\theta) - \widehat{l}_{nh}(\theta_0)\} \\ &= n^{-1} \sum_{i=1}^{[nt]} \log \frac{\widehat{f}_{nh} \circ \varphi_{\eta_1}(Y_i)}{\widehat{f}_{nh} \circ \varphi_{\eta_0}(Y_i)} + n^{-1} \sum_{i=[nt]+1}^{n} \log \frac{\widehat{f}_{nh} \circ \varphi_{\eta_2}(Y_i)}{\widehat{f}_{nh} \circ \varphi_{\eta_0}(Y_i)}\end{aligned}$$

converges a.s. uniformly under P_0 to the process $X(\theta)$ limit of the process $X_n(\theta)$ defined with a known density f in the previous section. By the concavity of the function X, maximum at θ_0, the estimators $\widehat{\eta}_{nh,t}$ and $\widehat{t}_{nh}$ are therefore a.s. consistent.

The process $\widehat{X}_{nh}$ has the mean $E_0\widehat{X}_{nh}(\theta) = X(\theta) + O(h^2)$ and the variance $Var_0\widehat{X}_{nh}(\theta) = O((nh)^{-1})$ due to the density estimation. Under the same conditions as in Section 3.4 and under Conditions 2.1, for every $\varepsilon > 0$, there exists a constant $\kappa_1 > 0$ such that for n large enough

$$E_0 \sup_{\rho(\theta,\theta_0) \le \varepsilon} n^{\frac{1}{2}} \widehat{X}_{nh}(\theta) \le \kappa_1 \varepsilon. \qquad (3.22)$$

The inequality (3.19) generalizes to the parameter θ. For $\varepsilon > 0$ sufficiently small, there exists a constant $\kappa_1 > 0$ such that for every θ

$$\sup_{\rho(\theta,\theta_0)\le\varepsilon} X(t) \le -\kappa_0\rho^2(\theta,\theta_0) \tag{3.23}$$

As in Theorem 3.2, the inequalities (3.23) and (3.22) provide the convergence rates of the estimator $\widehat{\theta}_n$.

Theorem 3.9. *Under the conditions of Section 3.4 and Conditions 2.1*

$$\overline{\lim}_{n,A\to\infty} P_0(n^{\frac{1}{2}}\|\widehat{\eta}_{n,h}-\eta_0\| > A, n|\widehat{t}_{n,h}-t_0| > A^2) = 0.$$

The derivatives of the estimated log-likelihood with respect to the parameters η_1 and η_2 define the process $\widehat{U}_{nh} = (\widehat{U}^T_{1nh}, \widehat{U}^T_{2nh})^T$ as

$$\widehat{U}_{1nh}(\theta) = n^{-\frac{1}{2}}\sum_{i=1}^{[nt]}\frac{\widehat{f}'_{nh,\eta_1}}{\widehat{f}_{nh}}(Y_i),$$

$$\widehat{U}_{2nh}(\theta) = n^{-\frac{1}{2}}\sum_{i=[nt]+1}^{n}\frac{\widehat{f}'_{nh,\eta_2}}{\widehat{f}_{nh}}(Y_i),$$

and the matrix $\widehat{I}_{nh}(\theta) = -n^{-1}\widehat{l}''_{nh,\eta}(\theta)$ such that $\widehat{I}_{nh}(\theta_0)$ converges a.s. under P_0 to the positive definite matrix $I_0 = -X''_\eta(\theta_0)$.

Under the conditions about the convergence of h tends to zero, $\widehat{U}_{nh}(\theta) - U_n(\theta)$ converges in probability under P_0 to zero, uniformly on bounded parameter intervals and the estimator of the density parameters maximizing $\widehat{l}_{nh}(\eta, t_0)$ is such that $n^{\frac{1}{2}}(\widehat{\eta}_{n,t_0} - \eta_0)$ converges weakly to a centered Gaussian process with variance I_0. At a fixed t, the estimator $\widehat{\eta}_{n,t}$ of η which maximizes $\widehat{l}_{nh}(\eta, t)$ has the expansion $n^{\frac{1}{2}}(\widehat{\eta}_{n,t} - \eta_{0,t}) = I^{-1}_{0,t}U_{nh}(\eta, t) + o_p(1)$, under P_0, where $\eta_{0,t} = E_0\widehat{\eta}_{n,t}$, and it converges weakly to a centered Gaussian variable with variance $I^{-1}_{0,t}$.

According to Theorem 3.9, we consider the parametrization $t_{n,u} = t_0 + n^{-1}u_n$, $\eta_{n,v} = \eta_0 + n^{-\frac{1}{2}}v_n$ and $\theta_n = (t_{n,u_n}, \eta^T_{n,v_n})^T$, with sequences $(u_n)_n$, and respectively $(v_n)_n$, converging to finite limits u, and respectively v. The process $\widehat{W}_{nh} = n\{\widehat{X}_{nh}(\theta_{n,u,v}) - X(\theta_{n,u,v})\}$ is a approximated by a sum

$$\widehat{W}_{nh}(u, v) = \widehat{W}_{1nh}(v) + \widehat{W}_{2nh}(u) + o_p(1)$$

where $\widehat{W}_{1nh}$ is the difference $n\{\widehat{X}_{nh}(\eta_{n,v}, t_0) - X(\eta_{n,v}, t_0)$ and $\widehat{W}_{2nh}(u)$ depends only on u and η_0

$$\widehat{W}_{1nh}(v) = \sum_{i=1}^{[nt_0]} \log \frac{\widehat{f}_{nh} \circ \varphi_{\eta_{1n,v_1}}(Y_i)}{\widehat{f}_{nh} \circ \varphi_{\eta_{01}}(Y_i)} + \sum_{i=[nt_0]+1}^{n} \log \frac{\widehat{f}_{nh} \circ \varphi_{\eta_{2n,v_2}}(Y_i)}{\widehat{f}_{nh} \circ \varphi_{\eta_{02}}(Y_i)}$$
$$- \sum_{i=1}^{[nt_0]} E_0 \log \frac{f_{\eta_{1n,v_1}}(Y_i)}{f_{\eta_{01}}(Y_i)} - \sum_{i=[nt_0]+1}^{n} E_0 \log \frac{f_{\eta_{2n,v_2}}(Y_i)}{f_{\eta_{02}}(Y_i)},$$

$$\widehat{W}_{2nh}(u) = 1_{\{t_{n,u}<t_0\}} \sum_{i=[nt_0+u]+1}^{[nt_0]} \Big\{\log \frac{\widehat{f}_{nh} \circ \varphi_{\eta_{02}}(Y_i)}{\widehat{f}_{nh} \circ \varphi_{\eta_{01}}(Y_i)} - E_0 \log \frac{f_{\eta_{02}}(Y_i)}{f_{\eta_{01}}(Y_i)}\Big\}$$
$$+1_{\{t_{n,u}>t_0\}} \sum_{i=[nt_0]+1}^{[nt_0+u]} \Big\{\log \frac{\widehat{f}_{nh} \circ \varphi_{\eta_{01}}(Y_i)}{\widehat{f}_{nh} \circ \varphi_{\eta_{02}}(Y_i)} - E_0 \log \frac{f_{\eta_{01}}(Y_i)}{f_{\eta_{02}}(Y_i)}\Big\}.$$

The asymptotic distribution of $\widehat{\eta}_{n,h}$ under P_0 is deduced from the asymptotic equivalence of the processes $\widehat{W}_{1nh}$ with the process defined with a known density, in the same way, its limit is given by Theorem 3.8.

Theorem 3.10. *The variable $n(\widehat{t}_n - t_0)$ converges weakly to the location u_0 of the maximum of an uncentered Gaussian process.*

Proof. Under the condition of convergence to zero of nh^4, the process

$$Z_{nh}(u, v) = n\{\widehat{X}_{nh}(\theta_{n,u,v}) - X(\theta_{n,u,v})\}$$

is a sum $Z_{nh}(u, v) = Z_{1nh}(v) + Z_{2nh}(u) + o_p(1)$. The variable $\widehat{u}_n = n(\widehat{t}_n - t_0)$ achieves the maximum of the process

$$Z_{2nh}(u) = 1_{\{t_{n,u}<t_0\}} n \sum_{i=[nt_0+u]+1}^{[nt_0]} \log \frac{\widehat{f}_{nh} \circ \varphi_{\eta_{02}}(Y_i)}{\widehat{f}_{nh} \circ \varphi_{\eta_{01}}(Y_i)}$$
$$+1_{\{t_{n,u}>t_0\}} n \sum_{i=[nt_0]+1}^{[nt_0+u]} \log \frac{\widehat{f}_{nh} \circ \varphi_{\eta_{01}}(Y_i)}{\widehat{f}_{nh} \circ \varphi_{\eta_{02}}(Y_i)} + o_p(1).$$

The process Z_{2nh} satisfies Billingsley's tightness criterion (15.21) and it converges weakly in $D([-A, A])$ by Proposition 1.1.

The mean and the variance of Z_{2nh} are $O(|u|)$ and they bounded on $\mathcal{U}_n^A$ the maximum of Z_{2nh} on $\mathcal{U}_n^A$ converges weakly to an uncentered Gaussian process with a finite variance function, Theorem 3.9 ends the proof. □

A test of the hypothesis H_0 of a density f against the alternative of a change at an unknown index k_0 is performed with the statistic

$$\widehat{T}_{n,h} = 2 \sup_{t\in]0,1[} \{\widehat{l}_{nh}(\widehat{\theta}_{n,t}) - \widehat{l}_{0nh}\} = 2\{\widehat{l}_{nh}(\widehat{\theta}_n) - \widehat{l}_{0nh}\},$$

where $\widehat{l}_{0nh} = \sup_{\eta\in\mathbb{R}} \sum_{i=1}^n \log \widehat{f}_{nh} \circ \varphi_{\widehat{\eta}_{0n}}(Y_i)$ is the maximum of the estimated log-likelihood with density f_{η_0} under the hypothesis H_0 and $\widehat{l}_{nh}$ is the empirical log-likelihood (3.21) under an alternative.

Under the conditions (2.1) and for a density f satisfying the integrability and differentiability conditions of Section 3.4, the asymptotic behaviour of $\widehat{T}_{nh}$ is the same as T_n in Propositions 3.7 and 3.8.

Proposition 3.9. *The statistic $\widehat{T}_{nh}$ converges weakly under H_0 to the variable T_0 as n tends to infinity. Under a fixed alternative, $\widehat{T}_{nh}$ diverges and under local alternatives P_{θ_n} contiguous to P_0, T_{nh} converges weakly to a variable $T_0 + S$ where S is a non-degenerated variable.*

The result under H_0 is a consequence of the uniform convergence in probability to zero of the variables $\widehat{U}_{nh}(\theta) - U_n(\theta)$ and $\widehat{I}_{nh}(\theta) - I_n(\theta)$. Under alternatives, the process $\widehat{X}_{nh}$ has an expansion similar to the expansion of X_n in the previous section.

Chapter 4

Change-points in parametric regressions

Abstract. This chapter studies regression models for a real variable with change of the parameters at unknown thresholds of the regressors or at an unknown sampling index. We first consider a linear regression and its empirical estimators with the least squares estimators of the change-point then the maximum likelihood estimators of parametric models. The convergence rates of the estimators are established. Least squares and likelihood ratio tests for the hypothesis of a constant mean are defined accordingly, the weak convergence of the estimators and the test statistics under the hypothesis and alternatives are proved.

4.1 Change-points in regressions

On a probability space $(\Omega, \mathcal{A}, P_0)$, we consider a vector of variables (X, Y) with finite mean and variance and such that the mean of Y conditionally on X is defined by a regression model

$$E_0(Y \mid X = x) = r(x)$$

where Y belongs to $\mathbb{R}$, $X = (X_1, \ldots, X_d)^T$ belongs to $\mathbb{R}^d$ and r is a real function on $\mathbb{R}^d$. The density of X is supposed to be continuous and the difference of Y and $r(X)$ is a random variable e such that $E_0(e \mid X = x)$ is zero. The conditional variance of Y is

$$\sigma_0(x) = E_0(Y^2 \mid X = x) - r^2(x) = E_0(e^2 \mid X = x),$$

we assume that it is a continuous function which does not depend on the parametrization of the regression function $r(x)$.

A regression of Y on X with a change-point when the components X_k of the variable X reach a threshold γ_k is determined by splitting each variable in a vector

$$(X_{1k}, X_{2k})^T = (X_k 1_{\{X_k \le \gamma_k\}}, X_k 1_{\{X_k > \gamma_k\}})^T,$$

for $k = 1, \ldots, d$. The parameter $\gamma = (\gamma_1, \ldots, \gamma_d)^T$ belongs to a bounded subset Γ of $\mathbb{R}^d$ and the observations of X are restricted to this interval. The indicators $1_{\{X_k \le \gamma_k\}}$ and $1_{\{X_k > \gamma_k\}}$ for the d components of the variable X and the threshold vector γ define $p = 2^d$ non-overlapping sets in $\mathbb{R}^d$ with indicators $I_k(X, \gamma)$, for $k = 1, \ldots, p$, and p regression functions

$$r_\eta(x, \gamma) = \sum_{k=1}^{p} r_k(x, \eta) I_k(x, \gamma) \tag{4.1}$$

where the p functions r_k differ.

For vectors a and b of the same dimension, the notation $a.\,b$ is for the vector with components $a_k b_k$. A linear model with a change of the regression parameter an unknown values of the variables and without discontinuity of the mean of Y is defined in a probability space $(\Omega, \mathcal{A}, P)$ by the equation

$$\begin{aligned} Y &= \mu_Y + \alpha^T\{(1-\delta_\gamma).X\} + \beta^T\{\delta_\gamma.X\} + e, \qquad (4.2)\\ \delta_{\gamma,k} &= 1_{\{X_k > \gamma_k\}},\ k = 1, \ldots, d, \end{aligned}$$

with d-dimensional vectors of parameters α and β, and an error variable e. If e has a Gaussian distribution independent of the regression variable X, it has a distribution $\sigma\mathcal{N}(0,1)$, otherwise its variance is a function of X. In a linear regression with a modified slope and mean after a change-point, the variable Y is

$$Y = \mu_Y +^T \alpha^T\{(1-\delta_\gamma).(X-\mu_1)\} + \beta^T\{\delta_\gamma.(X-\mu_2)\} + e, \tag{4.3}$$

where $\mu_Y = EY$, $\mu_1 = E\{(1-\delta_\gamma).X\}$ and $\mu_2 = E(\delta_\gamma.X)$.

The vector of the change-points γ has the dimension d, the regression parameter $\zeta = (\mu^T, \alpha^T, \beta^T)^T$ has the dimension $D = 4d+1$, with a vector of means $\mu = (\mu_Y, \mu_1^T, \mu_2^T)^T$ of dimension $2d+1$, and the vector of all parameter is denoted $\theta = (\zeta^T, \gamma^T)^T$, it belongs to a space Θ such that

$$E_0 \sup_\Theta r_\theta^2(X) < \infty. \tag{4.4}$$

The true value of the parameter vector is denoted θ_0.

These models have the form of parametric regressions

$$Y = r_\theta(X) + e = (1-\delta_\gamma)^T r_{1\zeta}(X) + \delta_\gamma^T r_{2\zeta}(X) + e \tag{4.5}$$

and the change-points do not depend on the observation sampling.

Other change-point models are chronological with a change at an unknown time γ in $]0,1[$ corresponding to the integer index $[n\gamma]$ with a sample of size n of the variable (X,Y), or to a random time T in a continuous sampling. A difference of consecutive observations under model (4.3) is modelled by (4.2) and both models will be considered. For a subset $\mathcal{S}_k$ of $\{1,\dots,d\}$, the change-point probabilities

$$p_{\gamma,\mathcal{S}_k} = \Pr([\cap_{i\in\mathcal{S}_k}\{X_i \le \gamma_i\}] \cap [\cap_{j\in\bar{\mathcal{S}}_k}\{X_j > \gamma_j\}]),$$

are mixture probabilities of two regression models with coefficients the components $\alpha_{\mathcal{S}_k}$ of α and, respectively $\beta_{\bar{\mathcal{S}}_k}$ of β, with the probabilities $p_{\gamma,\mathcal{S}_k}$

$$E(Y \mid X) = \sum_{\mathcal{S}_k} p_{\gamma,\mathcal{S}_k}\{r_{1,\mathcal{S}_k}(X;\mu,\alpha) + r_{2,\mathcal{S}_k}(X;\mu,\beta)\},$$

where the parameters are $\mu_1 = E\{X.(1-\delta_\gamma)\}$ and $\mu_1 = E\{X.\delta_\gamma\}$ in model (4.3), they are $\mu_1 = \mu_2 = EX$ in model (4.2). For a n-sample, the notation $Y_{(n)}$ is for a n-dimensional vector and $X_{(n)}$ is a matrix of dimension $n \times d$, and $\bar{X}_n = n^{-1}\sum_{i=1}^n X_i$ is a d-dimensional vector. The maximum likelihood estimators are first defined assuming that the change-point occurs at a known value γ, for the means they are

$$\begin{aligned}
\widehat{\mu}_{Yn} &= \bar{Y}_n, \\
\widehat{\mu}_{\gamma,1n} &= n^{-1}\sum_{i=1}^n X_i.(1-\delta_{\gamma,i}), \\
\widehat{\mu}_{\gamma,2n} &= n^{-1}\sum_{i=1}^n X_i.\delta_{\gamma,i},
\end{aligned}$$

in the model with a change-point of mean or $\widehat{\mu}_{\gamma,1n} = \widehat{\mu}_{\gamma,1n} = \bar{X}_n$ in the model without change-point of mean. The estimators of the parameters α_k and β_k, $k = 1,\dots,d$, are

$$\begin{aligned}
\widehat{\alpha}_{\gamma,n,k} &= [\{(X_{(n),k} - \widehat{\mu}_{\gamma,1n,k}).(1-\delta_{\gamma,k})\}^T (X - \widehat{\mu}_{\gamma,1n,k})]^{-1} \\
&\qquad \{(X -_{(n),k} \widehat{\mu}_{\gamma,1n,k}).(1-\delta_{\gamma,k}\}^T (Y_{(n),k} - \bar{Y}_{n,k}), \\
\widehat{\beta}_{\gamma,n,k} &= \{(X - \widehat{\mu}_{\gamma,2n,k}).\delta_{\gamma,k})^T (X - \widehat{\mu}_{\gamma,2n,k})\}^{-1} \\
&\qquad ((X - \widehat{\mu}_{\gamma,2n,k}).\delta_{\gamma,k})^T (Y_{(n),k} - \bar{Y}_{n,k}),
\end{aligned}$$

they are independent of the empirical estimator of the variance σ_0^2 of $Y - E_0(Y \mid X)$

$$\widehat{\sigma}^2_{\gamma,n} = n^{-1} \sum_i \{Y_i - \bar{Y}_n - (1 - \delta_{\gamma,i}).(X_i - \widehat{\mu}_{\gamma,1n})^T \widehat{\alpha}_{\gamma,n} - \delta_{\gamma,i}.(X_i - \widehat{\mu}_{\gamma,2n})^T \widehat{\beta}_{\gamma,n})\}^2.$$

At the true parameter value, σ_0^2 is estimated by $n^{-1} \sum_i e_i^2$.

Let $D_n(\gamma) = n^{-1} \sum_{i=1}^n \delta_{\gamma i}$, when n tends to infinity, $D_n(\gamma)$ converges to the probability vector p_γ with components

$$p_{k,\gamma} = \Pr(X_k > \gamma_k),\ k = 1, \ldots, d,$$

belonging to $]0,1[^d$. The true value $p_0 = p_{\gamma_0}$ of the change-point probability is supposed to belong to $]0,1[^d$ and under the probability $P_0 = P_{\theta_0}$, the convergence rate of $D_n(\gamma)$ and the estimators of the regression coefficients is $n^{-\frac{1}{2}}$. Furthermore, under P_0, the variables $n^{\frac{1}{2}}(\widehat{\alpha}_n - \alpha_0)$ and $n^{\frac{1}{2}}(\widehat{\beta}_n - \beta_0)$ converge weakly to centered Gaussian variables with variances the limits of $[n^{-1}\{(X - \mu_1).(1 - \delta_{\gamma_0})\}^T (X - \mu_1)]^{-1} \sigma_0^2$ and, respectively $n^{-1}\{(X - \mu_2).\delta_{\gamma_0})^T (X - \mu_2)\}^{-1} \sigma_0^2$. The estimator $\widehat{\sigma}_n^2$ has a $\sigma_0^2 \chi^2(n-d)$ distribution in the model with a Gaussian error. An asymptotic expansion when n tends to infinity provides an approximation of the distribution of $\widehat{\sigma}_n^2$ by a $\sigma_0^2 \chi^2(d)$ variable, this approximation of Wilk–Shapiro type statistics (1965), presented in a more general setting by Sen (0202), is valuable as n tends to infinity and for every distribution of the error e.

For $\gamma \neq \gamma_0$, the estimators of the means μ_1 and μ_2 and the coefficients α and β are biased under P_0. Under P_θ and when $D_n(\gamma)$ converges to 0, if there exists a constant $a > 0$ such that $D_n(\gamma, a) = n^{-a} \sum_{i=1}^n \delta_{\gamma i}$ converges to a limit $p_{\gamma,a} \neq 0$, then $n^{a/2}(\widehat{\beta}_{\gamma,n} - \beta_\gamma)$ converges weakly to a centred Gaussian variable with variance the limit of $[n^{-a}\{(X - \mu_1).(1 - \delta_\gamma)\}^T (X - \mu_1)]^{-1} \sigma_\gamma^2$ and the behaviour of the other estimators is unchanged. When $D_n(\gamma)$ converges to 1 and $n^{-a} \sum_{i=1}^n (1 - \delta_{\gamma i})$ converges to $p'_{\gamma,a} \neq 0$, $n^{a/2}(\widehat{\beta}_{\gamma,n} - \beta_\gamma)$ converges weakly under P_θ to a centred Gaussian variable with variance the limit of $n^{-a}\{(X - \mu_2).\delta_\gamma)^T (X - \mu_2)\}^{-1} \sigma_\gamma^2$, and the behaviour of the other estimators is not modified.

In model (4.2) where the mean of the variable X is constant, the behaviour of the other estimators is identical to those of model (4.3). The change-point parameter γ is estimated by minimization of the variance estimator

$$\widehat{\gamma}_n = \arg\inf_\gamma \widehat{\sigma}^2_{\gamma,n},$$

this minimization over all components of the parameter γ enables to estimate all changes in the same procedure. If several changes occur for the same variable, the consecutive minima are searched iteratively, starting from the previous one. The estimators of α, β and σ^2 are deduced by plugging $\widehat{\gamma}_n$ in the estimators defined for fixed γ and the mean square error of estimation of the regression is the empirical variance $\widehat{\sigma}_n^2 = \widehat{\sigma}^2_{\widehat{\gamma}_n,n}$.

4.2 Convergences in linear regressions

In model (4.3) and under the probability distribution P_0 of the observations, the parameter values are denoted γ_0, $\mu_0 = (\mu_{0Y}, \mu_{01}^T, \mu_{02}^T)^T$, α_0, β_0 and σ_0^2, and $\widehat{\sigma}^2_{\gamma_0 n}$ is the empirical residual variance in the true model. We assume that $E_0(\|X\|^2)$ is finite. For a n-sample, let us denote the norm defined as the empirical variance under a probability distribution P_θ by

$$\|Y - r_\theta(X)\|_n^2 = n^{-1}\sum_{i=1}^n \{Y_i - r_\theta(X_i)\}^2.$$

For parameter vectors $\theta = (\zeta^T, \gamma^T)^T$ and $\theta' = (\zeta'^T, \gamma'^T)^T$, we consider the semi-norm $\rho(\theta, \theta') = (|\gamma - \gamma'| + \|\zeta - \zeta'\|_d^2)^{\frac{1}{2}}$ and a neighborhood $V_\varepsilon(\theta_0)$ of θ_0 for ρ, with radius ε.

Lemma 4.1. *For $\varepsilon > 0$ sufficiently small, there exists a constant κ such that for n sufficiently large*

$$E_0 \sup_{\theta \in V_\varepsilon(\theta_0)} \|r_\theta(X) - r_{\theta_0}(X)\|_n^2 \le \kappa\varepsilon^2.$$

Proof. The norm of a real vector $(\alpha_k)_{k=1,\dots,d}$ of $\mathbb{R}^d$ is $\|\alpha\|_d^2 = \sum_{k=1}^d \alpha_k^2$ and for a matrix $(X_{ik})_{i=1,\dots,n,k=1,\dots,d}$, it is

$$\|X\|_n^2 = n^{-1}\sum_{i=1}^n \|X_i\|_d^2.$$

By linearity of the function r_θ, we have

$$E \sup_{\theta \in V_\varepsilon(\theta_0)} \|r_\theta(X) - r_{\theta_0}(X)\|_n^2 \le 3[(\mu_Y - \mu_{0Y})^2$$
$$+E_0(\|X\|_n^2) \sup_{\theta \in V_\varepsilon(\theta_0)} \{\|\alpha - \alpha_0\|_d^2 + \|\beta - \beta_0\|_d^2\}]$$

where $\|\alpha.X\|_n^2 \le \|X\|_n^2 \, \|\alpha\|_d^2$. □

Proposition 4.1. *Under P_0, the estimators $\widehat{\theta}_n$, $r_{\widehat{\theta}_n}$ and $\widehat{\sigma}^2_n$ are a.s. consistent.*

Proof. For every γ, the difference between $\widehat{\sigma}^2_{\gamma,n}$ and $\widehat{\sigma}^2_{\gamma_0,n}$ satisfies

$$\begin{aligned}0 &\leq \widehat{\sigma}^2_{\gamma,n} - \widehat{\sigma}^2_{\gamma_0,n} \leq \|r_{\widehat{\zeta}_{\gamma,n},\gamma}(X) - r_{\widehat{\zeta}_{\gamma_0,n},\gamma_0}(X)\|^2_n \\ &\leq 2E_0 \sup_{|\gamma-\gamma_0|\leq\varepsilon^2} [\|X\|^2_n\{\|\widehat{\alpha}_{\gamma,n} - \widehat{\alpha}_{\gamma_0,n}\|^2_d + \|\widehat{\beta}_{\gamma,n} - \widehat{\beta}_{\gamma_0,n}\|^2_d\}]\end{aligned}$$

where $\widehat{\zeta}_{n,\gamma_0}$ is an a.s. consistent estimator of ζ, and the minimum is achieved when $\widehat{\gamma}_n$ is a.s. in a neighborhood of γ_0, it follows that $\widehat{\theta}_n$ is a.s. consistent and the convergence of the other estimators follows. □

Let

$$l_n(\theta) = \|Y - r_\theta(X)\|^2_n - \|Y - r_{\theta_0}(X)\|^2_n, \tag{4.6}$$

in a model with a Gaussian error, the logarithm of the likelihood is $-l_n(\theta)$ up to a positive multiplicative constant, $l_n(\theta) \geq 0$ and it is minimum at θ_0 where it it zero. The process $l_n(\theta)$ converges a.s. uniform on Θ to its mean

$$l(\theta) = E_0\{|Y - r_\theta(X)|^2 - |Y - r_{\theta_0}(X)|^2\},$$

we have $E_0\{Yr_{\theta_0}(X)\} = E_0 r^2_{\theta_0}(X)$ and

$$E_0\{Yr_\theta(X)\} = E_0 r^2_{\theta_0}(X) + E_0[Y\{r_\theta(X) - r_{\theta_0}(X)\}],$$

it follows that for every θ in $V_\varepsilon(\theta_0)$

$$\begin{aligned}E_0 \sup_{V_\varepsilon(\theta_0)} \{Yr_\theta(X)\} &\leq E_0 r^2_{\theta_0}(X) + (E_0 Y^2)^{\frac{1}{2}} E_0\Big[\sup_{V_\varepsilon(\theta_0)} \{r_\theta(X) - r_{\theta_0}(X)\}^2\Big]^{\frac{1}{2}} \\ &\leq E_0 r^2_{\theta_0}(X) + \kappa\varepsilon\, E_0\|r_{\theta_0}(X)\|_2\end{aligned}$$

where the constant κ depends on the norms $L^2(P_0)$ of the variable X.

The variations of $\widehat{\alpha}_{\gamma,n}$ and $\widehat{\beta}_{\gamma,n}$ according to γ are approximated by functions $\widetilde{\alpha}_{\gamma,n}$ and $\widetilde{\beta}_{\gamma,n}$

$$\begin{aligned}\widehat{\alpha}_{\gamma,n} - \widehat{\alpha}_{\gamma_0,n} &= \widetilde{\alpha}^{-1}_{\gamma,1n}\widetilde{\alpha}_{\gamma,2n} - \widetilde{\alpha}^{-1}_{\gamma_0,1n}\widetilde{\alpha}_{\gamma_0,2n} + O_p(|\gamma - \gamma_0|), \\ \widetilde{\alpha}_{\gamma,1n} &= \int_{-\infty}^{\gamma} (x - \mu_1)^T(x - \mu_1)\, dF_{n,X}(x), \\ \widetilde{\alpha}_{\gamma,2n} &= \int_{-\infty}^{\gamma} (x - \mu_1)^T(y - \mu_Y)\, dF_{n,XY}(x, y), \\ \widehat{\beta}_{\gamma,n} - \widehat{\beta}_{\gamma_0,n} &= \widetilde{\beta}^{-1}_{\gamma,1n}\widetilde{\beta}_{\gamma,2n} - \widetilde{\beta}^{-1}_{\gamma_0,1n}\widetilde{\beta}_{\gamma_0,2n} + O_p(|\gamma - \gamma_0|),\end{aligned}$$

$$\widetilde{\beta}_{\gamma,1n} = \int_{\gamma}^{\infty} (y-\mu_2)^T (y-\mu_2)\, dF_{n,Y}(y),$$

$$\widetilde{\beta}_{\gamma,2n} = \int_{\gamma}^{\infty} (y-\mu_2)^T (y-\mu_2)\, dF_{n,XY}(x,y),$$

with the empirical distribution function of the variable (X, Y) and its marginal distribution functions. These approximations imply that the means of $\widehat{\alpha}_{\gamma,n} - \widehat{\alpha}_{\gamma_0,n}$ and $\widehat{\beta}_{\gamma,n} - \widehat{\beta}_{\gamma_0,n}$ are $O(\rho(\theta,\theta_0))$.

For the weak convergence of the change-point estimator in the linear regression, we consider the process

$$W_n(\theta) = n^{\frac{1}{2}}\{l_n(\theta) - l(\theta)\},\ \theta \in \Theta. \tag{4.7}$$

We assume the condition

$$E_0 \sup_{\theta \in V_\varepsilon(\theta_0)} \{Y - r_\theta(X)\}^4 < \infty. \tag{4.8}$$

Proposition 4.2. *For every $\varepsilon > 0$ sufficiently small, under the condition (4.8) there exist constants κ_0 and κ_1 such that for every θ belonging to $V_\varepsilon(\theta_0)$, as n tends to infinity*

$$l(\theta) \geq \kappa_0 \rho^2(\theta,\theta_0)$$

$$E_0 \sup\nolimits_{\theta \in V_\varepsilon(\theta_0)} |W_n(\theta)| \leq \kappa_1 \rho(\theta,\theta_0).$$

Proof. The function l is a sum of integrals of positive differences of linear functions of the parameters at θ and respectively θ_0 on the intervals $]-\infty, \gamma_0 \wedge \gamma]$ and $]\gamma \vee \gamma_0, \infty[$, where they are $O(\|\zeta - \zeta_0\|^2)$, and integrals of linear functions of ζ_0 on the intervals $]\gamma_0, \gamma]$ and $]\gamma, \gamma_0]$, according to the sign of $\gamma - \gamma_0$, which are $O(|\gamma - \gamma_0|)$.

Using the notation of equation (4.1) where $I_k(X,\gamma)$ and $I_{0k}(X)$ are the indicator variables for γ and γ_0, the function $l(\theta)$ is written as

$$E_0[\{Y - \sum_k r(X;\zeta_{0k}) I_{0k}(X)\}^2] - E_0[\{Y - \sum_k r(X;\zeta_k) I_k(X,\gamma)\}^2]$$

$$= \sum_k E_0[\{r(X;\zeta_{0k}) - r(X;\zeta_k)\}\{r(X;\zeta_{0k}) + r(X;\zeta_k) - 2Y\} I_{0k}(X)]$$

$$+ \sum_k E_0[\{I_k(X,\gamma) - I_{0k}(X)\} r(X;\zeta_k)\{r(X;\zeta_k) - 2Y\}],$$

$$= \sum_k E_0[\{r(X;\zeta_{0k}) - r(X;\zeta_k)\}^2 I_{0k}(X)]$$

$$+ \sum_k E_0[\{I_k(X,\gamma) - I_{0k}(X)\} r(X;\zeta_k)\{r(X;\zeta_k) - 2r(X;\zeta_{0k})\}].$$

This implies

$$E_0[\{Y - r(X;\zeta_{0k})I_{0k}\}^2] - E_0[\{Y - r(X;\zeta_k)I_k\}^2]$$
$$= E_0[\{r(X;\zeta_{0k}) - r(X;\zeta_k)\}^2 I_k] + E_0\{(I_k - I_{0k})r(X;\zeta_{0k})\}, \quad (4.9)$$

it is a $O(\|\zeta_k - \zeta_{0k}\|^2) + O(|\gamma - \gamma_0|)$ and $l(\theta)$ is strictly positive for θ different from θ_0 which ensures the bound for $l(\theta)$.

The second bound is proved writing

$$n^{\frac{1}{2}}\{\|Y - r_\theta(X)\|_n^2 - E_0|Y - r_\theta(X)|^2\} = \int |y - r_\theta(x)|^2 \, d\nu_n(x,y)$$

with the empirical process ν_n of the variable (X,Y), and

$$W_n(\theta) = \int \{|y - r_\theta(x)|^2 - |y - r_{\theta_0}(x)|^2\} \, d\nu_n(x,y)$$

then $E_0 W_n^2(\theta) = \int \{|y - r_\theta(x)|^2 - |y - r_{\theta_0}(x)|^2\}^2 \, d\{F(1-F)\}(x,y)$ and it is a $O(\rho^2(\theta,\theta_0))$ due to the indicator variables. □

The convergence rate of the estimators is deduced from Proposition 4.2 following the arguments of Ibragimov and Has'minskii (1981) for the weak convergence of the parameter estimators in the change-point models satisfying the bounds of Proposition 4.2. Let

$$\mathcal{U}_n = \{u = (u_\gamma, u_\zeta^T)^T : u_\gamma = n(\gamma - \gamma_0), u_\zeta = n^{\frac{1}{2}}(\zeta - \zeta_0); \gamma \neq \gamma_0, \zeta \in G\}$$

provided with the semi-norm $\|u\| = (|u_\gamma| + \|u_\zeta\|^2)^{\frac{1}{2}}$.

For $u = (u_\gamma, u_\zeta^T)^T \in \mathcal{U}_n$, let $\theta_{n,u} = (\gamma_{n,u}, \zeta_{n,u}^T)^T$, with $\gamma_{n,u} = \gamma_0 + n^{-1}u_\gamma$ and $\zeta_{n,u} = \zeta_0 + n^{-\frac{1}{2}}u_\zeta$, the norm in $\mathcal{U}_n$ is $\|u\| = n^{\frac{1}{2}}\rho(\theta_{n,u},\theta_0)$ and a sequence of ε-neighbourhoods of θ_0 defines neighbourhoods of zero as

$$\mathcal{U}_{n,\varepsilon} = \{u \in \mathcal{U}_n : \rho(\theta_{n,u},\theta_0) \leq n^{\frac{1}{2}}\varepsilon\}.$$

Theorem 4.1. *Under the condition of Proposition 4.2, for every $\varepsilon > 0$*

$$\overline{\lim}_{n,A\to\infty} P_0(\inf_{u\in\mathcal{U}_{n,\varepsilon}, \|u\|>A} l_n(\theta_{n,u}) \leq 0) = 0,$$
$$\overline{\lim}_{n,A\to\infty} P_0(n|\widehat{\gamma}_n - \gamma_0| > A) = 0,$$
$$\overline{\lim}_{n,A\to\infty} P_0(n^{\frac{1}{2}}\|\widehat{\zeta}_n - \zeta_0\| > A) = 0.$$

By construction of the estimators of the regression parameters ζ, $n^{\frac{1}{2}}(\widehat{\zeta}_n - \zeta_0)$ converges weakly to a Gaussian centred distribution with covariance given section 4.1. Let $r_\theta(X) = (1-\delta_\gamma).\{\mu + (X-\mu_1)^T\alpha\} + \delta_\gamma.\{\mu + (X-\mu_2)^T\beta\}$ be the regression function of models (4.2) and (4.3), also denoted $r_\theta = (1-\delta_\gamma).r_{\zeta_1} + \delta_\gamma.r_{\zeta_2}$. The process $r_\theta(X)$ is estimated by

$$r_{\widehat{\theta}_n}(X) = (1-\delta_{\widehat{\gamma}_n}).r_{\widehat{\zeta}_{1n}}(X) + \delta_{\widehat{\gamma}_n}.r_{\widehat{\zeta}_{2n}}(X).$$

For every x, let $\delta_0(x) = \delta_{\gamma_0}(x)$.

Theorem 4.2. *Under the condition (4.8), the process $G_n = n^{\frac{1}{2}}(r_{\widehat{\theta}_n} - r_{\theta_0})$ converges weakly under P_0 to a centred Gaussian process G, with a finite variance function.*

Proof. The process G_n develops as

$$G_n(x) = n^{\frac{1}{2}}\{r_{\widehat{\zeta}_n,\gamma_0}(x) - r_{\theta_0}(x)\} + n^{\frac{1}{2}}\{r_{\widehat{\theta}_n}(x) - r_{\widehat{\zeta}_n,\gamma_0}(x)\}$$

denoted $G_{1n}(x) + G_{2n}(x)$. The first term is a sum of two processes on the intervals $]-\infty, \gamma_0]$ and $]\gamma_0, +\infty[$ respectively, according to the sign of $x - \gamma_0$. We have

$$\begin{aligned} G_{1n}(x) &= n^{\frac{1}{2}}(\widehat{\alpha}_n - \alpha_0)\{(1-\delta_0).(x - \widehat{\mu}_{1n})\} \\ &\quad + n^{\frac{1}{2}}(\widehat{\beta}_n - \beta_0)\{\delta_0.(x - \widehat{\mu}_{2n})\} \end{aligned}$$

and it converges weakly to a centred Gaussian process by the weak convergence of the estimators of the regression coefficients. The second term splits according to the intervals defined by γ_0 and $\widehat{\gamma}_n$

$$G_{2n}(x) = n^{\frac{1}{2}}\widehat{\alpha}_n^T\{(\delta_0 - \delta_{\widehat{\gamma}_n}).(x - \widehat{\mu}_{1n})\} - n^{\frac{1}{2}}\widehat{\beta}_n^T\{(\delta_0 - \delta_{\widehat{\gamma}_n}).(x - \widehat{\mu}_{2n})\}$$

where

$$n^{\frac{1}{2}}P_0(\gamma_0 < x \le \widehat{\gamma}_n) = f_X(\gamma_0)n^{\frac{1}{2}}\|\widehat{\gamma}_n - \gamma_0\|$$

is a $O(n^{-\frac{1}{2}})$ by Theorem 4.2, and $E_0 n^{\frac{1}{2}}\bar{\eta}_n(X) = O(n^{-\frac{1}{2}})$. By the convergence of the estimators of the regression coefficients, the process G_{2n} is a $O_p(n^{-\frac{1}{2}})$ and it converges to zero in probability. □

The weak convergence of the normalized estimators relies on an expansion of $nl_n(\theta)$ given by (4.6), for θ in a neighborhood of θ_0 determined by the convergence rate of Theorem 4.1. Let

$$\widetilde{X} = (1, (1-\delta_\gamma)^T, \delta_\gamma^T, \{(1-\delta_\gamma).(X - \mu_1)\}^T, \{\delta_\gamma.(X - \mu_2)\}^T)^T,$$

the regression function is written as $r_\theta(X) = \zeta^T \widetilde{X}$, its first derivative $\dot{r}_{\theta,\zeta}(X)$ with respect to ζ is the vector $\widetilde{X}$ and the first two derivatives of the process l_n with respect to ζ are

$$\begin{aligned} \dot{l}_{n,\zeta}(\theta) &= 2\|\{Y_i - r_\theta(X_i)\}\dot{r}_{\theta,\zeta}(X_i)\|_n, \\ \ddot{l}_{n,\zeta}(\theta) &= -2\dot{r}^2_{\theta,\zeta}(X_i)\|_n. \end{aligned}$$

As $\widehat{\zeta}_n$ minimizes the process l_n, it satisfies the expansion

$$n^{\frac{1}{2}}(\widehat{\zeta}_n - \zeta_0) = -\ddot{l}_{n,\zeta}^{-1}(\zeta_0, \widehat{\gamma}_n)n^{\frac{1}{2}}\dot{l}_{n,\zeta}(\zeta_0, \widehat{\gamma}_n) + o_p(1) \quad (4.10)$$

where $\ddot{l}_n(\zeta_0, \widehat{\gamma}_n)$ converges in probability under P_0 to $V_0 = E_0 \dot{r}_{\theta_0}^{\otimes 2}(X)$. The asymptotic variance of $n^{\frac{1}{2}}(\widehat{\zeta}_n - \zeta_0)$ is therefore V_0^{-1} and it converges weakly to a centered Gaussian variable.

Let $\theta_{n,u} = (\gamma_0^T + n^{-1}u_\gamma^T, \zeta_0^T + n^{-\frac{1}{2}}u_\zeta^T)^T$, for every $u = (u_\gamma^T, u_\zeta^T)^T$ belonging to $\mathcal{U}_n^A = \{u \in \mathcal{U}_n; \|u\|^2 \leq A\}$, for $A > 0$. Let $u = (u_\gamma^T, u_\zeta^T)^T$ where $n^{-1}u_\gamma$ is the first component of $(\theta_{n,u} - \theta_0)$ and $n^{-\frac{1}{2}}u_\zeta$ is the vector of the other components corresponding to the regression parameters, we denote

$$\widetilde{l}_n(u) = \sum_{i=1}^n [\{Y_i - r_{\zeta_{n,u},\gamma_0}(X_i)\}^2 - \{Y_i - r_{\theta_0}(X_i)\}^2] \tag{4.11}$$

$$Q_n(u) = \sum_{i=1}^n \{Y_i - r_{\theta_{n,u}}(X_i)\}^2 - \{Y_i - r_{\zeta_{n,u},\gamma_0}(X_i)\}^2].$$

By definition, the process $\widetilde{l}_n$ does not depend on u_γ and the process Q_n is the difference of l_n and $\widetilde{l}_n$.

According to Section 1.4, let Q be the Gaussian process with mean function

$$\mu(u) = f_X(\gamma_0)[u_{\gamma_0}^T\{r_{1\zeta_0}(\gamma_0) - r_{2\zeta_0}(\gamma_0)\}]^2$$

and with a finite variance function limit of the variance of the process Q_n.

Theorem 4.3. *Under the condition (4.8), the process Q_n converges weakly in $\mathcal{U}_n^A$ to the process Q, as n and A tend to infinity.*

Proof. The conditional mean zero of the error variable e implies

$$\begin{aligned} &E_0[Y(\delta_\gamma - \delta_0)^T\{r_{1\zeta}(X) - r_{2\zeta}(X)\}] \\ &\quad = E_0[r_0(X)(\delta_\gamma - \delta_0)^T\{r_{1\zeta}(X) - r_{2\zeta}(X)\}], \end{aligned}$$

it follows that for $\gamma_0 < \gamma_{n,u}$

$$\begin{aligned} &E_0[\{Y - (\delta_\gamma - \delta_0)^T r_{1\zeta}(X)\}^2 - \{Y - (\delta_\gamma - \delta_0)^T r_{2\zeta}(X)\}^2] \\ &= E_0([(\delta_0 - \delta_\gamma)^T\{r_{1\zeta}(X) - r_{2\zeta}(X)\}]^2) \end{aligned}$$

and the expectation of the difference of the errors in the interval $]\gamma_{n,u}, \gamma_0]$, for $\gamma_{n,u} < \gamma_0$ tending to γ_0, has the same form. The expectation under P_0 of the process Q_n converges therefore to $E_0 Q$ such that

$$\begin{aligned} E_0 Q(u) &= \lim_{n\to\infty} n E_0([(\delta_{\gamma_{n,u}} - \delta_{\gamma_0})^T\{r_{1\zeta_{n,u}}(X) - r_{2\zeta_{n,u}}(X)\}]^2) \\ &= f_X(\gamma_0)[u_{\gamma_0}^T\{r_{1\zeta_0}(\gamma_0) - r_{2\zeta_0}(\gamma_0)\}]^2. \end{aligned}$$

The variance of $Q_n(u)$ is $E_0\{Q_n^2(u)\} - E_0^2 Q_n(u_\gamma)$, and $E_0\{Q_n^2(u)\}$ has the expansion

$$\begin{aligned}
&nE_0[\{Y - (\delta_{\gamma_{n,u}} - \delta_0)^T r_{1\zeta_{n,u}}(X)\}^2 - \{Y - (\delta_{\gamma_{n,u}} - \delta_0)^T r_{2\zeta_{n,u}}(X)\}^2]^2 \\
&= nE_0([(\delta_0 - \delta_{\gamma_{n,u}})^T \{r_{1\zeta_{n,u}}(X) - r_{2\zeta_{n,u}}(X)\}] \\
&\quad .[(\delta_0 - \delta_{\gamma_{n,u}})^T \{r_{1\zeta_{n,u}}(X) + r_{2\zeta_{n,u}}(X) - 2Y\}]^2),
\end{aligned}$$

using an approximation of means of functions between γ_0 and $\gamma_{n,u}$ by their value at γ_0, the variance of $Q_n(u)$ is bounded in $\mathcal{U}_n^A$. The weak convergence of Q_n on $D([-A, A])$ to an uncentered Gaussian process follows from Proposition 1.1 as the limit of sums of independent and identically distributed random variables with convergent moments and Q_n satisfies the tightness criterion in $D([-A, A])$, as n tends to infinity. Its weak convergence as A tends to infinity is a consequence of Theorem 4.1 and its limiting distribution is bounded in probability. □

By the expression of $l_n(\theta)$ as the sum of $\widetilde{l}_n(u_\zeta)+Q_n(u)$ defined by (4.11), the variable $u_{\widehat{\gamma}_n} = n(\widehat{\gamma}_n - \gamma_0)$ is such that

$$u_{\widehat{\gamma}_n} = \arg \inf_{u_\gamma \in \mathbb{R}^d} Q_n(u_{\widehat{\zeta}_n}, u_\gamma) + o_p(1),$$

this minimization over all components of the parameter γ provides the estimators of the changes for all regression variables in the same procedure.

Theorem 4.4. *Under the condition (4.8), the estimator $u_{\widehat{\gamma}_n}$ of the change-point γ_0 is asymptotically independent of the estimator of the regression parameters $\widehat{\zeta}_n$ and it converges weakly to a stopping time u_0 where the Gaussian process Q achieves its minimum, u_0 is bounded in probability.*

Proof. Theorem 4.1 implies $u_{\widehat{\gamma}_n}$ is a $O_p(1)$. By continuity of the minimum, the process Q achieves its minimum at $\lim_{n\to\infty} u_{\widehat{\gamma}_n}$ which is therefore bounded in probability. The covariance between the processes Q_n and $\widetilde{l}_n$ tends to zero since they depend on observations on intervals between $\gamma_{n,u}$ and γ_0 and for every $\varepsilon > 0$

$$P_0(\|\delta_{\gamma_{n,u}} - \delta_0\| > \varepsilon) \le \varepsilon^{-1} E_0 \|\delta_{\gamma_{n,u}} - \delta_0\| = \varepsilon^{-1} O(n^{-1}),$$

the processes Q_n and $\widetilde{l}_n$ are therefore asymptotically independent which entails the asymptotic independence of the estimators $\widehat{\gamma}_n$ and $\widehat{\zeta}_n$. □

4.3 Test of linear regressions without change

We consider the model defined by (4.1) and (4.3) where the variable X has a finite support. A test statistic for a linear regression model without change-point relies on the same criterion as the construction of the estimators

$$T_n = n \inf_{\gamma \in \Gamma} \{\widehat{\sigma}^2_{\gamma,n} - \widehat{\sigma}^2_{0n}\}$$

where $\widehat{\sigma}^2_{\gamma,n}$ is the variance estimator in the model with change-points and $\widehat{\sigma}^2_{0n}$ is the estimator of the variance under H_0, in the model without change, then γ_0 is the finite end-point of the support of the variable X, $\delta_{\gamma_0} \equiv 0$ and

$$T_n = n\{l_n(\widehat{\theta}_n) - l_{0n}(\widehat{\zeta}_{0n})\}.$$

With a Gaussian error e, $-nl_n(\gamma)$ is the logarithm of the likelihood ratio, up to an additive term $\log \widehat{\sigma}_{0n} - \log \widehat{\sigma}^2_n(\gamma)$ of smaller order than $nl_n(\gamma)$.

Under an alternative, $T_n = n\{\widehat{\sigma}^2_n(\widehat{\gamma}_n) - \widehat{\sigma}^2_{0n}\}$ and the variable $\widehat{\sigma}^2_n - \widehat{\sigma}^2_{0n}$ is asymptotically equivalent to the log-likelihood ratio test statistic when γ_0 is known, in a model with a Gaussian error. The parameter under H_0 is still denoted θ_0, ζ_0 is the value of the parameter vector ζ_{01} under H_0

Under the alternative, the combination of the indicators $1_{\{X_k \leq \gamma_k\}}$ and $1_{\{X_k > \gamma_k\}}$, for the d components of the variable X and the threshold vector γ, define p non-overlapping sets in $\mathbb{R}^d$ with indicators $I_l(\gamma)$, $l = 1, \ldots, p$, and p real regression functions such that

$$r_\theta(x) = \sum_{l=1}^{p} r(x, \zeta_l) I_l(\gamma). \tag{4.12}$$

The parameters $\zeta_2, \ldots, \zeta_p$ are zero under H_0.

Proposition 4.3. *If I_X is bounded and for variables X and Y satisfying the condition (4.8), the statistic T_n converges weakly under H_0 to a $\chi^2_{d(p-1)}$ variable T_0.*

Proof. Under H_0, the indicator vector $1_{\{X > \gamma_0\}}$ is zero and the difference $\{Y - r_\theta(X)\}^2 - \{Y - r_{\theta_0}(X)\}^2$ reduces to

$$[Y - 1_k^T r_{1\theta}(X) - \delta_\gamma^T \{r_{2\zeta}(X) - r_{1\zeta}(X)\}]^2 - \{Y - 1_k^T r_0(X)\}^2$$

where 1_k is the k-dimensional vector with components 1. At θ_n converging to θ_0 and ζ_{0n} converging to ζ_0, the difference

$$\begin{aligned}&\{Y - r_{\theta_n}(X)\}^2 - \{Y - r(X; \zeta_{0n})\}^2 \\ &= [Y - 1_k^T r(X; \zeta_{1n}) - \delta_{\gamma_n}^T \{r(X; \zeta_{2n}) - r(X; \zeta_{1n})\}]^2 - \{Y - 1_k^T r(X; \zeta_0)\}^2,\end{aligned}$$

it is the sum $\{Y - 1_k^T r(X;\zeta_{1n})\}^2 - \{Y - 1_k^T r(X;\zeta_{0n})\}^2 + r_n$. The first term is a $O(\|\zeta_{1n} - \zeta_0\|_2^2)$ and the remainder term is

$$\begin{aligned} r_n &= [\delta_{\gamma_n}^T \{r(X;\zeta_{2n}) - r(X;\zeta_{1n})\}]^2 \\ &\quad -2\{Y - 1_k^T r(X;\zeta_{1n})\}[\delta_{\gamma_n}^T \{r(X;\zeta_{2n}) - r(X;\zeta_{1n})\}]^2, \end{aligned}$$

it is a $o(\|\gamma_n - \gamma_0\|_1)$. The main term is minimum with the value zero as $\zeta_1 = \zeta_0$ and $\gamma = \gamma_0$, for every ζ_2. It follows that the estimators $\widehat{\zeta}_{1n}$ and $\widehat{\zeta}_{0n}$, and respectively, $\widehat{\gamma}_n$ converge a.s. under P_0 to ζ_0, and respectively, γ_0.

The difference $nl_n(\widehat{\theta}_n) - nl_{0n}(\widehat{\zeta}_{0n})$ is the sum of $T_{0n} = nl_{0n}(\widehat{\zeta}_{1n}) - nl_{0n}(\widehat{\zeta}_{0n})$ and $Q_n(\widehat{\theta}_n) = n\|Y - r_{\widehat{\theta}_n}(X)\|_n^2 - n\|Y - r(X;\widehat{\zeta}_{1n})\|_n^2$. The variable T_{0n} develops as

$$\begin{aligned} T_{0n} &= n\{\|Y - 1_k^T r(X;\widehat{\zeta}_{1n})\|_n^2 - \|Y - 1_k^T r(X;\widehat{\zeta}_{0n})\|_n^2\} \\ &= \sum_{i=1}^n [1_k^T \{r(X_i;\widehat{\zeta}_{1n}) - r(X_i;\widehat{\zeta}_{0n})\}\{1_k^T r(X_i;\widehat{\zeta}_{1n}) + 1_k^T r(X_i;\widehat{\zeta}_{0n}) - 2Y_i\}], \end{aligned}$$

The first order derivative of l_{0n} with respect to ζ_1 is

$$n\dot{l}_{0n,\zeta_1}(\widehat{\zeta}_{1n}) = -2\sum_{i=1}^n 1_k^T \dot{r}_\zeta(X_i;\widehat{\zeta}_{1n})\{Y_i - 1_k^T r(X_i;\widehat{\zeta}_{1n})\},$$

$\dot{l}_{0n,\zeta_1}(\widehat{\zeta}_{0n}) = 0$ and the second order derivative $\ddot{l}_{0n,\zeta_1}(\zeta_0)$ converges to a positive definite matrix I_{01}.

The weak convergence of $n^{\frac{1}{2}}(\widehat{\zeta}_{1n} - \zeta_0)$ and $n^{\frac{1}{2}}(\widehat{\zeta}_{0n} - \zeta_0)$ to the same limit, for $\delta_{\gamma_0} \equiv 0$ and $\delta_{\widehat{\gamma}_n} = 0(n^{-1})$, implies that $n^{\frac{1}{2}}(\widehat{\zeta}_{1n} - \widehat{\zeta}_{0n})$ converges to zero in probability under H_0. A first order asymptotic expansion of $\dot{l}_{0n,\zeta_1}(\widehat{\zeta}_{0n})$ entails that $\dot{l}_{0n,\zeta_1}(\widehat{\zeta}_{1n})$ also converges to zero in probability. By a second order asymptotic expansion of T_{0n} as

$$2T_{0n} = \dot{l}_{0n,\zeta_1}(\widehat{\zeta}_{1n}) I_{01}^{-1} \dot{l}_{0n,\zeta_1}(\widehat{\zeta}_{1n}),$$

it converges in probability to zero under H_0.

The process $Q_n(\theta) = \sum_{i=1}^n \{Y_i - \sum_{l=1}^p r(X_i;\zeta_l) I_l(\gamma)\}^2 - \sum_{i=1}^n \{Y_i - r(X_i;\zeta_1) I_1(\gamma)\}^2$ has first order derivatives

$$\dot{Q}_{n,l}(\theta) = -2\sum_{i=1}^n \dot{r}_{\zeta_l}(X_i;\zeta_l) I_l(\gamma) \Big[Y_i - \sum_{l=1}^p r(X_i;\zeta_l) I_l(\gamma)\Big],$$

for $l = 2, \ldots, p$. The variable $\dot{Q}_n(\eta, \widehat{\gamma}_n) = 0$ is asymptotically equivalent under H_0 to a centered Gaussian variable with positive definite variance I_{02}, from the convergence rate of $\widehat{\gamma}_n$. Under H_0, the variables $\widetilde{\zeta}_n = (\widehat{\zeta}_{2n}^T, \ldots, \widehat{\zeta}_{pn}^T)^T$ converges in probability to a limit $\widetilde{\zeta}_0$ and

$$\widetilde{\zeta}_n - \widetilde{\zeta}_0 = I_{02}^{-1} \dot{Q}_n(\widetilde{\zeta}_0, \widehat{\gamma}_n) + o_p(1).$$

The asymptotic expansion of Q_n as

$$Q_n(\widehat{\theta}_n) = \dot{Q}_n^T(\widetilde{\zeta}_0, \widehat{\gamma}_n) I_{02}^{-1} \dot{Q}_n(\widetilde{\zeta}_0, \widehat{\gamma}_n) + o_p(1)$$

entails its convergence in weakly to a $\chi^2_{d(p-1)}$ variable under H_0, the degree of the χ^2 variable corresponds to the $p-1$ terms of the partition $I_2, \ldots, I_p$, with d dimensional regression parameters. □

Under alternatives of a probability distribution P_θ with parameter θ such that γ is different from γ_0, the expression of the statistic depends under P_θ on the limit of the process $\widehat{Q}_n$. Let $P_n = P_{\theta_n}$ be the probability distributions of local alternatives with a sequence of parameters $(\theta_n)_n$ converging to the parameter θ_0 of the hypothesis, the parameter γ_n converges to the upper bound of the support of the variable X, ζ_{1n} converges to ζ_0 and ζ_{2n} converges to an arbitrary ζ_{02} different from ζ_0.

Proposition 4.4. *Under a fixed alternative P_θ, the statistic T_n tends to infinity. Under local alternatives $P_{\theta_{n,u}}$, the statistic T_n converges weakly to an uncentered $\chi^2_{d(p-1)}$ variable.*

Proof. Under an alternative P_θ, the estimator $Q_n(\widehat{\zeta}_n, \gamma)$ of the process Q_n defined by (4.11) tends to infinity for every $\gamma < \gamma_0$ and the statistic diverges.

Local alternatives are defined by a sequence of parameters $(\theta_n)_n$ converging to θ_0 with the rates $n^{-\frac{1}{2}}$ for the regression parameters and n^{-1} for the change-point parameter $\gamma_n = X_{n:n} - n^{-1}u_n$, where u_n converges to a non-zero limit u. The regression parameters under P_{θ_n} are $\zeta_n = \zeta_0 + n^{\frac{1}{2}} u_{n\zeta}$ where $u_{n\zeta}$ converges to a non-zero limit u_ζ. The estimators $\widehat{\zeta}_{1n}$ and $\widehat{\zeta}_{0n}$ converge a.s. to ζ_0, the variable $n^{\frac{1}{2}}(\widehat{\zeta}_{1n} - \zeta_0)$ converges weakly under P_{θ_n} to an uncentered Gaussian variable with mean u_{ζ_1} and with the same variance as $n^{\frac{1}{2}}(\widehat{\zeta}_{0n} - \zeta_0)$ hence $n^{\frac{1}{2}}(\widehat{\zeta}_{1n} - \widehat{\zeta}_{0n})$ converges in probability to u_{ζ_1} under P_n and

$$T_{0n} = u_{\zeta_1}^T I_{01}^{-1} u_{\zeta_1} + o_p(1).$$

Under P_n, the variable $n(\delta_{\gamma_n} - \delta_{\widehat{\gamma}_n})$ converges weakly to a variable U and $n(1 - \delta_{\widehat{\gamma}_n})$ converges weakly to $U + u$. The variable $Q_n(u_{\widehat{\theta}_n}) = 0$ is written in the same way as under H_0 and the conditional mean of Y under P_{θ_n} is $r_{\theta_n}(X)$, its difference with $\sum_{l=1}^p r(X_i; \widehat{\zeta}_{ln}) I_l(\widehat{\gamma}_n)$ is a $O_p(n^{-\frac{1}{2}})$. The variance of $\dot{Q}_n(u_{\widehat{\theta}_n})$ converges to a positive definite matrix I_1 depending on $U + u$ and the limit of $Q_n(u_{\widehat{\theta}_n})$ under P_{θ_n} is the same as under P_0. □

4.4 Change-points in parametric models

Let (X, Y) be a variable such that under a probability distribution P_θ

$$E_\theta(Y \mid X) = r_\eta(X, \gamma) = (1 - \delta_\gamma)^T r_{1\eta}(X) + \delta_\gamma^T r_{2\eta}(X) \tag{4.13}$$

where $\theta = (\eta^T, \gamma^T)^T$ and η belongs to a parameter sub-space $\mathcal{H}$ of $\mathbb{R}^s$, $s \geq 1$, with d-dimensional vectors of regression functions $r_{1\eta}$ and $r_{2\eta}$. The combination of the indicators $1_{\{X_k \leq \gamma_k\}}$ and $1_{\{X_k > \gamma_k\}}$, for the d components of the variable X and the threshold vector γ, define p non-overlapping sets in $\mathbb{R}^d$ with indicators $I_l(\gamma)$, $l = 1, \ldots, p$, and p real regression functions that define the conditional mean of Y by (4.12)

$$r_\eta(x, \gamma) = \sum_{l=1}^{p} r_l(x, \eta) I_l(\gamma).$$

The known parametric functions $r_l(x, \eta)$ belong to class $C^2(\mathcal{H})$, for every value x of a regression variable X in $\mathbb{R}^d$. In the model (4.12), the parameter may be a set of distinct parameters $\eta = (\eta_1^T, \ldots, \eta_p^T)^T$ with a common regression function r

$$r_\eta(x, \gamma) = \sum_{l=1}^{p} r(x, \eta_l) I_l(\gamma) \tag{4.14}$$

or (4.12) may be defined by p distinct real regression functions depending on a common parameter η. Let

$$r_0 = r_{\eta_0}(\gamma_0) = (1 - \delta_0)^T r_{01} + \delta_0 r_{02}$$

be the true regression function with a change-point at $\theta_0 = (\eta_0^T, \gamma_0^T)^T$ under the probability P_0. We assume that the condition (4.8) is fulfilled.

For a n-sample $(X^T, Y)^T$ with distribution function F_0 under P_0 and every value γ of the change-point parameter, the mean squares estimators of the parameters minimize

$$l_n(\eta, \gamma) = \|Y - r_\eta(X, \gamma)\|_n^2 - \|Y - r_0(X)\|_n^2,$$

they are asymptotically equivalent to the maximum likelihood estimators in the model with a Gaussian error independent of X. At a fixed γ, the estimator $\widehat{\eta}_{\gamma,n}$ is solution of the estimating equation $\dot{l}_{n,\eta}(\eta, \gamma) = 0$ or

$$\|\dot{r}_\eta(X, \gamma)(Y - r_\eta(X, \gamma))\|_n = 0,$$

and the matrix

$$I_n(\eta, \gamma) = -\ddot{l}_n(\eta, \gamma) = \|\dot{r}_\eta^{\otimes 2}(X, \gamma)\|_n - \|\ddot{r}_\eta(X, \gamma)(Y - r_\eta(X, \gamma))\|_n$$

converges a.s. under P_0 to its expectation $I_0(\eta, \gamma)$ such that

$$I_0 = I_0(\eta_0, \gamma_0) = E_0 \dot{r}_{\eta_0}^{\otimes 2}(X, \gamma_0)$$

is a positive definite matrix. The variance of the error at $\widehat{\eta}_{n,\gamma}$ is estimated by

$$\widehat{\sigma}_{\gamma,n}^2 = \left\| Y - \sum_{l=1}^{p} r(x, \widehat{\eta}_{\gamma,ln}) I_l(\gamma) \right\|_n^2 \tag{4.15}$$

and the change-point parameter γ is estimated by minimization of the estimator of variance

$$\widehat{\gamma}_n = \arg\inf_{\gamma} \widehat{\sigma}_{\gamma,n}^2.$$

The estimator of η is estimated by $\widehat{\eta}_n = \widehat{\eta}_{n,\widehat{\gamma}_n}$.

Let $\theta = (\eta^T, \gamma^T)^T$, the process $l_n(\theta)$ converges a.s. under P_0, to the function $l(\theta) = E_0[\{Y - r_\eta(X, \gamma)\}^2 - \{Y - r_0(X)\}]^2$ as n tends to infinity. For every γ, let η_γ be solution of the equation $\dot{l}_\eta(\eta, \gamma) = 0$, where

$$\dot{l}_\eta(\eta, \gamma) = E_0\left[\dot{r}_\eta(X, \gamma)\{Y - r_\eta(X, \gamma)\}\right]$$

then $\dot{l}_\eta(\eta_0, \gamma_0) = 0$ which is equivalent to

$$E_0([\dot{r}_{l,\eta}(X, \eta_0)\{Y - r_l(X, \eta_0)\}]I_l) = 0, \quad l = 1, \ldots, p,$$

and $I(\eta, \gamma) = -\ddot{l}_\eta(\eta, \gamma)$ is the a.s. limit under P_0 of the empirical information matrix $I_n(\eta, \gamma)$, let $I_0(\eta) = I(\eta, \gamma_0)$.

Let $W_n(\theta) = n^{\frac{1}{2}}(l_n - l)(\theta)$.

Proposition 4.5. *For every $\varepsilon > 0$ sufficiently small, there exist constants κ_0 and κ_1 such that for every θ belonging to $V_\varepsilon(\theta_0)$*

$$l(\theta) \geq \kappa_0 \rho^2(\theta, \theta_0)$$

$$E_0 \sup_{\theta \in V_\varepsilon(\theta_0)} |W_n(\theta)| \leq \kappa_1 \rho(\theta, \theta_0),$$

as n tends to infinity.

Proof. The proof is similar to the proof of Proposition 4.2 with a second order expansion of $E_0[\{r_l(X, \eta_0) - r_l(X; \eta_l)\}^2 I_l]$ for $l = 1, \ldots, p$ and for η in an ε- neighborhood of η_0

$$\begin{aligned}
&(\eta_l - \eta_0)^T E_0[\dot{r}_l(X; \eta_0)\{r_l(X, \eta_l) - r_l(X; \eta_0)\} I_l] \\
&-\frac{1}{2}(\eta_l - \eta_0)^T I_0 (\eta_l - \eta_0) + o(\|\eta_l - \eta_0\|^2) \\
&= -\frac{1}{2}(\eta_l - \eta_0)^T I_0 (\eta_l - \eta_0) + o(\|\eta_l - \eta_0\|^2).
\end{aligned}$$

The second inequality is proved in the same way as in Proposition 4.2. □

The convergence rates of $\widehat{\eta}_n$ and $\widehat{\gamma}_n$ are deduced from Proposition 4.5, like in Theorem 3.2.

Theorem 4.5. *For $\varepsilon > 0$ sufficiently small*

$$\overline{\lim}_{n,A\to\infty} P_0(\sup_{\theta\in V_\varepsilon(\theta_0),\|u_{n,\theta}\|>A} X_n(\gamma) \geq 0) = 0 \,,$$
$$\overline{\lim}_{n,A\to\infty} P_0\{n^{\frac{1}{2}}\rho(\widehat{\theta}_n, \theta_0) > A) = 0.$$

Proposition 4.6. *The estimator $\widehat{\eta}_n$ is such that $n^{\frac{1}{2}}(\widehat{\eta}_n - \eta_0)$ converges weakly under P_0 to a centred Gaussian variable with variance I_0^{-1}.*

Proof. By Theorem 4.5 and a first order expansion (3.5) of the process $\dot{l}_{n,\eta}$, we have

$$\begin{aligned} n^{\frac{1}{2}}\{\dot{l}_{n,\eta}(\widehat{\theta}_n) - \dot{l}_{n,\eta}(\eta_0, \widehat{\gamma}_n)\} &= -n^{\frac{1}{2}}\dot{l}_{n,\eta}(\eta_0, \widehat{\gamma}_n) \\ &= n^{\frac{1}{2}}(\widehat{\eta}_n - \eta_0)\ddot{l}_{n,\eta}(\eta_0, \widehat{\gamma}_n) + o_p(1), \end{aligned}$$

under P_0 the variables $-\ddot{l}_{n,\eta}(\theta_0)$ and $\ddot{l}_{n,\eta}(\eta_0, \widehat{\gamma}_n)$ converge a.s. to the matrix I_0, $n^{\frac{1}{2}}\dot{l}_{n,\eta}(\theta_0)$ is a centered and asymptotically Gaussian variable with variance I_0. The variable $\dot{l}_{n,\eta}(\eta_0, \gamma)$ has the mean

$$\begin{aligned} E_0\dot{l}_{n,\eta}(\eta_0, \gamma) &= -2n^{-1}\sum_{i=1}^n E_0[\dot{r}_{\eta_0}(X_i, \gamma)\{Y_i - r_{\eta_0}(X_i, \gamma)\}] \\ &= -2n^{-1}\sum_{i=1}^n E_0[\dot{r}_{\eta_0}(X_i, \gamma)\{r_{\eta_0}(X_i, \gamma_0) - r_{\eta_0}(X_i, \gamma)\}], \end{aligned}$$

and the variance of $n^{\frac{1}{2}}\dot{l}_{n,\eta}(\eta_0, \gamma)$ is $4n^{-1}\sum_{i=1}^n E_0[\dot{r}_{\eta_0}(X_i, \gamma)\{Y_i - r_0(X_i)\}]^2$, it converges to a finite limit $V_\gamma = 4\sigma_0^2 E_0[\dot{r}_{\eta_0}^{\otimes 2}(X, \gamma)]$. As $\widehat{\gamma}_n$ converges in probability to γ_0 with the rate n^{-1}, $\delta_{\widehat{\gamma}_n} - \delta_{\gamma_0} = O_p(n^{-1})$ and the variables $n^{\frac{1}{2}}\dot{l}_{n,\eta}(\eta_0, \widehat{\gamma}_n)$ and $n^{\frac{1}{2}}\dot{l}_{n,\eta}(\eta_0, \gamma_0)$ converge weakly under P_0 to a centered variable with variance $I_0 = V_{\gamma_0}$. Then

$$n^{\frac{1}{2}}(\widehat{\eta}_n - \eta_0) = n^{\frac{1}{2}}\dot{l}_{n,\eta}^T(\eta_0, \widehat{\gamma}_n)I_0^{-1} + o_p(1),$$

the result follows. □

Proposition 4.6 and a first order expansion of the regression function ensure the weak convergence of the process $n^{\frac{1}{2}}(r_{\widehat{\eta}_n}(\widehat{\gamma}_n) - r_0)$, like in Theorem 4.2. Let $u = (u_\eta^T, u_\gamma^T)^T$ and let $\eta_{nu} = \eta_0 + n^{-\frac{1}{2}}u_\eta$ and $\gamma_{nu} = \gamma_0 + n^{-1}u_\gamma$, the process $nl_n(\theta_{nu})$ is the sum of the processes

$$\widetilde{l}_n(u_\zeta) = \sum_{i=1}^n [\{Y_i - r_{\zeta_{n,u},\gamma_0}(X_i)\}^2 - \{Y_i - r_{\theta_0}(X_i)\}^2] \tag{4.16}$$

$$Q_n(u) = \sum_{i=1}^n [\{Y_i - r_{\theta_{n,u}}(X_i)\}^2 - \{Y_i - r_{\zeta_{n,u},\gamma_0}(X_i)\}^2]$$

where the process $\widetilde{l}_n$ does not depend on u_γ and the process Q_n is the difference of l_n and $\widetilde{l}_n$. The asymptotic behaviour of the process Q_n is established like in Theorem 4.3 then the behaviour of the estimators is deduced like in Theorem 4.4.

Theorem 4.6. *The process Q_n converges weakly to a Gaussian process Q with mean function*

$$\lambda_0(u_\zeta) = f_X(\gamma_0)[u_\gamma^T\{r(\gamma_0, \zeta_{01}) - r(\gamma_0, \zeta_{02})\}]^2$$

and with a finite variance function.

Theorem 4.7. *The estimator $u_{\widehat{\gamma}_n} = n(\widehat{\gamma}_n - \gamma_0)$ is asymptotically independent of the estimator of the regression parameters $\widehat{\eta}_n$ and it converges weakly to the variable u_0 where the process Q achieves its minimum and u_0 is finite in probability.*

A test statistic for the hypothesis a regression model without change-point is defined as

$$T_n = 2n \inf_{\gamma\in\Gamma}\{\widehat{\sigma}^2_{\gamma,n} - \widehat{\sigma}^2_{0n}\}$$

where $\widehat{\sigma}^2_{\gamma,n}$ is the variance estimator in the model with change-points and $\widehat{\sigma}^2_{0n}$ is the variance estimator in the model without change-point, it has the same asymptotic distributions as in Section 4.3 for a linear regression.

More generally, the variance σ^2 of the variable Y is supposed to be a strictly positive parametric function of the variable X, in the change-point model where the error e has the mean zero and the conditional variance

$$E_{\eta,\gamma}(e^2|X) = \sigma^2_\eta(X,\gamma),$$

with a change-point at the same point as in the mean. The maximum likelihood estimation in the model with a Gaussian error is based on the process

$$l_n(\eta,\gamma) = \frac{1}{2}[\|\sigma_\eta^{-2}(X,\gamma)\{Y - r_\eta(X,\gamma)\}\|_n^2 \\ -\|\sigma_{\eta_0}^{-2}(X)\{Y - r_{\eta_0}(X)\}\|_n^2] - n^{-1}\sum_{i=1}^n \log\frac{\sigma_\eta(X_i,\gamma)}{\sigma_0(X_i)}.$$

The estimator $\widehat{\eta}_{n,\gamma}$ of η minimizing l_n at any fixed γ is solution of the equation

$$\left\|\frac{\dot{r}_\eta(X,\gamma)}{\sigma^2_\eta(X,\gamma)}\{Y - r_\eta(X,\gamma)\}\right\|_n - \left\|\frac{\dot{\sigma}_\eta(X,\gamma)}{\sigma^2_\eta(X,\gamma)}[2\{Y - r_\eta(X,\gamma)\}^2 - 1]\right\|_n = 0.$$

This estimator is asymptotically Gaussian and centred at γ_0, moreover the estimator of γ that minimizes $\widehat{l}_n(\gamma) = l_n(\widehat{\eta}_{n,\gamma}, \gamma)$ is consistent, because $\widehat{\eta}_{n,\gamma}$ converges uniformly and the limit of $\widehat{l}_n$ is zero only at γ_0. The asymptotic distributions of the estimators are similar to the limits of the previous estimators.

4.5 Maximum likelihood in parametric models

Let f be the conditional density of the error given X in the model with a parametric regression (4.14) with distinct parameters η_1 and η_2 in a space $\mathcal{H}$ and a change-point γ in $\mathbb{R}^d$, it is written as

$$f(y - r_\eta(X,\gamma)) = f(Y - (1-\delta_\gamma)^T r_1(X,\eta) - \delta_\gamma^T r_2(X,\eta)).$$

The maximum likelihood estimator of the parameter θ achieves the maximum of

$$l_n(\theta) = \| \log f(Y_i - r_\eta(X_i,\gamma)) \|_n$$

with a n-sample of (X,Y). At a fixed γ, $\widehat{\eta}_{n,\gamma}$ maximizes the function $l_n(\eta,\gamma)$ and $\dot{l}_n(\widehat{\eta}_{n,\gamma}) = 0$, then $\widehat{\gamma}_n$ maximizes $\widehat{l}_n(\gamma) = l_n(\widehat{\eta}_{n,\gamma}, \gamma)$ and $\widehat{\eta}_n = \widehat{\eta}_{n,\widehat{\gamma}_n}$.

We consider a density f in $C^2(\mathbb{R})$ and a regression function r such that for every γ, the function $\eta \mapsto r(\eta,\gamma)$ belongs to $C^2(\mathcal{H})$, the first two derivatives of l_n with respect to η are

$$\dot{l}_{n,\eta}(\theta) = -n^{-1} \sum_{i=1}^n \frac{f'}{f}(Y_i - r_\eta(X_i,\gamma))\dot{r}_\eta(X_i,\gamma)$$

with components

$$\dot{l}_{n,\eta_1}(\theta) = -n^{-1} \sum_{i=1}^n \frac{f'}{f}(Y_i - r_\eta(X_i,\gamma))(1-\delta_\gamma)^T \dot{r}_{\eta_1}(X_i,\gamma),$$

$$\dot{l}_{n,\eta_2}(\theta) = -n^{-1} \sum_{i=1}^n \frac{f'}{f}(Y_i - r_\eta(X_i,\gamma))\delta_\gamma^T \dot{r}_{\eta_2}(X_i,\gamma)\},$$

$$\ddot{l}_n(\theta) = n^{-1} \sum_{i=1}^n \frac{ff'' - f'^2}{f^2}(Y_i - r_\eta(X_i,\gamma))\dot{r}_\eta^2(X_i,\gamma)$$
$$-n^{-1} \sum_{i=1}^n \frac{f'}{f}(Y_i - r_\eta(X_i,\gamma))\ddot{r}_\eta(X_i,\gamma).$$

The process l_n and its derivatives converge a.s. uniformly under P_0 to their expectation functions

$$\begin{aligned} l(\theta) &= E_0 \log f(Y - r_\eta(X,\gamma)) \\ &= E_0 \log f(Y - (1-\delta_\gamma)^T r_1(X,\eta) - \delta_\gamma^T r_2(X,\eta)) \end{aligned}$$

$\dot{l}(\theta) = E_0\dot{l}_n(\theta)$ such that $\dot{l}(\theta_0) = 0$ and respectively $\ddot{l}(\theta) = E_0\ddot{l}_n(\theta)$. We assume that the information matrix

$$I_0 = -\ddot{l}(\theta_0) = E_0\Big\{\frac{f'^2}{f^2}(Y - r_{\eta_0}(X,\gamma_0))\dot{r}^2_{\eta_0}(X,\gamma_0)\Big\}$$

is bounded. It is positive definite and diagonal by blocks for distinct parameters η_1 and η_2. The function l is a locally concave function of the parameter η with a maximum at θ_0, the estimator $\widehat{\theta}_n$ is therefore a.s. consistent under P_0 (cf. Theorem 3.1).

The logarithm of the likelihood ratio of the sample under P_θ and P_0 defines the process

$$S_n(\theta) = l_n(\theta) - l_n(\theta_0),$$

it splits as a sum $S_n = \widetilde{S}_n + Q_n$ with the processes

$$\widetilde{S}_n(\eta) = n^{-1}\sum_{i=1}^n \log\frac{f(Y_i - r_\eta(X_i,\gamma_0))}{f(Y_i - r_0(X_i))},$$

$$Q_n(\theta) = n^{-1}\sum_{i=1}^n \log\frac{f(Y_i - r_\eta(X_i,\gamma))}{f(Y_i - r_\eta(X_i,\gamma_0))}. \tag{4.17}$$

For γ in an ε^2-neighborhood $V_\varepsilon(\gamma_0)$ of γ_0

$$\begin{aligned} & f(Y_i - r_\eta(X_i,\gamma)) - f(Y_i - r_\eta(X_i,\gamma_0)) \\ &= -\{r_\eta(X_i,\gamma) - r_\eta(X_i,\gamma_0)\}f'(Y_i - r_\eta(X_i,\gamma_0)) \\ &\quad +O(\{r_\eta(X_i,\gamma) - r_\eta(X_i,\gamma_0)\}^2) \end{aligned}$$

where

$$r_\eta(X_i,\gamma) - r_\eta(X_i,\gamma_0) = (\delta_\gamma - \delta_{\gamma_0})^T\{r_2(X_i,\eta) - r_1(X_i,\eta)\}$$

and $E_0 \sup_{\gamma\in V_\varepsilon(\gamma_0)} \|\delta_\gamma - \delta_{\gamma_0}\| = O(\varepsilon^2)$. For every θ in an ε-neighborhood $V_\varepsilon(\theta_0)$ of θ_0 for the semi-norm $\rho(\theta,\theta_0) = (\|\eta - \eta_0\|_2^2 + \|\gamma - \gamma_0\|_1)^{\frac{1}{2}}$, then

$$E_0 \sup_{\theta\in V_\varepsilon(\theta_0)} Q_n(\theta) = O(\varepsilon^2).$$

Under P_0 and in $V_\varepsilon(\theta_0)$, the process Q_n has the approximation

$$Q_n(\theta) = n^{-1}\sum_{i=1}^n\Big\{\log\frac{f(Y_i - r_{\eta_0}(X_i,\gamma))}{f(Y_i - r_{\eta_0}(X_i,\gamma_0))}\Big\}\{1 + o_p(1)\}$$

which is denoted $Q_n(\theta) = \widetilde{Q}_n(\gamma)\{1 + o_p(1)\}$, where the main term $\widetilde{Q}_n$ of Q_n depends only on the change-point parameter γ and it is a $O_p(\varepsilon^2)$.

The processes $\widetilde{S}_n$ and $\widetilde{Q}_n$ converge a.s. under P_0 to the functions $\widetilde{S}(\eta) = E_0\widetilde{S}_n(\eta)$ and respectively $\widetilde{Q}(\gamma) = E_0\widetilde{Q}_n(\eta)$.

Let $W_n(\theta) = n^{\frac{1}{2}}\{S_n(\theta) - E_0S_n(\theta)\}$. We assume the condition

$$E_0\Big\{\sup_{\theta\in V_\varepsilon(\theta_0)} \log^2 f(Y - r_\theta(X))\Big\} < \infty. \tag{4.18}$$

Proposition 4.7. *Under the condition (4.18), for every $\varepsilon > 0$ sufficiently small, there exist constants κ_0 and κ_1 such that for every θ belonging to $V_\varepsilon(\theta_0)$, as n tends to infinity*

$$E_0S_n(\theta) \le -\kappa_0\rho^2(\theta,\theta_0)$$
$$E_0 \sup\nolimits_{\theta\in V_\varepsilon(\theta_0)} |W_n(\theta)| \le \kappa_1\varepsilon.$$

Proof. To prove the first inequality for the function $E_0S_n(\theta) = \widetilde{S}(\eta) + \widetilde{Q}(\gamma)\{1+o(1)\}$, it is sufficient to prove it for the functions $\widetilde{S}$ and $\widetilde{Q}$. For the function $\widetilde{S}$ such that $\widetilde{S}'(\eta_0) = 0$, it is obtained by a second order expansion in $V_\varepsilon(\theta_0)$, they imply

$$\begin{aligned}\widetilde{S}(\eta) &= \frac{1}{2}(\eta-\eta_0)^T\widetilde{S}''(\eta_0)(\eta-\eta_0) + o(\|\eta-\eta_0\|^2)\\ &= -\frac{1}{2}(\eta-\eta_0)^T I_0(\eta-\eta_0) + o(\|\eta-\eta_0\|^2),\end{aligned} \tag{4.19}$$

it follows that $E_0\widetilde{Q}_n(\theta) = O(\varepsilon^2)$ in $V_\varepsilon(\theta_0)$ which provides the inequality for the function $\widetilde{Q}$.

The process W_n is asymptotically centered and the second inequality is a consequence of the Cauchy–Schwarz inequality and of the bound for the variance of

$$\sup_{\theta\in V_\varepsilon(\theta_0)} W_n(\theta) = \sup_{\theta\in V_\varepsilon(\eta_0)} [n^{\frac{1}{2}}\{\widetilde{S}_n(\theta) - \widetilde{S}(\eta)\} + n^{\frac{1}{2}}\{\widetilde{Q}_n(\gamma) - \widetilde{Q}(\gamma)\}] + O_p(\varepsilon^2),$$

using a second order expansion of the first process and the equality

$$E_0 \sup_{\gamma\in V_\varepsilon(\gamma_0)} (\delta_\gamma - \delta_{\gamma_0}) = O(\varepsilon^2)$$

for the second process. □

The convergence rates of $\widehat{\eta}_n$ and $\widehat{\gamma}_n$ are deduced from Proposition 4.7, by the arguments of Theorem 3.2.

Theorem 4.8. *Under the condition (4.18), for $\varepsilon > 0$ sufficiently small*

$$\overline{\lim}_{n,A\to\infty} P_0(\sup\nolimits_{\theta\in V_\varepsilon(\theta_0),\|u_{n,\theta}\|>A} S_n(\gamma) \ge 0) = 0,$$
$$\overline{\lim}_{n,A\to\infty} P_0(n^{\frac{1}{2}}\rho(\widehat{\theta}_n,\theta_0) > A) = 0.$$

Let $u = (u_\eta^T, u_\gamma^T)^T$ and let $\eta_{n,u} = \eta_0 + n^{-\frac{1}{2}} u_\eta$ and $\gamma_{n,u} = \gamma_0 + n^{-1} u_\gamma$.

Theorem 4.9. *For every $A > 0$, in $\mathcal{U}_n^A$ the process $nS_n(\theta_{n,u})$ has the uniform approximation*

$$nS_n(\theta_{n,u}) = n\widetilde{Q}_n(\gamma_{n,u}) + n\widetilde{S}_n(\eta_{n,u}) + o_p(1).$$

Proof. By (4.17) and (4.19), the process S_n is such that $nS_n(\theta_{n,u})$ is the sum $n\widetilde{Q}_n(\gamma_{n,u}) + n\widetilde{S}_n(\eta_{n,u}) + r_n$ where $n\widetilde{S}_n(\eta_{n,u})$ is a uniform $O_p(1)$ and $n\widetilde{Q}_n(\gamma_{n,u})$ converges a.s. uniformly in $\mathcal{U}_n^A$ to its means which is a $O(1)$. The remainder term r_n is a $o_p(n\widetilde{Q}_n(\gamma_{n,u}))$ it is then a $o_p(1)$. □

The estimator $\widehat{\eta}_n$ maximizes the process $\widetilde{S}_n$ and $\widehat{\gamma}_n$ maximizes the process $\widetilde{Q}_n$.

Proposition 4.8. *Under the condition (4.18), the estimator $\widehat{\eta}_n$ is such that $n^{\frac{1}{2}}(\widehat{\eta}_n - \eta_0)$ converges weakly under P_0 to a centred Gaussian variable with variance I_0^{-1}.*

Proof. By Theorem 4.8 and a first order expansion of $\widetilde{S}'_n$ in a neighborhood of η_0, the estimator satisfies

$$n^{\frac{1}{2}}\{\widetilde{S}'_n(\widehat{\eta}_n) - \widetilde{S}'_n(\eta_0)\} = n^{\frac{1}{2}}(\widehat{\eta}_n - \eta_0)\widetilde{S}''_n(\theta_0) + o_p(1)$$

with $\widetilde{S}'_n(\widehat{\eta}_n) = 0$, therefore $n^{\frac{1}{2}}(\widehat{\eta}_n - \eta_0) = I_0^{-1}\widetilde{S}''_n(\theta_0) + o_p(1)$. The variable $\widetilde{S}'_n(\eta_0)$ is centered and asymptotically Gaussian, and the asymptotic variance of $n^{\frac{1}{2}}\widetilde{S}'_n(\theta_0)$ is I_0, the result follows. □

Theorem 4.10. *Under the condition (4.18) and P_0, the process $\widetilde{Q}_n$ converges weakly in $\mathcal{U}_n^A$ to a Gaussian process Q with mean $u_\gamma^T \lambda_0$ where*

$$\lambda_0 = \{r_1(\gamma_0, \eta_0) - r_2(\gamma_0, \eta_0)\} \int f'(y - r_{\eta_0}(\gamma_0, \gamma_0))\, dy$$

and with a finite variance function.

Proof. Under P_0, the process $\widetilde{Q}_n$ has the mean

$$\begin{aligned}
E_0\widetilde{Q}_n(\theta_{n,u}) &= E_0\Big[\log\Big\{1 + \frac{f(Y_i - r_{\eta_0}(X_i, \gamma_{n,u}) - f(Y_i - r_{\eta_0}(X_i, \gamma_0)))}{f(Y_i - r_{\eta_0}(X_i, \gamma_0))}\Big\}\Big] \\
&= E_0\Big\{\frac{f(Y_i - r_{\eta_0}(X_i, \gamma_{n,u}) - f(Y_i - r_{\eta_0}(X_i, \gamma_0)))}{f(Y_i - r_{\eta_0}(X_i, \gamma_0))}\Big\}\{1 + o(1)\} \\
&= -E_0\Big[\{\dot{r}_{\eta_0}(X_i, \gamma_{n,u}) - \dot{r}_{\eta_0}(X_i, \gamma_0)\}\frac{f'}{f}(Y_i - r_{\eta_0}(X_i, \gamma_0))\Big] \\
&\quad + r_n,
\end{aligned}$$

the main term of $nE_0\widetilde{Q}_n(\theta_{n,u})$ is

$$n\int (\delta_{\gamma_{n,u}} - \delta_{\gamma_0})^T\{\dot r_1(x,\eta_0) - \dot r_2(x,\eta_0)\}f'(y - r_{\eta_0}(x,\gamma_0))\,dx\,dy$$

$$= u_\gamma^T\{\dot r_1(\gamma_0,\eta_0) - \dot r_2(\gamma_0,\eta_0)\}\int f'(y - r_{\eta_0}(\gamma_0,\gamma_0))\,dy$$

and $nr_n = O(n|\dot r_{\eta_0}(X_i,\gamma_{n,u}) - \dot r_{\eta_0}(X_i,\gamma_0)|^2) = o(1)$. The variance of the process $n\widetilde{Q}_n(\gamma_{n,u})$ is a 0(1). The limiting distribution of $n\widetilde{Q}_n$ is the limit of a sequence of sums of independent and identically distributed random variables with convergent first two moments. □

By Theorem 4.8, the variable $u_{\widehat{\gamma}_n} = n(\widehat{\gamma}_n - \gamma_0)$ is such that

$$u_{\widehat{\gamma}_n} = \arg\max_{\mathbb{R}^d} n\widetilde{Q}_n(u_\gamma) + o_p(1)$$

where the minimization deals with all components of the parameter γ.

Theorem 4.11. *Under the condition (4.18), the estimator $u_{\widehat{\gamma}_n}$ of the change-point γ_0 is asymptotically independent of the estimator of the regression parameters $\widehat{\eta}_n$, it converges weakly under P_0 to the point u_0 where the process $\widetilde{Q}$ achieves its maximum and u_0 is finite with a probability converging to one.*

The proof relies on Theorems 4.8 and 4.10, like for Theorem 3.3.

4.6 Likelihood ratio test in parametric models

We consider the log-likelihood ratio test for the hypothesis H_0 of an unknown regression function $r_0 = r(\eta_0)$ without change of parameter against the alternative of a change at an unknown breakpoint γ according to model (4.13) with two distinct regression parameters η_1 and η_2. The statistic of the likelihood ratio test for H_0 is

$$T_n = 2n \sup_{X_{n:1}<\gamma\leq X_{n:n}} \{\widehat{l}_n(\gamma) - \widehat{l}_{0n}\},$$

it is defined by the logarithm of the estimated density $\widehat{l}_{0n} = l_n(\widehat{\eta}_{0n})$ of the sample under H_0, with a regression parameter η_0 under the probability P_0 of the sample space under the hypothesis, and $\widehat{l}_n(\gamma) = l_n(\widehat{\eta}_{n,\gamma},\gamma)$, with the process l_n of Section 4.5.

Using the notation of (4.14), the process S_n is written as

$$\begin{aligned} S_n(\theta) &= \int_{\mathbb{R}^{d+1}} \log \frac{f(y - r_\eta(x,\gamma))}{f(y - r_{\eta_0}(x))} \, d\widehat{F}_n(x,y) \\ &= \int_{\mathbb{R}^{d+1}} \sum_{k=1}^{p} I_k(x,\gamma) \log \frac{f(y - r(x,\eta_k))}{f(y - r_{\eta_0}(x))} \, d\widehat{F}_n(x,y), \end{aligned} \tag{4.20}$$

$\widehat{S}_n(\gamma) = S_n(\widehat{\theta}_{n,\gamma})$ and $T_n = 2n \sup_\gamma \widehat{S}_n(\gamma)$. Under the alternative, the p distinct parameter vectors η_k and the vector γ define a set of $d(p+1)$ parameters. Under H_0 and denoting $I_1(x,\tau) = 1_{\{x \le \tau\}}$, θ_0 reduces to a d-dimensional vector $\eta_{01} = \eta_0$ and $\gamma_0 = \tau$, the end-point of the support of the variable S, the other components of θ_0 are arbitrary and mutually distinct.

Under H_0 and under the differentiability and integrability conditions for the density f and the regression function, the estimator $\widehat{\theta}_n$ maximizing S_n in the model (4.13) is such that $\lim_n S_n(\theta_0) = 0$, $\lim_n S_n(\theta) < 0$ for θ distinct of θ_0 and $S'_{n,\eta}(\widehat{\theta}_n) = 0$. The first derivative of S_n with respect to η converges a.s. under H_0 to its expectation S'_η which is zero at θ_0 and

$$S'_{n,\eta}(\theta_0) = -\int_{\mathbb{R}^{d+1}} \dot{r}_{\eta_0}(x,\gamma_0) \frac{f'(y - r_{\eta_0}(x,\gamma_0))}{f(y - r_{\eta_0}(x,\gamma_0))} \, d\widehat{F}_n(x,y).$$

Let

$$U_n(\theta) = n^{\frac{1}{2}} S'_{n,\eta}(\theta),$$

under P_0, its expectation at θ_0 is asymptotically equivalent to

$$n^{\frac{1}{2}} \int_{\mathbb{R}^{d+1}} \dot{r}_{\eta_0}(x,\gamma_0) f'(y - r_{\eta_0}(x,\gamma_0)) \, dx \, dy = 0.$$

Let η_1 be the regression parameter on the set $\cap_{i=1,\ldots,n}\{X_i \le \gamma_i\}$ and let U_{1n} be the vector corresponding to the derivative of S_n with respect to η_1, its variance is a positive definite $d \times d$-dimensional matrix I_{01}, then the variable U_n converges weakly under H_0 to a centered Gaussian variable with variance I_0. The second order derivative of the variable $\lim_n S_n$ is $-I_0$ at θ_0 hence $\lim_n S_n$ is a concave function of η_1 at η_0 and the estimators $\widehat{\eta}_{0n}$ and $\widehat{\eta}_{1n}$ converges a.s. to η_0, and $\widehat{\gamma}_n$ converges a.s. to $\gamma_0 = \tau$, under H_0.

The derivatives of S_n with respect to the other components of η converge a.s. to zero at θ_0, due to the restriction of the integral on sets where at least one component of the change-point is the end-point of the support of S. By a normalization with n, for $k = 2, \ldots, p$, the first two derivatives of S_n with respect to η_k converge to non-zero limits and the estimators $\widehat{\eta}_{kn}$ converge a.s. to the components η_{0k} of θ_0.

A first order expansion of U_n, as n tends to infinity, yields

$$U_{1n}(\eta_0, \widehat{\eta}_{2n}, \ldots, \widehat{\eta}_{pn}, \widehat{\gamma}_n) = -n^{\frac{1}{2}}(\widehat{\eta}_{1n} - \eta_0)^T \\ .n^{-1}S''_{1n,\eta}(\eta_0, \widehat{\eta}_{2n}, \ldots, \widehat{\eta}_{pn}, \widehat{\gamma}_n) + o_p(1)$$

and $n^{\frac{1}{2}}(\widehat{\eta}_{1n} - \eta_0)$ converges weakly to a centered Gaussian variable with variance I_{01}^{-1}.

Under the conditions of Section 4.5, the asymptotic behaviour of the statistic T_n is established by the arguments of Proposition 3.7 for the log-likelihood ratio test statistic of parametric densities.

Proposition 4.9. *In model (4.13) with a regression variable X on a finite support, the statistic T_n converges weakly under H_0 to a $\chi^2_{d(p-1)}$ variable T_0.*

Proof. Under the condition of a finite bound τ, the convergences of Section 4.5 are satisfied and the convergence rate of $\widehat{\gamma}_n$ is n^{-1}, then $U_{1n}(\widehat{\eta}_{0n}, \widehat{\eta}_{2n}, \ldots, \widehat{\eta}_{pn}, \widehat{\gamma}_n)$ is asymptotically equivalent to $U_{0n}(\widehat{\eta}_{0n}) = 0$, the difference $U_{1n}(\widehat{\theta}_n) - U_{0n}(\widehat{\eta}_{1n})$ is a $o_p(1)$ hence $U_{0n}(\widehat{\eta}_{1n}) = o_p(1)$.

A first order expansion in a neighborhood of μ_0 implies

$$n^{\frac{1}{2}}(\widehat{\eta}_{1n} - \widehat{\eta}_{0n}) = I_{01}^{-1}U_{1n}(\widehat{\eta}_{0n}\widehat{\eta}_{2n}, \ldots, \widehat{\eta}_{pn}, \widehat{\gamma}_n) + o_p(1)$$

with a positive definite matrix I_{01}. It follows that $n^{\frac{1}{2}}(\widehat{\mu}_{1n} - \widehat{\mu}_{0n})$ converges in probability to zero under H_0.

A second order asymptotic expansion of the process S_n implies

$$\begin{aligned} n\{S_n(\widehat{\theta}_n) - S_{0n}(\widehat{\theta}_{0n})\} &= n\{S_n(\widehat{\eta}_{0n}, \widehat{\eta}_{2n}, \ldots, \widehat{\eta}_{pn}, \widehat{\gamma}_n) - S_{0n}(\widehat{\theta}_{0n})\} \\ &\quad + n^{\frac{1}{2}}(\widehat{\eta}_{1n} - \widehat{\eta}_{0n})^T U_{1n}(\widehat{\eta}_{0n}, \widehat{\eta}_{2n}, \ldots, \widehat{\eta}_{pn}, \widehat{\gamma}_n) \\ &\quad - \frac{n}{2}(\widehat{\eta}_{1n} - \widehat{\eta}_{0n})^T I_{01}(\widehat{\eta}_{1n} - \widehat{\eta}_{0n}) + o_p(1) \\ &= n\{S_n(\widehat{\eta}_{0n}, \widehat{\eta}_{2n}, \ldots, \widehat{\eta}_{pn}, \widehat{\gamma}_n) - S_{0n}(\widehat{\theta}_{0n})\} + o_p(1). \end{aligned}$$

The variable $\widehat{W}_n = n\{S_n(\widehat{\eta}_{0n}, \widehat{\eta}_{2n}, \ldots, \widehat{\eta}_{pn}, \widehat{\gamma}_n) - S_{0n}(\widehat{\theta}_{0n})\}$ is

$$n \int \Big\{ \sum_{k=2}^{p} I_k(x, \widehat{\gamma}_n) \log \frac{f(y - r_k(x, \widehat{\eta}_n))}{f(y - r(x, \widehat{\eta}_{0n}))} \Big\} \, d\widehat{F}_n(x, y)$$

with indicators I_k such that there exist observations between at least one component of the vectors $\widehat{\gamma}_n$ and τ, the other components are below $\widehat{\gamma}_n$.

At $\widehat{\gamma}_n$ converging to τ, the process

$$W_n(\eta_2, \ldots, \eta_p, \gamma) = n\{S_n(\eta_0, \eta_2, \ldots, \eta_p, \gamma) - S_{0n}(\eta_0)\}$$

is asymptotically equivalent to a sum for $\min_{j=1,\ldots,d} \widehat{u}_{nj}$ observations and this minimum converges weakly to a limit determined by Theorem 4.10. The first derivative $\dot{W}_{nk}$ of W_n with respect to η_k is such that $\dot{W}_{nk}(\widehat{\eta}_{2n}, \ldots, \widehat{\eta}_{pn}, \widehat{\gamma}_n) = 0$ and $\dot{W}_{nk}(\eta_{02}, \ldots, \eta_{0p}, \gamma_0)$ is asymptotically equivalent under H_0 to a centered Gaussian variable with finite variance I_{02} such that $-I_0$ is the limit in probability of the second order derivative $\ddot{W}_n(\theta_0)$ with respect to the parameter vector $\widetilde{\eta} = (\eta_2, \ldots, \eta_p)$.

Under H_0, the variables $\widehat{\eta}_{nk}$ converge to limits η_{0k}, for $k = 2, \ldots, p$, the vector $\widetilde{\eta}_0 = (\eta_{02}^T, \ldots, \eta_{0p}^T)^T$ and its estimator $\widehat{\widetilde{\eta}}_n$ satisfy

$$\widehat{\widetilde{\eta}}_n - \widetilde{\eta}_0 = I_{02}^{-1} \dot{W}_n(\widetilde{\eta}_0, \widehat{\gamma}_n) + o_p(1).$$

The variable $\widehat{W}_n$ has the expansion

$$\widehat{W}_n = \dot{W}_n^T(\widetilde{\eta}_0, \widehat{\gamma}_n) I_{02}^{-1} \dot{W}_n^T(\widetilde{\eta}_0, \widehat{\gamma}_n) + o_p(1)$$

and it converges weakly to a $\chi^2_{d(p-1)}$ variable under H_0. □

Under the alternative K of a model with a change point at γ between $Y_{n:1}$ and $Y_{n:n}$ and a regression parameter vector η with distinct components $\eta_1, \eta_2, \ldots, \eta_p$. the process $n^{-1} l_{0n}$ converges in probability under P_θ to the function $l_0(\theta)$ and the log-likelihood ratio process under $P_{\theta'}$, relatively to P_θ, is $S_n(\theta'; \theta) = S_n(\theta') - S_n(\theta)$, it has the expectation

$$\begin{aligned} S_\theta(\theta') = & \int \log \frac{f(y - r_{\eta'}(x, \gamma'))}{f(y - r_\eta(x, \gamma))} \, dF_\theta \\ & + \int \log \frac{f(y - r_\eta(x, \gamma))}{f(y - r_{\eta_0}(x))} \, dF_\theta, \end{aligned} \tag{4.21}$$

where the second term of the sum is a constant with respect to the parameter θ' and the first term is maximum at $\theta' = \theta$. Under P_θ, the estimator $\widehat{\eta}_n$ converges a.s. to η and $\widehat{\gamma}_n$ converges a.s. to γ. At θ such that the integral $S''_\theta(\theta)$ is not singular, the maximum likelihood estimator of η has the convergence rate $n^{-\frac{1}{2}}$ and the estimator of the change-point has the convergence rate n^{-1}, with limiting distributions given by Proposition 4.8 and Theorem 4.11.

Proposition 4.10. *Under a fixed alternative P_θ, the statistic T_n tends to infinity. Under local alternatives $P_{\theta_{n,u}}$ and if the support of X is finite, the statistic T_n converges weakly to $T_0 + v^T I_0^{-1} v$.*

Proof. Under P_θ, the estimator $\widehat{\theta}_n$ maximize the log-likelihood ratio process $S_n(\theta';\theta)$ which is zero as $\theta' = \theta$, and $\widehat{\theta}_n$ converges a.s. to θ. The test statistic T_n is the value at the estimators of the process $S_n(\theta;\theta_0)$ and for every θ different from θ_0, $S_n(\widehat{\theta}_n;\widehat{\theta}_{0n})$ converges a.s. under P_θ to $S(\theta;\theta_0)$ which is different from zero, so $T_n = nS_n(\widehat{\theta}_n;\widehat{\theta}_{0n})$ diverges.

Let $P_n = P_{\theta_n}$ be local alternatives with densities f_{θ_n} in $C^2(\mathcal{H})$ converging to the density $f_0 = f_{\theta_0}$ of the hypothesis H_0, with parameters γ_n converging to τ and η_{1n} converging to η_0 and η_{2n} converging to a vector η_{02}. If the change-point parameter belongs to a bounded set, $\gamma_n = X_{n:n} - n^{-1}u_n$, with a sequence $(u_n)_n$ converging to a non-zero limit u, and $\eta_{1n} = \eta_0 + n^{-\frac{1}{2}}v_{1n}$, with a sequence $(v_{1n})_n$ converging to non-zero limits v_1.

The estimators $\widehat{\eta}_{1n}$ and $\widehat{\eta}_{0n}$ converge a.s. to η_0 under P_n and the variables $n^{\frac{1}{2}}(\widehat{\eta}_{1n} - \eta_{1n})$ and $n^{\frac{1}{2}}(\widehat{\eta}_{0n} - \eta_0)$ converge weakly to the same Gaussian limit, with mean zero and variance I_{01}, then the variable $n^{\frac{1}{2}}(\widehat{\eta}_{1n} - \widehat{\eta}_{0n})$ converges to u_1 in probability under P_n. By the same approximation as in the proof of Proposition 4.9, the convergence of the estimators imply

$$n\{S_n(\widehat{\theta}_n) - S_n(\widehat{\eta}_{0n}, \widehat{\eta}_{2n}, \ldots, \widehat{\eta}_{pn}, \widehat{\gamma}_n)\} = u_1^T I_0^{-1} u_1 + o_p(1)$$

and the test statistic has the expansion

$$n\{l_n(\widehat{\theta}_n) - l_n(\widehat{\theta}_{0n})\} = \widehat{W}_n + u_1^T I_0^{-1} u_1 + o_p(1).$$

The first derivative $\dot{W}_n$ of the process

$$W_n(\eta_2, \ldots, \eta_p, \gamma) = n\{S_n(\eta_0, \eta_2, \ldots, \eta_p, \gamma) - S_{0n}(\eta_0)\}$$

with respect to $(\eta_2, \ldots, \eta_p)$ is zero at $(\widehat{\eta}_{2n}, \ldots, \widehat{\eta}_{pn}, \widehat{\gamma}_n)$ and the variable $\dot{W}_n(\eta_{2n}, \ldots, \eta_{pn}, \gamma_n)$ is asymptotically equivalent under P_n to a centered Gaussian variable with finite variance I_{02} such that $-I_{20}$ is the limit in probability of the second order derivative $\ddot{W}_n(\eta_{2n}, \ldots, \eta_{pn}, \gamma_n)$, as n tends to infinity.

Under P_n, the estimator of the vector $\widetilde{\eta} = (\eta_2, \ldots, \eta_p)$ satisfies

$$\widehat{\widetilde{\eta}}_n - \widetilde{\eta}_n = I_{02}^{-1}\dot{W}_n(\widetilde{\eta}_0, \widehat{\gamma}_n) + o_p(1)$$

so $\widehat{\widetilde{\eta}}_n - \widetilde{\eta}_n$ converges weakly to a centered Gaussian variable and $\widetilde{\eta}_n - \widetilde{\eta}_0$ converges to zero. The variable $\widehat{W}_n$ has the expansion

$$\widehat{W}_n = \dot{W}_n^T(\widetilde{\eta}_0, \widehat{\gamma}_n) I_{02}^{-1} \dot{W}_n^T(\widetilde{\eta}_0, \widehat{\gamma}_n) + o_p(1)$$

and it converges weakly to a $\chi^2_{d(p-1)}$ variable under P_n. □

4.7 Chronological changes in parametric models

Consider a regression model for (X, Y) with a random sampling vector T independent of the regression variable X. Under a probability distribution $P_{\eta,t}$, the regression with a change-point parameter t of $[0,1]$ has the conditional expectation

$$r_\eta(X,t) = E_{\eta,t}(Y|X,T) = E_\eta(Y|X,T=t)$$

and the regression parameter of Y_i on X_i is modified at $k = [nt]$, they are η_1 for $i \leq k$ and η_2 for $i > k$. Under the probability distribution P_0 of the observations, the parameter values are t_0 and η_0, with distinct components η_{01} and η_{02}, the regression function is $r_0(x) = r_{\eta_0}(x, t_0)$.

For a variable Y such that E_0Y^2 is finite, the mean square error estimators are defined by minimization of the mean square error

$$l_n(\eta, t) = \|Y - r_\eta(X,t)\|_n^2 - \|Y - r_0(X)\|_n^2,$$

at a fixed value t of the sampling time in $[0,1]$. For every t, the variance of the error is estimated by

$$\widehat{\sigma}^2_{n,t} = n^{-1} \sum_{i=1}^n \{Y_i - r(X_i, \widehat{\eta}_{n,t}, t)\}^2,$$

the change-point parameter t is estimated by minimization of the empirical variance $\widehat{\sigma}^2_{n,t}$

$$\widehat{t}_n = \arg\inf_t \widehat{\sigma}^2_{n,t}$$

and the estimator of η_j is estimated by $\widehat{\eta}_n = \widehat{\eta}_{n,\widehat{t}_n}$. A change-point at the index $[nt]$ determines the empirical sub-distribution functions of the sample

$$\widehat{F}_{1n}(x,y,t) = n^{-1} \sum_{i=1}^n 1_{\{X_i \leq x, Y_i \leq y, i \leq [nt]\}},$$

$$\widehat{F}_{2n}(x,y,t) = n^{-1} \sum_{i=1}^n 1_{\{X_i \leq x, Y_i \leq y, i > [nt]\}},$$

they converge a.s. under P_0 to F_{01}(x,y,t), and respectively $F_{02}(x,y,t)$. Let $\widehat{F}_n = \widehat{F}_{1n} + \widehat{F}_{2n}$, the process l_n is written as

$$\begin{aligned} l_n(\theta) = & \int \{y - r(x, \eta_1, t)\}^2 \, d\widehat{F}_{1n}(x,y,t) + \int \{y - r(x, \eta_2, t)\}^2 \, d\widehat{F}_{2n}(x,y,t) \\ & - \int \{y - r(x, \eta_{01}, t_0)\}^2 \, d\widehat{F}_{1n}(x,y,t_0) \\ & - \int \{y - r(x, \eta_{02}, t_0)\}^2 \, d\widehat{F}_{2n}(x,y,t_0). \end{aligned}$$

Under P_0, for a variable Y such that $E_0[\{Y - r_\eta(X,t)\}^2]$ is finite on $\mathcal{H}^2 \times [0,1]$, the process l_n converges a.s. uniformly to the function

$$l(\eta,t) = \int \{y - r_\eta(x,t)\}^2 \, dF(x,y,t) - \int \{y - r_0(x)\}^2 \, dF(x,y,t_0).$$

Let $\rho^2(\theta,\theta_0) = \sum_{j=1}^p \rho^2(\theta,\theta_{0,j})$ and let $V_\varepsilon(\theta_0)$ be an ε-neighborhood of θ_0, we assume that the condition (4.18) is fulfilled.

Lemma 4.2. *For ε small enough, there exists a constant $\kappa_0 > 0$ such that for every $\theta = (\eta,t)$ in $\mathcal{H}^2 \times [0,1]$*

$$\inf_{\rho(\theta,\theta_0)\le\varepsilon} l(\theta) \ge \kappa_0 \rho^2(\theta,\theta_0).$$

Proof. The function l is the sum of the positive function

$$l_1(\eta) = l(\eta,t_0) = \int \{r_\eta(x,t_0) - r_0(x)\}^2 \, dF(x,y,t_0)$$

and

$$\begin{aligned} l_2(\eta,t) &= l(\eta,t) - l(\eta,t_0) \\ &= \int \{y - r_\eta(x,t)\}^2 \, dF(x,y,t) - \int \{y - r_\eta(x,t_0)\}^2 \, dF(x,y,t_0). \end{aligned}$$

By a second order expansion of $l_1(\eta)$ for η in a neighborhood $V_\varepsilon(\eta_0)$ of η_0 and by the equality $l'_{1,\eta}(\eta_0) = 0$, we have

$$l_1(\eta) = \frac{1}{2}(\eta - \eta_0)^T J_0 (\eta - \eta_0)\{1 + o(1)\}$$

with the positive definite matrix $J_0 = l''_{1,\eta}(\eta_0)$ and the lemma is fulfilled for the function l_1, for every positive and sufficiently small ε. In the same way, the function $l_2(\theta)$ is a $O_p(\rho^2(\theta,\theta_0))$ for every η. □

Like the function l, the process $W_n = n^{\frac{1}{2}}(l_n - l)$ is the sum of the processes $W_{kn} = n^{\frac{1}{2}}(l_{kn} - l_k)$, for $k = 1, 2$, where

$$\begin{aligned} l_{1n}(\eta) &= \|Y - r_\eta(X,t_0)\|_n^2 - \|Y - r_0(X)\|_n^2, \\ l_{2n}(\theta) &= \|Y - r_\eta(X,t)\|_n^2 - \|Y - r_\eta(X,t_0)\|_n^2 \end{aligned} \tag{4.22}$$

and the functions l_k are their a.s. limits under P_0.

Lemma 4.3. *For every $\varepsilon > 0$ sufficiently small, there exists a constant κ_1 such that*

$$E_0 \sup_{\rho(\theta,\theta_0)\le\varepsilon} |W_n(\theta)| \le \kappa_1 \varepsilon.$$

Proof. The processes W_{kn} are centered and by first order expansions with respect to η in a neighborhood of η_0, there exists a constant κ such that their variance satisfy the inequality

$$E_0 \sup_{\rho(\theta,\theta_0)\leq\varepsilon} |W_{kn}(t)|^2 \leq \kappa\rho^2(\theta,\theta_0),$$

the result is deduced from the Cauchy–Schwarz inequality. □

Let $\theta_{n,u} = \theta_0 + n^{-1}u$, u in $\mathcal{U}_n = \{u = n(\theta - \theta_0), \theta \in \mathcal{H}^2 \times [0,1]\}$. By the same arguments as in Section 2.1, the inequalities of Lemmas 4.2 and 4.3 provide the convergence rates of the estimators.

Theorem 4.12.

$$\overline{\lim}_{n,A\to\infty} P_0(n\rho^2(\widehat{\theta}_n,\theta_0) > A) = 0.$$

Let $W'_{1n,\eta}$ be the first derivative of W_{1n} with respect to η, at t_0. The weak convergence of $n^{\frac{1}{2}}(\widehat{\eta}_n - \eta_0)$ to a centered Gaussian variable with variance J_0^{-1} is deduced from a first order asymptotic expansion of $W'_{1n,\eta}$, at η_0, according to the classical arguments and the convergence rate of $\widehat{t}_n$ is n^{-1}.

Theorem 4.13. *The process $nl_{2n}(t_{n,u})$ converges weakly in $\mathcal{U}_n^A$ to a centered Gaussian process L_2 under P_0, with a finite variance function.*

Proof. Let u in $\mathcal{U}_n^A$ with components u_η and u_t, the process $l_{2n}(\theta)$ has the mean

$$E_0 l_{2n}(\theta) = E_0[\{r_\eta(X,t) - r_\eta(X,t_0)\}\{r_\eta(X,t) + r_\eta(X,t_0) - 2r_0(X)\}]$$

such that $nE_0 l_{2n}(\theta_{n,u}) = o(1)$ and the variance of $nl_{2n}(\theta_{n,u})$ is asymptotically equivalent to $E_0([\{r_\eta(X,t) - r_\eta(X,t_0)\}\{r_\eta(X,t) + r_\eta(X,t_0) - 2Y\}]^2)$ then the variance of the process $nl_{2n}(t_{n,u})$ converges to a finite limit as n tends to infinity. It converges weakly to an uncentered Gaussian process, by the arguments of Section 1.4. □

The estimator $\widehat{t}_n$ of the change-point that minimizes the process l_{2n} is such that $\widehat{u}_n = n(\widehat{t}_n - t_0)$ converges weakly to U_0, the location of the minimum of the process L_2.

4.8 Test of parametric models without change

A test for a parametric regression model without change of parameter relies on the statistic

$$T_n = 2n \inf_{t\in[0,1]} \{\widehat{\sigma}^2_{t,n} - \widehat{\sigma}^2_{0n}\}$$

where $\widehat{\sigma}^2_{t,n}$ is the variance estimator in the model with a change of regression parameters at t and $\widehat{\sigma}^2_{0n}$ is the variance estimator in the model without change, where $t_0 = 1$ is known so

$$T_n = 2n\{l_n(\widehat{\theta}_n) - l_{0n}(\widehat{\eta}_{0n})\}.$$

With a Gaussian error e, $-nl_n(t)$ is the logarithm of the likelihood ratio, up to an additive term $\log \widehat{\sigma}_{0n} - \log \widehat{\sigma}^2_n(t)$ of smaller order than $nl_n(t)$.

Under alternatives, there exists t in $]0,1[$ and distinct regression parameters η_1 and η_2 such that

$$r_\eta(X_i, t) = r(X_i, \eta_1)1_{\{i \le [nt]\}} + r(X_i, \eta_2)1_{\{i > [nt]\}}$$

and $T_n = 2n\{\widehat{\sigma}^2_n(\widehat{t}_n) - \widehat{\sigma}^2_{0n}\}$. The process $l_n(\theta, \eta_0)$ is the sum of the processes $l_{1n}(\eta_1, \eta_0) = l_n(\eta_1, t_0, \eta_0)$ and $l_{2n}(\theta, \eta_0) = l_n(\theta, \eta_0) - l_{1n}(\eta_1, \eta_0)$ given by (4.22). Under H_0, the parameter is still denoted θ_0 with $t_0 = 1$, η_{01} is the value of the regression parameter and ζ_{02} is unspecified.

Proposition 4.11. *The statistic T_n converges weakly under H_0 to a χ^2_d variable T_0.*

Proof. Under H_0, the indicator $1_{\{i > [nt_0]\}}$ is zero and Theorem 4.12 applies. Let $r_\eta(X, t_0) = r(X, \eta_1)$, the difference $\|Y - r_\eta(X,t)\|^2_n - \|Y - r_{\eta_0}(X)\|^2_n$ is the sum of the processes

$$l_{1n}(\eta_1, \eta_0) = \|Y - r(X, \eta_1)\|^2_n - \|Y - r(X, \eta_0)\|^2_n,$$
$$l_{2n}(\theta, \eta_0) = \|Y - r_\eta(X, t)\|^2_n - \|Y - r(X, \eta_1)\|^2_n.$$

Due to the convergence rate of $\widehat{t}_n$, the regression parameter estimators $\widehat{\eta}_{0n}$ and $\widehat{\eta}_{1n}$ have the same asymptotic behaviour and $n^{\frac{1}{2}}(\widehat{\eta}_{0n} - \widehat{\eta}_{1n})$ converges in probability to zero. The process nl_{1n} develops as

$$nl_{1n}(\eta_1, \eta_0) = \sum_{i=1}^n \{r(X_i, \eta_0) - r(X_i, \eta_1)\}\{2Y_i - r(X_i, \eta_0) - r(X_i, \eta_1)\},$$

its expectation $nE_0\{r(X_i, \eta_0) - r(X_i, \eta_1)\}^2$ and its variance converge to zero at $(\widehat{\eta}_{1n}, \widehat{\eta}_{0n})$. The first derivative $n^{\frac{1}{2}}\dot{l}_{1n}(\widehat{\eta}_{1n}) = -2n^{\frac{1}{2}} \sum_{i=1}^n \dot{r}_{\eta_1}(X_i, \widehat{\eta}_{1n})\{Y_i - r(X_i, \widehat{\eta}_{1n})\}$, with respect to η_1, is zero at $\widehat{\eta}_{1n}$ and it converges to zero at $\widehat{\eta}_{0n}$, its variance converges in probability to a positive definite matrix J_{01}. By a second order expansion

$$2nl_{1n}(\widehat{\eta}_{1n}, \widehat{\eta}_{0n}) = n^{\frac{1}{2}}(\widehat{\eta}_{0n} - \widehat{\eta}_{1n})^T J_{01}^{-1} n^{\frac{1}{2}}(\widehat{\eta}_{0n} - \widehat{\eta}_{1n}) + o_p(1)$$

and it converges in probability to zero.

By (4.22), the process nl_{2n} is a sum on $\widehat{u}_n = n[1 - \widehat{t}_n]$ observations indexed by $i > [n\widehat{t}_n]$

$$nl_{2n}(\theta) = \sum_{i=[nt]+1}^{n} [\{Y_i - r(X_i, \eta_2)\}^2 - \{Y_i - r(X_i, \eta_1)\}^2.$$

Its first derivative with respect to η_2 is

$$n\dot{l}_{2n,\eta_2}(\theta) = 2 \sum_{i=[nt]+1}^{n} \dot{r}_{\eta_2}\{Y_i - r(X_i, \eta_2)\},$$

it is zero at $\widehat{\theta}_n$ and at θ_0, its expectation and its variance are finite in a neighborhood of θ_0 where it has the expansion $\widehat{\eta}_{2n} - \eta_{02} = I_{02}^{-1} n\dot{l}_{2n,\eta_2}(\eta_{02}, \widehat{t}_n) + o_p(1)$. Then $l_{2n}(\widehat{\theta}_n)$ has the expansion

$$2nl_{2n}(\widehat{\theta}_n) = n\dot{l}_{2n,\eta_2}(\eta_{02}, \widehat{t}_n) I_{02}^{-1} n\dot{l}_{2n,\eta_2}(\eta_{02}, \widehat{t}_n) + o_p(1)$$

and it converges weakly to a χ_d^2 variable. □

Under alternatives with parameter θ, the process $\|Y - r(X;\zeta_0)\|_n^2$ is uncentered therefore the variable $l_{1n}(\widehat{\zeta}_{1n}, \widehat{\zeta}_{0n})$ is uncentered. Let P_{θ_n} be the probability distribution for local alternatives with a sequence of parameters $(\theta_n)_n$ converging to the parameter θ_0 of the hypothesis, the parameter t_n converges to one, ζ_{1n} converges to ζ_0 and ζ_{2n} converges to ζ_{02} different from ζ_0.

Proposition 4.12. *Under a fixed alternative P_θ, the statistic T_n tends to infinity. Under local alternatives P_{θ_n}, the statistic T_n converges weakly to $T_0 + a^2$ with a non-null constant a.*

Proof. Under P_θ, the variable $nl_{1n}(\widehat{\eta}_{1n}, \widehat{t}_n, \widehat{\eta}_{0n})$ is uncentered like the process $\|Y - r(X;\zeta_0)\|_n^2$ and it diverges, the asymptotic behaviour of the variable $nl_{2n}(\widehat{\theta}_n, \widehat{\eta}_{0n}))$ is the same as under P_0, hence the test statistic diverges.

Local alternatives are defined by a sequence of parameters $(\theta_n)_n$ such that the regression parameters converge to η_0 with the rate $n^{-\frac{1}{2}}$ and the change-point parameter for the regression parameters and n^{-1} for the change-point parameter converge to one with the rate n^{-1}, $t_n = t_0 - n^{-1}u_n$, where u_n converges to a non-zero limit u. The variable $n^{\frac{1}{2}}(\widehat{\eta}_{0n} - \widehat{\eta}_{1n})$ converges in probability to a non-null limit v_1 and the variance of the first derivative of nl_{1n} with respect to η_1, at $(\widehat{\eta}_{0n}, \widehat{t}_n, \widehat{\eta}_{0n})$, converges to a positive definite matrix I_0, then

$$nl_{1n}(\widehat{\eta}_{1n}, \widehat{t}_n, \widehat{\eta}_{0n}) = v_1^T I_0^{-1} v_1 + o_p(1).$$

The variable $l_{2n}(\theta_n)$ is a sum over $\widehat{u}_n + \bar{u}_n$ observations, its asymptotic behaviour is similar under P_n and P_0 and the limit of $2nl_{2n}(\widehat{\theta}_n)$ is a χ^2_d variable. □

4.9 Maximum likelihood for chronological changes

Let us consider a model with a change of parameter in a parametric regression of a random variable Y on a random vector X at a sampling index, the sample splits in two independent sub-samples of independent and identically distributed observations. The conditional density $f_\theta(Y_i \mid X_i)$ of Y_i given X_i is defined in

$$f(Y_i - r_\eta(X_i, t)) = f(Y_i - r(X_i, \eta_1))1_{\{i \le [nt]\}} + f(Y_i - r(X_i, \eta_2))1_{\{i > [nt]\}}.$$

The log-likelihood of the sample with a change of parameter at t in $]0, 1[$ is

$$l_n(\theta) = \sum_{i=1}^{[nt]} \log f(Y_i - r(X_i, \eta_1)) + \sum_{i=1+[nt]}^{n} \log f(Y_i - r(X_i, \eta_1)) \qquad (4.23)$$

at the vector parameter $\theta = (\eta^T, t)^T$ in $\mathcal{H}^{2d} \times]0, 1[$ and the parameter value under P_0 is θ_0. The map $\eta \mapsto f_\eta(y \mid x)$ is supposed to belongs to $C^2(\mathcal{H} \times \mathcal{H})$, uniformly with respect (x, y) in $\mathbb{R}^{d+1}$. The maximum likelihood estimators $\widehat{\eta}_{n,t}$ of η at an arbitrary t are solutions of the estimating equations

$$\sum_{i=1}^{[nt]} \frac{f'_{\eta_1}}{f_{\eta_1}}(Y_i \mid X_i) = 0, \qquad \sum_{i=1+[nt]}^{n} \frac{f'_{\eta_2}}{f_{\eta_2}}(Y_i \mid X_i) = 0$$

and the change-point is estimated by

$$\widehat{t}_n = \arg\max_t l_n(\widehat{\eta}_{n,t}, t),$$

then $\widehat{\eta}_{jn} = \widehat{\eta}_{jn,\widehat{t}_n}$, for $j = 1, 2$.

The process $Z_n(\theta) = n^{-1}\{l_n(\theta) - l_n(\theta_0)\}$ is the sum

$$Z_n(\theta) = n^{-1} \sum_{i=1}^{[nt]\wedge[nt_0]} \log \frac{f_{\eta_1}(Y_i \mid X_i)}{f_{\eta_{01}}(Y_i \mid X_i)} + n^{-1} \sum_{i=[nt]\vee[nt_0]+1}^{n} \log \frac{f_{\eta_2}(Y_i \mid X_i)}{f_{\eta_{02}}(Y_i \mid X_i)}$$

$$+1_{\{t<t_0\}}n^{-1}\sum_{i=[nt]+1}^{[nt_0]}\log\frac{f_{\eta_2}(Y_i\mid X_i)}{f_{\eta_{01}}(Y_i\mid X_i)}$$

$$+1_{\{t>t_0\}}n^{-1}\sum_{i=[nt_0]+1}^{[nt]}\log\frac{f_{\eta_1}(Y_i\mid X_i)}{f_{\eta_{02}}(Y_i\mid X_i)},$$

it converges a.s. uniformly under P_0 to the function

$$\begin{aligned}Z(\theta) &= (t\wedge t_0)E_0\Big\{\log\frac{f_{\eta_1}(Y_i\mid X_i)}{f_{\eta_{01}}(Y_i\mid X_i)}1_{\{i\le k\wedge k_0\}}\Big\}\\ &+(1-t\vee t_0)E_0\Big\{\log\frac{f_{\eta_2}(Y_i\mid X_i)}{f_{\eta_{02}}(Y_i\mid X_i)}1_{\{i>k\vee k_0\}}\Big\}\\ &+1_{\{k<k_0\}}(t_0-t)E_0\Big\{\log\frac{f_{\eta_2}(Y_i\mid X_i)}{f_{\eta_{01}}(Y_i\mid X_i)}1_{\{k<i\le k_0\}}\Big\}\\ &+1_{\{k>k_0\}}(t-t_0)E_0\Big\{\log\frac{f_{\eta_1}(Y_i\mid X_i)}{f_{\eta_{02}}(Y_i\mid X_i)}\Big\}.\end{aligned}\tag{4.24}$$

Proposition 4.13. *Under the condition of a family of densities such that the map $\eta\mapsto f_\eta$ belongs to $C^2(\mathcal{H}\times\mathcal{H})$ uniformly on $\mathbb{R}^{d+1}$, and its second order derivatives is uniformly bounded on $\mathcal{H}$, the estimator $\widehat{\theta}_n$ converges a.s. to θ_0 under P_0.*

Proof. Under P_0, the function Z is such that $Z'_\eta(\theta_0)=0$ and the matrix $I_0=-Z''_\eta(\theta_0)$ is definite positive, then the function $\eta\mapsto Z(\eta,t)$ is concave in a neighborhood of η_0 with distinct components η_{01} and η_{02}. By the uniform convergence of Z_n to Z, $\widehat{\eta}_n$ is a.s. consistent. Moreover, Z is maximum at θ_0 where it is zero, then $\widehat{t}_n$ converges a.s. to t_0. □

Lemma 4.4. *If $E_0\log f_\theta(Y\mid X)$ is finite in a neighborhood $V_\varepsilon(\theta_0)$ of θ_0, there exists a constant $\kappa_0>0$ such that for every θ in $V_\varepsilon(\theta_0)$*

$$Z(\theta)\ge-\kappa_0\rho^2(\theta,\theta_0).$$

Proof. The function $Z(\theta)$ is a sum $Z_1(\eta)+\cdots+Z_4(\theta)$ defined by (4.24) and it is negative under P_0. The functions $Z_3(\theta)$ and $Z_4(\theta)$ are bounded and there exists a constant κ such that they are larger than $-\kappa(|t-t_0|)$.

By second order expansions of Z_1 and Z_2 with respect to η in a neighborhood of η_0 and by the equality $Z'_{j,\eta_j}(\theta_0)=0$, with $Z''_{j,\eta_j}(\theta)\le0$ in a neighborhood of η_0, there exists a constant $\kappa'>0$ such that in a neighborhood of η_0, $Z_j(\theta)\ge-\kappa'\|\eta_j-\eta_0\|^2$. □

The variance of the process $W_n = n^{\frac{1}{2}}(Z_n - Z)$ has a similar uniform upper bound under the condition that $E_0 \sup_{\eta \in V_\varepsilon(\eta_0)} \log^2 f_\theta(Y \mid X)$ is finite.

Lemma 4.5. *There exists a constant $\kappa_1 > 0$ such that for n large enough*
$$E_0 \sup_{\rho(\theta,\theta_0)\leq\varepsilon} W_n(\theta) \leq \kappa_1\varepsilon.$$

The convergence rates of the components of $\widehat{\theta}_n$ are deduced from Lemmas 4.4 and 4.5.

Theorem 4.14. $\overline{\lim}_{n,A\to\infty} P_0(n\rho^2(\widehat{\theta}_n,\theta_0) > A) = 0$.

Theorem 4.15. *The variable $n^{\frac{1}{2}}(\widehat{\eta}_n - \eta_0)$ is asymptotically independent of $\widehat{\eta}_n$ and converges weakly to a centered Gaussian variable with variance I_0^{-1} and, for t_0 in $]0,1[$, $n(\widehat{t}_n - t_0)$ converges weakly to the location U of the maximum of an uncentered Gaussian process.*

Proof. Under P_0, the process nZ_n is the sum $nZ_n(\theta) = nZ_{1n}(\eta) + nZ_{2n}(\theta)$ with $Z_{1n}(\eta) = Z_n(\eta, t_0)$ and $Z_{2n}(\theta) = Z_n(\theta) - Z_{1n}(\eta)$

$$nZ_{1n}(\eta) = \sum_{i=1}^{[nt_0]} \log \frac{f_{\eta_1}(Y_i \mid X_i)}{f_{\eta_{01}}(Y_i \mid X_i)} + \sum_{i=[nt_0]+1}^{n} \log \frac{f_{\eta_2}(Y_i \mid X_i)}{f_{\eta_{02}}(Y_i \mid X_i)},$$

$$nZ_{2n}(\theta) = 1_{\{t<t_0\}} \sum_{i=[nt]+1}^{[nt_0]} \log \frac{f_{\eta_2}(Y_i \mid X_i)}{f_{\eta_1}(Y_i \mid X_i)}$$
$$+1_{\{t>t_0\}} \sum_{i=[nt_0]+1}^{[nt]} \log \frac{f_{\eta_1}(Y_i \mid X_i)}{f_{\eta_2}(Y_i \mid X_i)}.$$

At $\theta_{n,u}$ in an neighborhood of θ_0, such that u belongs to $\mathcal{U}_n^A$ for A sufficiently large, the process nZ_{2n} is expanded as
$$nZ_{2n}(\theta_{n,u}) = nZ_{2n}(\eta_0, t_{n,u})\{1 + o_p(1)\}$$
and the process nZ_{2n} is asymptotically independent of the parameter η, the estimators $\widehat{\eta}$ and $\widehat{t}$ are therefore asymptotically independent.

The process

$$Z_2(\eta_0, t_{n,u}) = 1_{\{t_{n,u}<t_0\}} \sum_{i=[nt_{n,u}]+1}^{[nt_0]} \log \frac{f_{\eta_{02}}(Y_i \mid X_i)}{f_{\eta_{01}}(Y_i \mid X_i)}$$
$$+1_{\{t_{n,u}>t_0\}} \sum_{i=[nt_0]+1}^{[nt_{n,u}]} \log \frac{f_{\eta_{01}}(Y_i \mid X_i)}{f_{\eta_{02}}(Y_i \mid X_i)},$$

has under P_0 the expectation

$$|u|(1_{\{[nt_{n,u}]<i\leq[nt_0]\}} - 1_{\{[nt_0]<i\leq[nt_{n,u}]\}})E_0 \log \frac{f_{\eta_{02}}(Y_i \mid X_i)}{f_{\eta_{01}}(Y_i \mid X_i)},$$

by the same argument its variance is also a $O(|u|)$, then for every u of $\mathcal{U}_n^A$, the variable $nZ_{2n}(\eta_0, t_{n,u})$ converges weakly to an uncentered Gaussian variable $Q_A(u)$. Donsker's theorem implies the tightness of the process $nZ_{2n}(\eta_0, t)$ so the process converges weakly $nZ_{2n}(\eta_0, t_{n,u})$ to a Gaussian process Q_A on $\mathcal{U}_n^A$, as n tends to infinity.

By Theorem 4.14, the maximum of the process nZ_n2n is achieved at $\widehat{u}_n = n(\widehat{t}_n - t_0)$ belonging to $\mathcal{U}_n^A$, it converges weakly on $\mathcal{U}_n^A$ to the location of the maximum of Q_A and it is bounded in probability as n and A tend to infinity.

As the process Z_{2n} is asymptotically independent of the regression parameter, the weak convergence of $n^{\frac{1}{2}}(\widehat{\eta}_n - \eta_0)$ to a centered Gaussian variable is deduced from a second order expansion of $n^{\frac{1}{2}} Z'_{1n}(\widehat{\eta}_n)$ and from the weak convergence of $n^{\frac{1}{2}} Z'_{1n}(\eta_0)$ to a centered Gaussian variable with variance I_0. □

4.10 Likelihood ratio test

The log-likelihood ratio test of the hypothesis H_0 of regression without change of parameter against the alternative of a change at an unknown index vector k_0, with distinct parameters η_1 and η_2, is performed with the statistic

$$T_n = 2\{l_n(\widehat{\theta}_n) - \widehat{l}_{0n}\},$$

where $\widehat{l}_{0n} = \sup_{\eta\in\mathcal{H}} \sum_{i=1}^n \log f_\eta(Y_i \mid X_i)$ is the maximum of the likelihood under the hypothesis H_0, it converges weakly under P_0 to a χ^2_d variable by the $l_{0n} = U_{0n}^T I_0^{-1} U_{0n} + o_p(1)$, with $U_{0n} = n^{-\frac{1}{2}} \sum_{i=1}^n (f_{\eta_0}^{-1} f'_{\eta_0})(Y_i \mid X_i)$ converging weakly to a centered Gaussian variable with variance I_0.

Under P_0, $t_0 = 1$ and the density is $f_{\eta_0}(Y_i \mid X_i) = f(Y_i - r(X_i, \eta_0))$, the log-likelihood ratio process Z_n is the sum

$$\begin{aligned} Z_n(\theta) = n^{-1} &\sum_{i=1}^{[nt]} \log \frac{f_{\eta_1}(Y_i \mid X_i)}{f_{\eta_0}(Y_i \mid X_i)} \\ &+n^{-1} \sum_{i=[nt]+1}^{n} \log \frac{f_{\eta_2}(Y_i \mid X_i)}{f_{\eta_0}(Y_i \mid X_i)} \end{aligned}$$

it converges a.s. uniformly under P_0 to the function

$$Z_0(\theta) = tE_0\Big\{\log \frac{f_{\eta_1}(Y_i \mid X_i)}{f_{\eta_0}(Y_i \mid X_i)} 1_{\{i\le k\}}\Big\} \\ +(1-t)E_0\Big\{\log \frac{f_{\eta_2}(Y_i \mid X_i)}{f_{\eta_0}(Y_i \mid X_i)} 1_{\{k<i\}}\Big\}$$

and the function Z_0 is maximum at η_{01} and t_0, for every η_2, therefore $\widehat{\eta}_{1n}$ converges a.s. to η_0 and $\widehat{t}_n$ converges a.s. to 1.

The first derivative of Z_n with respect to η has the components

$$\dot{Z}_{n,\eta_1}(\eta_1, t) = n^{-1} \sum_{i=1}^{[nt]} \frac{\dot{f}_{\eta_1}(Y_i \mid X_i)}{f_{\eta_1}(Y_i \mid X_i)},$$

$$\dot{Z}_{n,\eta_2}(\eta_2, t) = n^{-1} \sum_{i=[nt]+1}^{n} \frac{\dot{f}_{\eta_2}(Y_i \mid X_i)}{f_{\eta_2}(Y_i \mid X_i)}$$

and the process $U_n(\theta) = n^{\frac{1}{2}} \dot{Z}_{n,\eta}(\theta)$ is zero at $\widehat{\theta}_n$, its second derivative $\ddot{Z}_{n,\eta}$ is such that the derivative $-\ddot{Z}_{n,\eta_1}(\theta_0)$ with respect to η_1 converges in probability under H_0 to a positive definite matrix I_{01}, limit of the variance of $U_{n1}(\theta_0)$ under H_0. The asymptotic behaviour of T_n is proved by the same arguments as in Section 2.7.

Proposition 4.14. *The statistic T_n converges weakly under H_0 to a χ_d^2 variable.*

Proof. Under H_0, Lemmas 4.4 and 4.5 are still true under H_0 and the convergence rate of $\widehat{t}_n$ to one is n^{-1} by Theorem 4.14, the convergence rate of $\widehat{\eta}_{1n}$ and $\widehat{\eta}_{0n}$ is $n^{-\frac{1}{2}}$. A first order asymptotic expansion of $U_{1n}(\widehat{\eta}_{1n}, \widehat{t}_n)$ with respect to the parameter η_1 entails

$$n^{\frac{1}{2}}(\widehat{\eta}_{1n} - \widehat{\eta}_{0n}) = I_{01}^{-1} U_{1n}(\widehat{\eta}_{0n}, \widehat{t}_n) + o_p(1).$$

The equalities $U_{1n}(\widehat{\eta}_{0n}, t_0) = U_{0n}(\widehat{\eta}_{0n}) = 0$ and $U_{1n}(\widehat{\eta}_{1n}, \widehat{t}_n) = 0$ imply that the variable $U_{1n}(\widehat{\eta}_{0n}, t_0) - U_{1n}(\widehat{\eta}_{0n}, \widehat{t}_n)$ converges in probability to zero, as a normalized sum of variables, its expectation and its variance converging to zero. It follows that the variables $U_{1n}(\widehat{\eta}_{0n}, \widehat{t}_n)$ and $n^{\frac{1}{2}}(\widehat{\eta}_{0n} - \widehat{\eta}_{1n})$ converge in probability to zero under P_0.

A second order asymptotic expansion of the process

$$nZ_{1n}(\theta) = \sum_{i=1}^{[nt]} \{\log f_{\eta_1}(Y_i \mid X_i) - \log f_{\eta_{01}}(Y_i \mid X_i)\}$$

at the estimator values has the approximation

$$2nZ_{1n} = U_{1n}^T(\widehat{\eta}_{0n}, \widehat{t}_n) I_{01}^{-1} U_{1n}(\widehat{\eta}_{0n}, \widehat{t}_n) + o_p(1)$$

and it converges in probability to zero under H_0.

The process

$$nZ_{2n}(\theta) = \sum_{i=[nt]+1}^{n} \{\log f_{\eta_2}(Y_i \mid X_i) - \log f_{\eta_0}(Y_i \mid X_i)\}$$

is asymptotically equivalent to a sum over $n - [n\widehat{t}_n]$ observations. The first derivative $\dot{W}_n$ of $W_n = nZ_n$ with respect to η_2 is

$$\dot{W}_n(\eta_2, t) = \sum_{i=[nt]+1}^{n} \frac{\dot{f}_{\eta_2}(Y_i \mid X_i)}{f_{\eta_2}(Y_i \mid X_i)}$$

and $\dot{W}_n(\widehat{\eta}_{2n}, \widehat{t}_n) = 0$, under H_0, its variance converges to a finite limit, the expectation of the process $\dot{W}_n$ is asymptotically equivalent to

$$\dot{W}_0(\eta_2, t) = (n - [nt]) E_0 \frac{\dot{f}_{\eta_2}(Y_i \mid X_i)}{f_{\eta_2}(Y_i \mid X_i)}$$

and it is zero at t_0. Let η_{02} be the limit of the estimator $\widehat{\eta}_{2n}$ under H_0. The variable $\dot{W}_n(\eta_{02}, \widehat{t}_n)$ is asymptotically equivalent to an centered Gaussian variable with a positive definite variance I_{02} and $-I_{02}$ is the limit in probability of the second derivative $\ddot{W}_n(\theta_0)$.

The estimator $\widehat{\eta}_{2n}$ satisfies

$$\widehat{\eta}_{2n} - \eta_{02} = I_{02}^{-1} \dot{W}_n(\eta_{02}, \widehat{t}_n) + o_p(1)$$

and the variable $\widehat{W}_n = W_n(\widehat{\eta}_{2n}, \widehat{t}_n)$ has the expansion

$$2\widehat{W}_n = \dot{W}_n^T(\eta_{02}, \widehat{t}_n) I_{02}^{-1} \dot{W}_n^T(\eta_{02}, \widehat{t}_n) + o_p(1),$$

it converges weakly to a χ_d^2 variable under H_0. □

Proposition 4.15. *Under fixed alternatives, the statistic T_n tends to infinity as n tends to infinity and under local alternatives P_{θ_n} contiguous to H_0, the statistic T_n converges weakly to $T_0 + c^2$ where c is non-null.*

Proof. Under the alternative of a probability P_θ with t in $]0,1[$ and distinct parameters η_1 and η_2, the process $Z_n(\theta') - Z_n(\theta)$ converges a.s.

uniformly under P_0 to the function

$$\begin{aligned}
Z_\theta(\theta',\theta) &= (t'\wedge t)E_\theta\Big\{\log\frac{f_{\eta'_1}(Y_i\mid X_i)}{f_{\eta_1}(Y_i\mid X_i)}1_{\{i\le k\wedge k'\}}\Big\}\\
&+(1-t\vee t')E_\theta\Big\{\log\frac{f_{\eta'_2}(Y_i\mid X_i)}{f_{\eta_2}(Y_i\mid X_i)}1_{\{i>k\vee k\}}\Big\}\\
&+(t-t')E_\theta\Big\{\log\frac{f_{\eta'_2}(Y_i\mid X_i)}{f_{\eta_1}(Y_i\mid X_i)}1_{\{k'<i\le k\}}\Big\}\\
&+(t'-t)E_\theta\Big\{\log\frac{f_{\eta'_1}(Y_i\mid X_i)}{f_{\eta_2}(Y_i\mid X_i)}1_{\{k<i\le k'\}}\Big\}.
\end{aligned}$$

where $k=[nt]$ and $k'=[nt']$.

The function Z_θ is maximum as $\theta'=\theta$ where it is zero. It follows that the maximum likelihood estimator $\widehat{\theta}_n$ converges a.s. under P_θ to θ. The process $Z_n(\theta')$ converges a.s. under P_θ to a strictly negative function $Z_\theta(\theta')$ depending on η_0, by a second order asymptotic expansion of T_n for $\widehat{\theta}_n$ in a neighborhood of θ, it diverges under P_θ.

Local alternatives $P_n=P_{\theta_n}$ are defined by parameters $t_n=t_0-n^{-1}u_n$ in $]0,1[$ converging to t_0 and by distinct parameters $\eta_{jn}=\eta_j+n^{-\frac12}v_{jn}$, for $j=1,2$, where $\eta_1=\eta_0$, u_n and v_{1n} and v_{2n} converge to non-null limits u, and respectively v_1, v_2, as n tends to infinity.

The test statistic is the sum $T_n=T_{1n}+T_{2n}$ with

$$\begin{aligned}
T_{1n} &= 2\sum_{i=1}^{[n\widehat{t}_n]}\log\frac{f_{\widehat{\eta}_{1n}}}{f_{\widehat{\eta}_{0n}}}(Y_i\mid X_i),\\
T_{2n} &= 2\sum_{i=[n\widehat{t}_n]+1}^{n}\log\frac{f_{\widehat{\eta}_{2n}}}{f_{\widehat{\eta}_{0n}}}(Y_i\mid X_i).
\end{aligned}$$

Under P_n, the variable $n(1-\widehat{t}_n)$ converges weakly to a limit $U+u$ where U is similar to its limit under P_0. The estimators $\widehat{\eta}_{1n}$ and $\widehat{\eta}_{0n}$ converges a.s. to η_0, the variables $n^{\frac12}(\widehat{\eta}_{1n}-\eta_{1n})$ and $n^{\frac12}(\widehat{\eta}_{0n}-\eta_0)$ converge weakly to the same centered Gaussian variable with variance I_{01}, therefore $n^{\frac12}(\widehat{\eta}_{1n}-\widehat{\eta}_{0n})$ converges in probability to zero under P_n.

By an asymptotic expansion of $Z'_{n,\eta_1}(\widehat{\eta}_{1n})$, we have

$$n^{\frac12}(\widehat{\eta}_{1n}-\widehat{\eta}_{0n})=I_{01}^{-1}Z'_{n,\eta_1}(\eta_0)=o_p(1),$$

second order expansion entails

$$T_{1n}=n^{\frac12}(\widehat{\eta}_{1n}-\widehat{\eta}_{0n})^TI_{01}^{-1}n^{\frac12}(\widehat{\eta}_{1n}-\widehat{\eta}_{0n})=o_p(1)$$

and it converges in probability under P_n to $v_1^T I_{01}^{-1} v_1$.

Under P_n, $W_n = nZ_{2n}(\widehat{\theta}_n)$ is a sum over $\widehat{u}_n + u_n$ observations and the first derivative of the process $W_n(\eta_2, \widehat{t}_n)$ with respect to η_2 is

$$\dot{W}_n(\eta_2, \widehat{t}_n) = \sum_{i=[n\widehat{t}_n]+1}^{[nt_n]} \frac{\dot{f}_{\eta_2}(Y_i \mid X_i)}{f_{\eta_2}(Y_i \mid X_i)} + \sum_{i=[nt_n]+1}^{n} \frac{\dot{f}_{\eta_2}(Y_i \mid X_i)}{f_{\eta_2}(Y_i \mid X_i)},$$

the second term is asymptotically equivalent to a finite sum T of centered variables with finite variance. The variable $\widehat{\eta}_{2n}$ has the same convergence rate as $\widehat{t}_n$ and the expansion $\widehat{\eta}_{2n} - \eta_{02} = I_{02}^{-1} \dot{W}_n(\eta_{02}, \widehat{t}_n) + o_p(1)$ therefore the first sum has an expansion similar to the expansion of l_{2n} under H_0, hence T_{2n} converges weakly to $T_0 + T$ under P_n. □

Chapter 5

Change-points for point processes

Abstract. This chapter studies estimation and testing methods for the intensities of parametric or functional Poisson models, for parametric or nonparametric hazard function under right-censoring and for point processes, with change-points at unknown time thresholds. We consider empirical estimators, least squares parametric estimators and nonparametric estimators, and maximum likelihood estimators for the intensity parameters and their change-points, the convergence rates are established. Tests for the hypothesis of functions without changes are defined by the same methods. The weak convergence of the estimators and the test statistics under the hypothesis and alternatives are proved.

In regular parametric models for a density, the estimators for a parametric or nonparametric hazard function under right-censoring behave like those of a density and regression curves in the i.i.d. case. We consider a class of hazard functions λ with a change-point at an unknown point $\gamma > 0$ under a probability distribution P_γ

$$\lambda(t, \gamma) = \{1 - \delta_\gamma(t)\}\lambda_1(t) + \delta_\gamma(t)\lambda_2(t). \tag{5.1}$$

The estimators of the hazard function on the sub-intervals defined by the change-points are asymptotically independent of the estimators of the change-points. An expansion of the log-likelihood with a change-point is proved and the asymptotic properties of the likelihood ratio test are deduced. We first study a real Poisson process with a change-point, where λ_1 and λ_2 are constants, then a counting process for the occurrence of right-censored variables and a counting process with multiplicative intensity, $\lambda_{n,k}(t) = \lambda_k(t)Y_{n,k}(t)$.

5.1 Change in the intensity of a Poisson process

Several change-point models for the multiplicative intensity of a Poisson process have been studied, their estimators are consistent and the likelihood ratio test for a single intensity has been studied (Matthews, Farewell and Pyke 1985, Loader 1991, Loader 1992). Here, other estimators and another formulation of the likelihood ratio test are provided for intensities of point processes. Consider a Poisson process N observed on an increasing interval $[0, T]$ and let

$$N_T(t) = T^{-1}N(Tt), 0 \leq t \leq 1.$$

The intensity λ_T of N_T is supposed to have a change-point at an unknown time γ in the interval $]0, 1[$

$$\lambda(t) = \lambda_1\delta_\gamma(t) + \lambda_2\{1 - \delta_\gamma(t)\}, \tag{5.2}$$

where λ_1 and λ_2 are distinct parameters belonging to an open and bounded real interval and $\delta_\gamma(t) = 1\{t \leq \gamma\}$. In the model with a change-point, the parameters are restricted by the constraints of distinct parameters λ_1 and λ_2, and γ different from 0 and 1, the true intensity under a probability P_0 is denoted $\lambda_0(t)$, with parameter values λ_{01}, λ_{02} and γ_0. Let θ be the parameter with components λ_1, λ_2 and γ. The expectation of the process N_T under P_θ is discontinuous at γ and piece-wise linear $E_\theta N_T(t) = \Lambda_\theta(t)$, with the cumulative hazard function

$$\Lambda_\theta(t) = \int_0^T \lambda_\theta(s)\,ds = \lambda_1 t\delta_\gamma(t) + \{\gamma\lambda_1 + (t - \gamma)\lambda_2\}\{1 - \delta_\gamma(t)\},\ 0 \leq t \leq 1.$$

At a fixed value γ of the change-point parameter, the maximum likelihood estimators of the parameters λ_1 and λ_2 are

$$\begin{aligned}
\widehat{\lambda}_{1,\gamma} &= \gamma^{-1}N_T(\gamma) = (\gamma T)^{-1}N(\gamma T),\\
\widehat{\lambda}_{2,\gamma} &= (1 - \gamma)^{-1}\{N_T(1) - N_T(\gamma)\}\\
&= (T - \gamma T)^{-1}\{N(T) - N(\gamma T)\}
\end{aligned}$$

and the intensity $\lambda_0(t)$ is estimated from the observation of the process on $[0, 1]$ by

$$\widehat{\lambda}_T(t, \gamma) = \widehat{\lambda}_{1,\gamma}\delta_\gamma(t) + \widehat{\lambda}_{2,\gamma}\{1 - \delta_\gamma(t)\}.$$

Under P_0, the mean of $\widehat{\lambda}_{2,\gamma}$ is

$$\lambda_{2,\gamma} = (1 - \gamma)^{-1}\{\lambda_{02}(1 - \gamma_0) + \lambda_{01}(\gamma_0 - \gamma)\}\{1 - \delta_\gamma(\gamma_0)\} + \lambda_{02}\delta_\gamma(\gamma_0)$$

and $\widehat{\lambda}_{2,\gamma}$ converges a.s. to $\lambda_{2,\gamma}$ as T tends to infinity, for every γ of $[0,1]$. The mean of $\widehat{\lambda}_{1,\gamma}$ is

$$\lambda_{1,\gamma} = \gamma^{-1}\{\gamma_0\lambda_{01} + (\gamma - \gamma_0)\lambda_{02}\}\delta_\gamma(\gamma_0)\} + \lambda_{01}\{1 - \delta_\gamma(\gamma_0)\}$$

and $\widehat{\lambda}_{1,\gamma}$ converges a.s. to $\lambda_{1,\gamma}$ as T tends to infinity. For every θ, we have $\gamma\lambda_{1,\gamma} + (1-\gamma)\lambda_{2,\gamma} = \gamma_0\lambda_{01} + (1-\gamma_0)\lambda_{02}$.

The variance of $N(t)$ is $\Lambda_0(t) = \Lambda_{\theta_0}(t)$ and the variances of the estimators are

$$\begin{aligned} V_{1,\gamma} = Var\widehat{\lambda}_{1,\gamma} &= \gamma^{-2}\Lambda_0(\gamma) = \gamma^{-1}[\delta_{\gamma_0}(\gamma)\lambda_{01} + \{1 - \delta_{\gamma_0}(\gamma)\}\lambda_{02}] \\ &+ \gamma_0\gamma^{-2}\{1 - \delta_{\gamma_0}(\gamma)\}(\lambda_{01} - \lambda_{02}) = \gamma^{-1}\lambda_{1,\gamma}, \\ V_{2,\gamma} = Var\widehat{\lambda}_{2,\gamma} &= (1-\gamma)^{-2}\delta_{\gamma_0}(\gamma)\{(\gamma_0 - \gamma)\lambda_{01} + (1-\gamma_0)\lambda_{02}\} \\ &+ (1-\gamma)^{-1}\{1 - \delta_{\gamma_0}(\gamma)\}\lambda_{02} = (1-\gamma)^{-1}\lambda_{2,\gamma}, \end{aligned}$$

the variance of $\widehat{\lambda}_T(t,\gamma)$ is $V_\gamma(t) = V_{1,\gamma}\delta_\gamma(t) + V_{2,\gamma}\{1 - \delta_\gamma(t)\}$, it is consistently estimated by plug-in with

$$\widehat{V}_\gamma(t) = \widehat{V}_{1,\gamma}\delta_\gamma(t) + \widehat{V}_{2,\gamma}\{1 - \delta_\gamma(t)\} = \gamma^{-1}\widehat{\lambda}_{1,\gamma}\delta_\gamma(t) + (1-\gamma)^{-1}\widehat{\lambda}_{2,\gamma}\{1 - \delta_\gamma(t)\}.$$

A mean square estimator of the parameter γ is then defined as

$$\widehat{t}_T = \arg\inf_{\gamma\in]0,T[}\sup_{t\in]0,T[}\widehat{V}_\gamma(t) = \arg\inf_{\gamma\in]0,T[}\max\{\gamma^{-1}\widehat{\lambda}_{1,\gamma}, (1-\gamma)^{-1}\widehat{\lambda}_{2,\gamma}\} \tag{5.3}$$

and for $j = 1, 2$, λ_j is estimated by $\widehat{\lambda}_{jT}$, the value of $\widehat{\lambda}_{j,\gamma}$ at $\widehat{t}_T$.

For a Poisson process N with intensity λ, the variable $\Lambda_T^{-\frac{1}{2}}\{N_T - \Lambda_T\}$ converges weakly to a normal variable as T tends to infinity. With a change point, the variance is modified and it follows that for every γ, the independent variables

$$(\gamma T)^{\frac{1}{2}}\frac{\widehat{\lambda}_{1,\gamma} - \lambda_{1,\gamma}}{\widehat{V}_{1,\gamma}^{\frac{1}{2}}}, \quad \{(1-\gamma)T\}^{\frac{1}{2}}\frac{\widehat{\lambda}_{2,\gamma} - \lambda_{2,\gamma}}{\widehat{V}_{2,\gamma}^{\frac{1}{2}}}$$

converge weakly under P_0 to normal variables as T tends to infinity.

Theorem 5.1. *The mean square error estimator of γ_0 is a.s. consistent as T tends to infinity.*

Proof. According to the location of γ with respect to γ_0, the maximum of $Var\widehat{\lambda}_{1,\gamma}$ and $Var\widehat{\lambda}_{2,\gamma}$ is the maximum of $\gamma^{-1}\lambda_{01}\delta_{\gamma_0}(\gamma)$, $\{\gamma^{-1}\lambda_{02} + \gamma_0\gamma^{-2}(\lambda_{01} - \lambda_{02})\}\{1 - \delta_{\gamma_0}(\gamma)\}$, $(1-\gamma)^{-2}\{(\gamma_0 - \gamma)\lambda_{01} + (1-\gamma_0)\lambda_{02}\}\delta_{\gamma_0}(\gamma)$

and $(1-\gamma)^{-1}\lambda_{02}\{1-\delta_{\gamma_0}(\gamma)\}$. The first and the last terms are minimum at γ_0, for the second term, $\gamma\lambda_{02}+\gamma_0(\lambda_{01}-\lambda_{02})$ is larger than $\gamma\lambda_{01}$ if and only if λ_{01} is smaller than λ_{02} for every $\gamma>\gamma_0$, and its minimum is at γ_0. For the third term, $(\gamma_0-\gamma)\lambda_{01}+(1-\gamma_0)\lambda_{02}$ is larger than $(1-\gamma)\lambda_{02}$ if and only if λ_{01} is larger than λ_{02}, for every $\gamma\leq\gamma_0$, and it reaches its minimum at γ_0. The a.s. uniform convergence of the process $\widehat{V}_\gamma$ to the function V_γ minimum at γ_0 implies $\widehat{t}_T$ is a.s. consistent. □

The log-likelihood of the independent Poisson variables $N_T(\gamma)$ and $N_T(1)-N_T(\gamma)$ is

$$l_T(\theta)=N_T(\gamma)\log\Lambda_\theta(\gamma)+\{N_T(1)-N_T(\gamma)\}\log\{\Lambda_\theta(1)-\Lambda_\theta(\gamma)\}-\Lambda_\theta(1),$$

up to an additive variable which does not depend on the parameters. Replacing the function Λ_θ by its estimator, it becomes

$$\widehat{l}_T(\gamma)=N_T(\gamma)\log N_T(\gamma)+\{N_T(1)-N_T(\gamma)\}\log\{N_T(1)-N_T(\gamma)\}-N_T(1) \tag{5.4}$$

The process $\widehat{l}_T(\gamma)$ reaches its maximum at

$$\widehat{\gamma}_T=\arg\sup_{\gamma\in]0,1[}\widehat{l}_T(\gamma) \tag{5.5}$$

and $\widehat{l}_T(\gamma)$ converges a.s. uniformly under P_0 to the function

$$\begin{aligned}l(\gamma)&=\gamma\lambda_{1,\gamma}\log(\gamma\lambda_{1,\gamma})+(1-\gamma)\lambda_{2,\gamma}\log\{(1-\gamma)\lambda_{2,\gamma}\}\\&\quad-\{\gamma_0\lambda_{01}+(1-\gamma_0)\lambda_{02}\}\end{aligned}$$

which is maximum at γ_0, this implies the consistency theorem.

Theorem 5.2. *The maximum likelihood estimator of γ_0 is a.s. consistent as T tends to infinity.*

The convergence rate of the maximum likelihood estimator $\widehat{\gamma}_T$ with γ_0 in $]0,1[$ is deduced from the next two lemmas.

Lemma 5.1. *For ε small enough, there exists a constant $\kappa_0>0$ such that for every γ*

$$\sup_{|\gamma-\gamma_0|\leq\varepsilon} l(\gamma)-l(\gamma_0)\leq-\kappa_0\varepsilon.$$

Proof. The differences $\gamma\lambda_{1,\gamma}-\gamma_0\lambda_{01}=(\gamma-\gamma_0)\{\lambda_{01}\delta_{\gamma_0}(\gamma)+\lambda_{02}\delta_\gamma(\gamma_0)\}$ and $(1-\gamma)\lambda_{2,\gamma}-(1-\gamma_0)\lambda_{02}=-(\gamma\lambda_{1,\gamma}-\gamma_0\lambda_{01})$ imply that $l(\gamma)-l(\gamma_0)$ is a $O(|\gamma-\gamma_0|)$, furthermore the function l is maximum at γ_0 and it is negative under P_0. □

Lemma 5.2. *For γ_0 in $]0,1[$ and for every $\varepsilon > 0$, there exists a constant $\kappa_1 > 0$ such that, as T tends to infinity*

$$0 \le E_0 \sup_{|\gamma-\gamma_0|\le\varepsilon} \{\widehat{l}_T(\gamma) - \widehat{l}_T(\gamma_0)\} \le \kappa_1 \varepsilon^{\frac{1}{2}}.$$

Proof. For every γ in $]0,1[$ and for $k = 1,2$, let $\widehat{\lambda}_{0k} = \widehat{\lambda}_{k,\gamma_0}$, then

$$\widehat{\lambda}_{1,\gamma} - \widehat{\lambda}_{01} = \frac{N_T(\gamma) - N_T(\gamma_0)}{\gamma_0} + \frac{\gamma_0 - \gamma}{\gamma_0\gamma} N_T(\gamma)$$

$$\widehat{\lambda}_{2,\gamma} - \widehat{\lambda}_{02} = \frac{N_T(\gamma_0) - N_T(\gamma)}{1-\gamma_0} + \frac{\gamma - \gamma_0}{(1-\gamma_0)(1-\gamma)}\{N_T(1) - N_T(\gamma)\}.$$

The mean of $E_0\{N_T(\gamma) - N_T(\gamma_0)\}^2 = E_0\{M_T(\gamma) - M_T(\gamma_0)\}^2 + \{\Lambda_0(\gamma) - \Lambda_0(\gamma_0)\}^2$ where $E_0\{M_T(\gamma) - M_T(\gamma_0)\}^2 = \Lambda_0(\gamma) - \Lambda_0(\gamma_0)$ is a $O(|\gamma-\gamma_0|)$ and the variance of $\widehat{\lambda}_{k,\gamma} - \widehat{\lambda}_{0k}$ is a $O(|\gamma-\gamma_0|)$. Let γ in an ε neighborhood of γ_0

$$\widehat{l}_T(\gamma) - \widehat{l}_T(\gamma_0) = \{N_T(\gamma) - N_T(\gamma_0)\} \log \frac{\widehat{\lambda}_{01}}{\widehat{\lambda}_{2,\gamma}} + N_T(\gamma_0) \log \frac{\widehat{\lambda}_{1,\gamma}}{\widehat{\lambda}_{01}}$$
$$+\{N_T(1) - N_T(\gamma_0)\} \log \frac{\widehat{\lambda}_{2,\gamma}}{\widehat{\lambda}_{02}}.$$

As T tends to infinity, we obtain an approximation of $\log(\widehat{\lambda}_{k,\gamma}\widehat{\lambda}_{0k}^{-1}) - (\widehat{\lambda}_{k,\gamma} - \widehat{\lambda}_{0k})\widehat{\lambda}_{0k}^{-1}$ by an expansion of the logarithm and

$$\begin{aligned}\widehat{l}_T(\gamma) - \widehat{l}_T(\gamma_0) &= \{N_T(\gamma) - N_T(\gamma_0)\} \log \frac{\widehat{\lambda}_{01}}{\widehat{\lambda}_{2,\gamma}} + \gamma_0(\widehat{\lambda}_{1,\gamma} - \widehat{\lambda}_{01})\{1 + o_p(1)\}\\ &\quad +(1-\gamma_0)(\widehat{\lambda}_{2,\gamma} - \widehat{\lambda}_{02})\{1 + o_p(1)\}\\ &= \{N_T(\gamma) - N_T(\gamma_0)\} \log \frac{\widehat{\lambda}_{01}}{\widehat{\lambda}_{2,\gamma}}\\ &\quad +\frac{\gamma-\gamma_0}{1-\gamma}\{N_T(1) - \widehat{\lambda}_{1,\gamma}\} + o_p(|\gamma-\gamma_0|). \qquad (5.6)\end{aligned}$$

which is bounded and $E_0\{\widehat{l}_T(\gamma) - \widehat{l}_T(\gamma_0)\}^2$ is a $O(|\gamma-\gamma_0|)$, the result follows from the Cauchy–Schwarz inequality. □

The process $X_T(\gamma) = \widehat{l}_T(\gamma) - \widehat{l}_T(\gamma_0)$ converges a.s. uniformly on $]0,1[$ to the function $X(\gamma) = l(\gamma) - l(\gamma_0)$ such that

$$X(\gamma) = \{\Lambda_\theta(\gamma) - \Lambda_\theta(\gamma_0)\} \log \frac{\lambda_{01}(1-\gamma)}{\Lambda_0(1) - \Lambda_0(\gamma)}$$
$$+\frac{\gamma_0 - \gamma}{\gamma(1-\gamma)}\{\Lambda_0(\gamma) - \gamma\Lambda_0(1)\} + o(|\gamma-\gamma_0|).$$

By the same proof as Theorem 2.2, we deduce from Lemmas 5.1–5.2 the convergence rate of $\widehat{\gamma}_T$

$$\overline{\lim}_{T,A\to\infty} P_0(T|\widehat{\gamma}_T - \gamma_0| > A) = 0. \tag{5.7}$$

Let $\mathcal{U}_T = \{u = T(\gamma - \gamma_0), \gamma \in]0,1[\}$ and let $\mathcal{U}_T^A = \{u \in \mathcal{U}_T : |u| < A\}$, for $A > 0$. By (5.7), the asymptotic behaviour of the variable $T(\widehat{\gamma}_T - \gamma_0)$ is deduced from the limiting distribution of the process $W_T = T^{\frac{1}{2}}(X_T - X)$ in the set $\mathcal{U}_T^A$, as A tends to infinity. The process W_T is written under P_0 according to the difference

$$M_T = N_T - \Lambda_0.$$

At $\gamma_{T,u} = \gamma_0 + T^{-1}u$, u in $\mathcal{U}_T^A$, the process $T^{\frac{1}{2}}\{M_T(\gamma_{T,u}) - M_T(\gamma_0)\}$ converges weakly under P_0 to a centered Gaussian process with independent increments and with variance $0(|u|)$.

Theorem 5.3. *Under P_0, the variable $T(\widehat{\gamma}_T - \gamma_0)$ is asymptotically independent of $\widehat{\lambda}_T$ and it converges weakly to the limit U of the location of the maximum U_A of an uncentered Gaussian process in $\mathcal{U}_n^A$, as A tends to infinity.*

Proof. The estimator $\widehat{u}_T = T(\widehat{\gamma}_T - \gamma_0)$ of $u = T(\gamma_{T,u} - \gamma_0)$ in $\mathcal{U}_T^A$ maximizes $X_T(\gamma_{T,u})$ and $P_0(|\widehat{u}_T| > A)$ converges to zero as T and A tend to infinity by (5.7), the limit of $\widehat{u}_T = T(\widehat{\gamma}_T - \gamma_0)$ is therefore bounded in probability under P_0. From the approximation (5.6) of the process X_T, the asymptotic mean and variance of $TX_T(\gamma_{T,u})$ have the order $|u|$ and the process $TX_T(\gamma_{T,u})$ converges weakly to an uncentered Gaussian process, $\widehat{u}_T$ converges weakly to the location of the maximum in U_A of this process in $\mathcal{U}_T^A$. By the consistency of the estimators of the intensities, $\widehat{\gamma}_T$ is asymptotically independent of $\widehat{\lambda}_{1T}$ and $\widehat{\lambda}_{2T}$. □

5.2 Likelihood ratio test for a Poisson process

The log-likelihood ratio test for the hypothesis H_0 of a Poisson process with a constant intensity λ_0 against the alternative of an intensity with a change at an unknown time γ according to the model (5.2) is performed with the statistic

$$S_T = 2T\{\sup_{0<\gamma<1} \widehat{l}_T(\gamma) - \widehat{l}_{0T}\}$$

where $\widehat{l}_T(\gamma)$ is defined by (5.4) and $\widehat{l}_{0T}$ estimates the logarithm of the likelihood l_{0T} of $N_T(1)$ under H_0, which is $l_{0T} = N_T(1)\log\lambda_0 - \lambda_0$, up to a variable which does not depend on the parameters. The estimator of λ_0 converges with the rate $T^{-\frac{1}{2}}$ and it is denoted $\widehat{\lambda}_{0T} = \lambda_0 + T^{-\frac{1}{2}} h_{0T}$. The limit of the statistic under H_0 depends on the limit of the variable $T(\widehat{\lambda_{1T}} - \lambda_0)$.

Proposition 5.1. *The statistic S_T converges weakly under H_0 to a variable T_0 with distribution $-\lambda_0\chi_1^2$.*

Proof. Under H_0, $\gamma_0 = 1$ and by (5.7) the maximum likelihood estimator of γ_0 in model (5.2) has the convergence rate T. The parameters under the alternative are written as $\gamma_{Tu} = 1 - T^{-1}u$ and $\lambda_{kT} = \lambda_k + T^{-\frac{1}{2}} h_{kT}$, for $k = 1, 2$, with $\lambda_1 = \lambda_0 = \lambda_2$. The statistic is the value at $\widehat{\theta}_T$ of the process $2s_T(\theta)$ where

$$\begin{aligned} s_T(\theta) &= TN_T(\gamma_{T,u}) \log \frac{\lambda_{1T}}{\lambda_{0T}} + T\{N_T(1) - N_T(\gamma_{T,u})\} \log \frac{\lambda_{2T}}{\lambda_{0T}} \\ &\quad + \gamma_{T,u} T^{\frac{1}{2}} (h_{1T} - \lambda_{0T}) + u(\lambda_{2T} - \lambda_{0T}) \\ &= T \int_0^{\gamma_{T,u}} \log \frac{\lambda_{1T}}{\lambda_{0T}} \, dM_T + T \int_{\gamma_{T,u}}^1 \log \frac{\lambda_{2T}}{\lambda_{0T}} \, dM_T \\ &\quad + T\gamma_{T,u}\lambda_0 \Big\{ \log \frac{\lambda_{1T}}{\lambda_{0T}} - \frac{\lambda_{1T} - \lambda_{0T}}{\lambda_0} \Big\} \\ &\quad + \lambda_0 u \Big\{ \log \frac{\lambda_{2T}}{\lambda_{0T}} - \frac{\lambda_{2T} - \lambda_{0T}}{\lambda_0} \Big\} \end{aligned}$$

where $\lambda_{1T} - \lambda_{0T} = T^{-\frac{1}{2}}(h_{1T} - h_{0T})$ tends to zero. The first term of this expansion is asymptotically equivalent to $(h_{1T} - h_{0T})\lambda_0^{-1} T^{\frac{1}{2}} M_T(\gamma_{T,u})$, at the maximum $\widehat{h}_{0T} - \widehat{h}_{1T} = T^{\frac{1}{2}}\{M_T(1) - \gamma_{T,u}^{-1} M_T(\gamma_{T,u})\} = T^{\frac{1}{2}}\{M_T(1) - M_T(\gamma_{T,u})\} + o_p(1)$ is asymptotically equivalent to a centered process with independent increments and independent of the centered process $T^{\frac{1}{2}} M_T(\gamma_{T,u})$, their product at $\widehat{u}_T$, converges in probability to its mean which is zero. The second term on the negligible interval $]\gamma_{T,u}, 1]$ is a $o_p(1)$. The third term including the logarithm of the ratio $\lambda_{1T}\lambda_{0T}^{-1}$ has the expansion $-\frac{1}{2}\lambda_0^{-2} h_1^2 + o(1)$ whereas the last term is a $o_p(1)$. At the maximum, the limit of the statistic is the limit of $-\lambda_0^{-1}\{T(\widehat{\lambda}_{1T} - \lambda_1)\}^2$, it is therefore proportional to a χ_1^2 variable. □

Under a fixed alternative with γ in $]0, 1[$ and distinct parameters λ_1 and λ_2, one of them is different from λ_0 and the test statistic S_T tends to infinity.

Proposition 5.2. *Under local alternatives with parameters θ_T, the statistic S_T converges weakly to $-T_0$.*

Proof. The parameters under a local alternative are $\gamma_{Tu} = 1 - T^{-1}u$ and for $k = 1, 2$, $\lambda_{kT} = \lambda_k + T^{-\frac{1}{2}} h_{kT}$ with $\lambda_1 = \lambda_0 = \lambda_2$. The local martingale M_T and the process s_T are now centered according to these values and s_T has the expansion

$$\begin{aligned} s_T(\theta) &= -T \int_0^{\gamma_{T,u}} \log \frac{\lambda_{1T}}{\lambda_{0T}} \, dM_T - T \int_{\gamma_{T,u}}^1 \log \frac{\lambda_{2T}}{\lambda_{0T}} \, dM_T \\ &\quad - T\gamma_{T,u}\lambda_{1T} \Big\{ \log \frac{\lambda_{1T}}{\lambda_{0T}} - \frac{\lambda_{1T} - \lambda_{0T}}{\lambda_{1T}} \Big\} \\ &\quad - \lambda_{2T} u \Big\{ \log \frac{\lambda_{2T}}{\lambda_{0T}} - \frac{\lambda_{2T} - \lambda_{0T}}{\lambda_{2T}} \Big\} \end{aligned}$$

where the the third term is asymptotically equivalent to $(2\lambda_{0T})^{-1}(\lambda_{1T} - \lambda_{0T})^2 + o(1)$ and the other terms converge to zero in probability. □

5.3 Parametric models for Poisson processes

Parametric models such as additive models for the logarithm of the intensity of Poisson processes are often used in data analysis. Let $N_T(t) = T^{-1}N(Tt)$ be a Poisson process on $[0, T]$, with a parametric intensity function $\lambda_{T\theta}$ satisfying an ergodic property: There exists a real function λ_θ on $[0, 1]$ such that

$$\lim_{T\to\infty} \sup_{t\in[0,1]} |\lambda_{T\theta}(t) - \lambda_\theta(t)| = 0, \tag{5.8}$$

for every θ. The function λ_θ is piecewise continuous with a discontinuity at an unknown time γ in the interval $]0, 1[$

$$\lambda_\theta(t) = \lambda_{\eta_1}(t)\delta_\gamma(t) + \lambda_{\eta_2}(t)\{1 - \delta_\gamma(t)\}. \tag{5.9}$$

The parameters η_1 and η_2 are distinct in an open parameter space $\mathcal{H}$ of $\mathbb{R}^d$, the function λ_η is $C_b^1(]0, 1[)$ for every η of $\mathcal{H}^2$ and the function $\eta \mapsto \lambda_\eta(t)$ is $C_b^2(\mathcal{H})$, with a bounded second order derivative on $\mathcal{H}$, uniformly for t on $]0, 1[$. Let $\theta = (\eta^T, \gamma)^T$ be the parameter vector with value θ_0 under the probability measure P_0 of the observations of the process N.

Let $\lambda_{01} = \lambda_{\eta_{01}}$ and $\lambda_{02} = \lambda_{\eta_{02}}$ denote the ergodic intensities on $[0, 1]$ under the probability measure P_0, let $\Lambda_{01}(t) = \int_0^t \lambda_{01}(s)\, ds$, for $t \le \gamma_0$ and let $\Lambda_{02}(t) = \Lambda_{02}(\gamma_0) + \int_{\gamma_0}^t \lambda_{0s}(s)\, ds$, for $t > \gamma_0$. The cumulative intensity of the Poisson process N_T under P_0 is

$$\Lambda_{0T}(t) = \Lambda_{01T}(t)\delta_{\gamma_0}(t) - \{\Lambda_{01T}(\gamma_0) + \Lambda_{02T}(t) - \Lambda_{02T}(\gamma_0)\}\{1 - \delta_{\gamma_0}(t)\},$$

it determines the local martingale

$$M_T(t) = N_T(t) - \Lambda_{0T}(t),$$

for t in $[0,1]$, such that $E_0 N_T(t) = \Lambda_{0T}(t) = T^{-1} E_0 N(Tt)$. By the ergodic property, there exist real functions λ_{01} and λ_{02} on $[0,1]$ such that λ_{0kT} converges uniformly to λ_{0k}, for $k = 1, 2$.

Under probability measure P_θ, the log-likelihood of the independent Poisson variables $N_T(\gamma)$ and $N_T(1) - N_T(\gamma)$ is defined, up to an additive term which does not depend on the parameters, as

$$\begin{aligned} l_T(\theta) &= N_T(\gamma) \log\{\Lambda_{T\eta_1}(\gamma)\} + \{N_T(1) - N_T(\gamma)\} \log\{\Lambda_{T\eta_2}(1) - \Lambda_{T\eta_2}(\gamma)\} \\ &\quad -\Lambda_{T\eta_1}(\gamma) - \Lambda_{T\eta_2}(1) + \Lambda_{T\eta_2}(\gamma). \end{aligned}$$

The maximum likelihood estimators $\widehat{\eta}_{T1,\gamma}$ and $\widehat{\eta}_{T2,\gamma}$ of the parameters η_1 and η_2 maximize the log-likelihood variable $l_T(\theta)$, at γ. They are solutions of the estimating equations

$$\begin{aligned} \dot{l}_{T,\eta_1}(\theta) &= N_T(\gamma) \frac{\dot{\Lambda}_{T\eta_1}(\gamma)}{\Lambda_{T\eta_1}(\gamma)} - \dot{\Lambda}_{T\eta_1}(\gamma) = 0, \\ \dot{l}_{T,\eta_2}(T\theta) &= \{N_T(1) - N_T(\gamma)\} \frac{\dot{\Lambda}_{T\eta_2}(1) - \dot{\Lambda}_{T\eta_2}(\gamma)}{\Lambda_{T\eta_2}(1) - \Lambda_{T\eta_2}(\gamma)} - \{\dot{\Lambda}_{T\eta_2}(1) - \dot{\Lambda}_{T\eta_2}(\gamma)\} = 0, \end{aligned}$$

equivalently

$$\begin{aligned} &\dot{\Lambda}_{T\widehat{\eta}_{1T}}(\gamma) = 0 \text{ or } \Lambda_{T\widehat{\eta}_{1T},\gamma}(\gamma) = N_T(\gamma), \\ &\dot{\Lambda}_{T\widehat{\eta}_{2T}}(\gamma) = 0 \text{ or } \Lambda_{T\widehat{\eta}_{2T},\gamma}(1) - \Lambda_{T\widehat{\eta}_{2T},\gamma}(\gamma) = N_T(1) - N_T(\gamma). \end{aligned}$$

Then $\widehat{\gamma}_T$ maximizes $l_T(\widehat{\eta}_\gamma, \gamma)$ and $\widehat{\eta}_{T,k} = \widehat{\eta}_{k,\widehat{\gamma}_T}$, for $k = 1, 2$.

The logarithm of the likelihood ratio of the sample under P_θ and P_0 defines the process $X_T(\theta) = l_T(\theta) - l_T(\theta_0)\}$ such that

$$\begin{aligned} X_T(\theta) &= N_T(\gamma_0) \log \frac{\Lambda_{T\eta_1}(\gamma)}{\Lambda_{T01}(\gamma_0)} + \{N_T(1) - N_T(\gamma_0)\} \log \frac{\Lambda_{T\eta_2}(1) - \Lambda_{T\eta_2}(\gamma)}{\Lambda_{T02}(1) - \Lambda_{T02}(\gamma_0)} \\ &\quad + \{N_T(\gamma_0) - N_T(\gamma)\} \log \frac{\Lambda_{T\eta_2}(1) - \Lambda_{T\eta_2}(\gamma)}{\Lambda_{T\eta_1}(\gamma)} \\ &\quad - (\Lambda_{T\eta_1} - \Lambda_{T01})(\gamma_0) - \{(\Lambda_{T\eta_2} - \Lambda_{T02})(1) - (\Lambda_{T\eta_2} - \Lambda_{T02})(\gamma_0)\} \\ &\quad - \{\Lambda_{T\eta_1}(\gamma) - \Lambda_{T\eta_2}(\gamma) - \Lambda_{T\eta_1}(\gamma_0) + \Lambda_{T\eta_2}(\gamma_0)\}. \end{aligned}$$

Under P_0, X_T converges a.s. uniformly on Θ to the function

$$X(\theta) = \Lambda_{01}(\gamma_0)\log\frac{\Lambda_{\eta_1}(\gamma)}{\Lambda_{01}(\gamma_0)} + \{\Lambda_{02}(1) - \Lambda_{02}(\gamma_0)\}\log\frac{\Lambda_{\eta_2}(1) - \Lambda_{\eta_2}(\gamma)}{\Lambda_{02}(1) - \Lambda_{02}(\gamma_0)}$$
$$+\{\Lambda_0(\gamma_0) - \Lambda_0(\gamma)\}\log\frac{\Lambda_{\eta_2}(1) - \Lambda_{\eta_2}(\gamma)}{\Lambda_{\eta_1}(\gamma)} - (\Lambda_{\eta_1} - \Lambda_{01})(\gamma_0)$$
$$-\{(\Lambda_{\eta_2} - \Lambda_{02})(1) - (\Lambda_{\eta_2} - \Lambda_{02})(\gamma_0)\}$$
$$-\{\Lambda_{\eta_1}(\gamma) - \Lambda_{\eta_2}(\gamma) - \Lambda_{\eta_1}(\gamma_0) + \Lambda_{\eta_2}(\gamma_0)\}.$$

Theorem 5.4. *Under P_0, the maximum likelihood estimator $\widehat{\theta}_T$ of θ_0 is a.s. consistent.*

Proof. The function X is concave with respect to Λ_1 and Λ_2 and it is zero at Λ_0 where it reaches its maximum. By the convergence of X_T to X, we have

$$0 \leq X_T(\widehat{\theta}_T) \leq \sup_\Theta |X_T(\theta) - X(\theta)| + X(\widehat{\theta}_T)$$

and $\lim_{T\to\infty} X(\widehat{\theta}_T) \geq 0$ but $0 = X(\theta_0) \geq X(\widehat{\theta}_T)$ which implies $\widehat{\theta}_T$ converges a.s. to θ_0. □

The function X is differentiable with respect to the parameter γ on the sets $\{\gamma < \gamma_0\}$ and $\{\gamma_0 < \gamma\}$, its derivative with respect to γ is

$$X'_\gamma(\theta) = \Lambda_{01}(\gamma_0)\frac{\lambda_{\eta_1}(\gamma)}{\Lambda_{\eta_1}(\gamma)} - (\lambda_{\eta_1} + \lambda_{\eta_2})(\gamma^+),$$
$$-\{\Lambda_{02}(1) - \Lambda_{20}(\gamma_0)\}\frac{\lambda_{\eta_2}(\gamma^+)}{\Lambda_{\eta_2}(1) - \Lambda_{\eta_2}(\gamma)}$$
$$-\lambda_0(\gamma)\log\frac{\Lambda_{\eta_2}(1) - \Lambda_{\eta_2}(\gamma)}{\Lambda_{\eta_1}(\gamma)}$$
$$-\{\Lambda_0(\gamma_0) - \Lambda_0(\gamma)\}\left\{\frac{\lambda_{\eta_2}(\gamma^+)}{\Lambda_{\eta_2}(1) - \Lambda_{\eta_2}(\gamma)} + \frac{\lambda_{\eta_1}(\gamma)}{\Lambda_{\eta_1}(\gamma)}\right\},$$

and the derivatives of $X(\theta)$ with respect to η are expressed by the means of the derivatives $\dot{\Lambda}_{T\eta_1}$ and $\dot{\Lambda}_{T\eta_2}$ of the function Λ with respect to the components η_1 and η_2 of η

$$\dot{X}_{\eta_1}(\theta) = \Lambda_{01}(\gamma)\frac{\dot{\Lambda}_{\eta_1}(\gamma)}{\Lambda_{\eta_1}(\gamma)} - \dot{\Lambda}_{\eta_1}(\gamma),$$

$$\dot{X}_{\eta_2}(\theta) = \{\Lambda_{02}(1) - \Lambda_0(\gamma)\}\frac{\dot{\Lambda}_{\eta_2}(1) - \dot{\Lambda}_{\eta_2}(\gamma)}{\Lambda_{\eta_2}(1) - \Lambda_{\eta_2}(\gamma)} - \{\dot{\Lambda}_{\eta_2}(1) - \dot{\Lambda}_{\eta_2}(\gamma)\},$$

$$\ddot{X}_{\gamma,\eta_1}(\theta) = \Lambda_0(\gamma)\frac{\Lambda_{\eta_1}(\gamma)\dot{\lambda}_{\eta_1}(\gamma) - \lambda_{\eta_1}(\gamma)\dot{\Lambda}_{\eta_1}(\gamma)}{\Lambda_{\eta_1}^2\gamma)} + \lambda_0(\gamma)\frac{\dot{\Lambda}_{\eta_1}(\gamma)}{\Lambda_{\eta_1}(\gamma)} - \dot{\lambda}_{\eta_1}(\gamma),$$

$$\ddot{X}_{\gamma,\eta_2}(\theta) = -\lambda_0(\gamma)\frac{\dot{\Lambda}_{\eta_2}(1) - \dot{\Lambda}_{\eta_2}(\gamma)}{\Lambda_{\eta_2}(1) - \Lambda_{\eta_2}(\gamma)} - \dot{\lambda}_{\eta_2}(\gamma^+)\frac{\Lambda_{02}(1) - \Lambda_0(\gamma)}{\Lambda_{\eta_2}(1) - \Lambda_{\eta_2}(\gamma)} + \dot{\lambda}_{\eta_2}(\gamma^+),$$

$$+\{\Lambda_{02}(1) - \Lambda_0(\gamma)\}\frac{\{\dot{\Lambda}_{\eta_2}(1) - \dot{\Lambda}_{\eta_2}(\gamma)\}\{\lambda_{\eta_2}(1) - \lambda_{\eta_2}(\gamma^+)\}}{\{\Lambda_{\eta_2}(1) - \Lambda_{\eta_2}(\gamma)\}^2},$$

where $\dot{X}_\eta(\theta_0) = 0$ and $I_0 = -\ddot{X}_\eta(\theta_0)$ is a definite positive matrix, with

$$\ddot{X}_{\eta_1}(\theta_0) = -\frac{\dot{\Lambda}_{01}^2(\gamma_0)}{\Lambda_{01}(\gamma_0)}, \tag{5.10}$$

$$\ddot{X}_{\eta_2}(\theta_0) = -\frac{\{\dot{\Lambda}_{02}(1) - \dot{\Lambda}_{02}(\gamma_0)\}^2}{\Lambda_{02}(1) - \Lambda_{02}(\gamma_0)}.$$

The function X then a concave function of η in a neighborhood of θ_0 where it reaches its maximum $X(\theta_0) = 0$, then $X(\eta_0, \gamma) < 0$ for every γ different from γ_0.

Let $\varepsilon > 0$ and let $V_\varepsilon(\theta_0)$ be an ε- neighborhood of θ_0 for the semi-norm

$$\rho(\theta, \theta_0) = (\|\eta - \eta_0\|_2^2 + |\gamma - \gamma_0|)^{\frac{1}{2}}.$$

Lemma 5.3. *For ε small enough, there exists a constant $\kappa_0 > 0$ such that for every θ in Θ*

$$X(\theta) \leq -\kappa_0\rho^2(\theta, \theta_0). \tag{5.11}$$

Proof. A second order expansion of X with respect to η in a neighborhood of θ_0 is written as

$$\begin{aligned} X(\theta) &= X(\eta_0, \gamma) + (\eta - \eta_0)^T\dot{X}_\eta(\eta_0, \gamma) - \frac{1}{2}(\eta - \eta_0)^T I(\eta_0, \gamma)(\eta - \eta_0) \\ &\quad + o(\|\eta - \eta_0\|^2) \\ &\leq (\eta - \eta_0)^T\dot{X}_\eta(\eta_0, \gamma) - \frac{1}{2}(\eta - \eta_0)^T I(\eta_0, \gamma)(\eta - \eta_0) + o(\|\eta - \eta_0\|^2). \end{aligned}$$

As $\Lambda_{01}(\gamma) = \Lambda_{\eta_{01}}(\gamma)$ and $\Lambda_{02}(\gamma) = \Lambda_{\eta_{02}}(\gamma)$, the first order derivative $\dot{X}_\eta(\eta_0, \gamma)$ is zero for every γ. □

The process $W_T = T^{\frac{1}{2}}(X_T - X)$ is written as according to the local martingale $M_T = N_T - \Lambda_T$ as

$$\begin{aligned} W_T(\theta) &= T^{\frac{1}{2}} M_T(\gamma_0) \log\frac{\Lambda_{T\eta_1}(\gamma)}{\Lambda_{T01}(\gamma_0)} \\ &\quad + T^{\frac{1}{2}}\{M_T(1) - M_T(\gamma_0)\}\log\frac{\Lambda_{T\eta_2}(1) - \Lambda_{T\eta_2}(\gamma)}{\Lambda_{T02}(1) - \Lambda_{T02}(\gamma_0)} \\ &\quad + T^{\frac{1}{2}}\{M_T(\gamma_0) - M_T(\gamma)\}\log\frac{\Lambda_{T\eta_2}(1) - \Lambda_{T\eta_2}(\gamma)}{\Lambda_{T\eta_1}(\gamma)}. \end{aligned}$$

The process $\nu_T - T^{\frac{1}{2}} M_T$ is centered and its variance is $E_0 \nu_T^2(t) = \Lambda_0(t)$ on $[0,1]$, by Rolando's theorem it converges weakly to a centered Gaussian process with independent increments and with variance $\Lambda_0(t)$ at t in $[0,1]$, as T tends to infinity. For ε sufficiently small, under the condition

$$\sup_{\theta \in V_\varepsilon(\theta_0)} \{\log^2 \Lambda_{T\eta_1} + \log^2 \Lambda_{T\eta_2}\} < \infty,$$

a first order expansion of the intensity (5.9) in $V_\varepsilon(\theta_0)$ implies the existence of a constant $\kappa_1 > 0$ such that $E_0 \sup_{\theta \in V_\varepsilon(\theta_0)} W_T^2(\theta) \le \kappa_1^2 \varepsilon^2$, as T tends to infinity, therefore

$$E_0 \sup_{\theta \in V_\varepsilon(\theta_0)} W_T(\theta) \le \kappa_1 \varepsilon. \tag{5.12}$$

Denoting now

$$\mathcal{U}_T = \{u_T = (T^{\frac{1}{2}}(\eta - \eta_0)^T, T(\gamma - \gamma_0))^T, \eta \in \mathcal{H}, \gamma \in]0,1[\}$$

and for u in $\mathcal{U}_T$, $\theta_{T,u}$ is the vector with components $\eta_0 + T^{-\frac{1}{2}} u_1$ and $\gamma_0 + T^{-1} u_2$, with $u = (u_1^T, u_2)^T$, and $\|u\| = T^{\frac{1}{2}} \rho(\theta_{T,u}, \theta_0)$, reversely $u_{T,\theta}$ denotes the vector of $\mathcal{U}_T$ such that $\theta = \theta_{T,u_{T,\theta}}$.

For $\varepsilon > 0$, let $\mathcal{U}_{T,\varepsilon} = \{u \in \mathcal{U}_T : \|u\| \le T^{\frac{1}{2}} \varepsilon\}$, there is equivalence between u belongs to $\mathcal{U}_{T,\varepsilon}$ and $\theta_{T,u}$ belongs to $V_\varepsilon(\theta_0)$.

Theorem 5.5. *For every $\varepsilon > 0$ sufficiently small, as T and A tend to infinity, the probability $P_0(\sup_{\theta \in V_\varepsilon(\theta_0) \|u_{T,\theta}\| > A} X_T(\gamma) \ge 0)$ converges to zero and*

$$\overline{\lim}_{T,A\to\infty} P_0\{T^{\frac{1}{2}} \rho(\widehat{\theta}_T, \theta_0) > A) = 0.$$

Proof. Let $\widehat{u}$ be the vector with components $T^{\frac{1}{2}}(\widehat{T\eta} - \eta_0)$ and $T(\widehat{\gamma}_T - \gamma_0)$, its norm is $\|\widehat{u}_T\| = T^{-\frac{1}{2}} \rho(\widehat{\theta}_T, \theta_0)$. For every $\eta > 0$ the consistency of the estimators implies that for $\varepsilon > 0$ sufficiently small

$$P_0\{\widehat{u}_T \in \mathcal{U}_{T,\varepsilon}\} = P_0\{\widehat{\theta}_T \in V_\varepsilon(\theta_0)\} > 1 - \eta,$$

therefore $P_0(\|\widehat{u}_T\| > A) \le P_0(\sup_{u \in \mathcal{U}_{T,\varepsilon}, \|u\| > A} l_T(\theta_{T,u}) \ge 0) + \eta$. Let g be an increasing function such that $\sum_{g(j) > A} g(j+1) g^{-2}(j)$ tends to zero as A tends to infinity, and let

$$H_{T,j} = \{u \in \mathcal{U}_{T,\varepsilon} : g(j) < \|u\| \le g(j+1)\}, \ j \in \mathbb{N}.$$

For every u belonging to $H_{T,j}$, $T^{-\frac{1}{2}} g(j) \le \rho(\theta_{T,u}, \theta_0) \le T^{-\frac{1}{2}} g(j+1)$ and the inequality 3.3 implies $X(\theta_{T,u}) \le -\kappa_0 T^{-1} g^2(j)$, with $X(\theta_0) = 0$. For every $\varepsilon \leq T^{-\frac{1}{2}}g(j+1)$, the sets $H_{T,j}$ split the probability as a sum

$$P_0\Big(\sup_{u\in\mathcal{U}_{T,\varepsilon},\|u\|>A} l_T(\theta_{T,u}) \geq 0\Big) \leq \sum_{g(j)>A} P_0\Big(\sup_{u\in H_{T,j}} l_T(\theta_{T,u}) \geq 0\Big)$$

$$\leq \sum_{2^j>A} P_0\Big(\sup_{u\in H_{T,j}} |W_T(\theta_{T,u})| \geq T^{-\frac{1}{2}}g^2(j)\kappa_0\Big)$$

$$\leq \frac{T^{\frac{1}{2}}}{\kappa_0}\sum_{g(j)>A} g^{-2}(j)E_0 \sup_{u\in H_{T,j}} |W_T(\theta_{T,u})| \leq \frac{\kappa_1}{\kappa_0}\sum_{g(j)>A}\frac{g(j+1)}{g^2(j)}$$

by the inequality (5.12), this bound tends to zero as A tends to infinity. □

Proposition 5.3. *Under P_0, the variable $T^{\frac{1}{2}}(\widehat{\eta}_T - \eta_0)$ converges weakly to a centered Gaussian variable with variance I_0^{-1} and the processes $T^{\frac{1}{2}}(\widehat{\lambda}_{kT} - \lambda_{0k})$ converges weakly to independent centered Gaussian processes with variances $\dot{\lambda}^T_{\eta_{0k}}(t)I_{0k}^{-1}\dot{\lambda}_{T\eta_{0k}}(t)$, for $k = 1, 2$.*

Proof. The equality $\dot{X}_{T,\eta}(\widehat{\theta}_T) = 0$, Theorems 5.4 and 5.5 and a first order expansion of $\dot{X}_{T,\eta}(\widehat{\theta}_T)$ for $\widehat{\theta}_T$ in a neighborhood of θ_0 yield

$$-T^{\frac{1}{2}}\dot{X}_{T,\eta}(\eta_0,\widehat{\gamma}_T) = T^{\frac{1}{2}}(\widehat{\eta}_T - \eta_0)^T\ddot{X}_{T,\eta}(\eta_0,\widehat{\gamma}_T) + o_p(1),$$

$$T^{\frac{1}{2}}(\widehat{\eta}_T - \eta_0) = T^{\frac{1}{2}}\dot{X}_{T,\eta}(\theta_0)^T I_0^{-1} + o_p(1) \tag{5.13}$$

where

$$\dot{X}_{T,\eta_1}(\theta_0) = M_T(\gamma_0)\frac{\dot{\Lambda}_{T\eta_{01}}(\gamma_0)}{\Lambda_{T01}(\gamma_0)}, \tag{5.14}$$

$$\dot{X}_{T,\eta_2}(\theta_0) = \{M_T(1) - M_T(\gamma_0)\}\frac{\dot{\Lambda}_{T02}(1) - \dot{\Lambda}_{T02}(\gamma_0)}{\Lambda_{T02}(1) - \Lambda_{T02}(\gamma_0)} \tag{5.15}$$

and the variance of $T^{\frac{1}{2}}\dot{X}_{T,\eta}(\theta_0)$ is I_0 according to (5.10), as the variance of $T^{\frac{1}{2}}M_T(t)$ is $\Lambda_{01}(t)$, for $t \leq \gamma_0$, and the variance of $T^{\frac{1}{2}}\{M_T(t) - M_T(\gamma_0)\}$ is $\Lambda_{T02}(t) - \Lambda_{T02}(\gamma_0)$, for $t > \gamma_0$. The process $T^{\frac{1}{2}}M_T$ converges weakly to a centered Gaussian process with independent increments, then the variable $T^{\frac{1}{2}}(\widehat{\eta}_T - \eta_0)$ converges weakly to a Gaussian variable with variance I_0^{-1}.

The estimators $T^{\frac{1}{2}}(\widehat{\eta}_{T1T} - \eta_{01})$ and $T^{\frac{1}{2}}(\widehat{\eta}_{2T} - \eta_{02})$ depend on the increments of M_T on disjoint intervals and the process $T^{\frac{1}{2}}M_T$ converges weakly to a Gaussian process with independent increments, therefore the estimators are asymptotically independent.

The estimators $\widehat{\lambda}_{kT}$ have a first order expansion

$$T^{\frac{1}{2}}\{\widehat{\lambda}_{Tk}(t) - \lambda_{0k}(t)\} = T^{\frac{1}{2}}(\widehat{\eta}_{Tk} - \eta_{0k})^T\dot{\lambda}_{T0k}(t) + o_p(1),$$

uniformly for t in $]0,1[$, and the processes $T^{\frac{1}{2}}\{\widehat{\lambda}_{Tk}(t) - \lambda_{0k}(t)\}$ converge weakly to independent centered Gaussian processes with variances $\dot{\lambda}_{T0k}^T(t) I_{0k}^{-1} \dot{\lambda}_{T0k}(t)$, for $k = 1, 2$. □

For every $A > 0$, let $\mathcal{U}_T^A = \{u \in \mathcal{U}_T : \|u\| \leq A\}$ and for u in $\mathcal{U}_T^A$, let $\eta_{T,u} = \eta_0 + T^{-\frac{1}{2}} u_\eta$ and $\gamma_{T,u} = \gamma_0 + T^{-1} u_\gamma$. We consider the process defined on $\mathcal{U}_T^A$ as $\widetilde{W}_T(u) = T^{\frac{1}{2}} W_T(\eta_{T,u_\eta}, \gamma_{T,u_\gamma})$, with $u = (u_\eta^T, u_\gamma)^T$ in $\mathcal{U}_T^A$ and $\theta_{T,u}$ in a neighborhood of θ_0.

The process $\widetilde{W}_T(u) = T(X_T - X)(\theta_{T,u})$, for u in $\mathcal{U}_T^A$, is written as

$$\begin{aligned}
\widetilde{W}_T(u) = TM_T(\gamma_0) \log \frac{\Lambda_{T\eta_{1,T,u}}(\gamma_{T,u})}{\Lambda_{T01}(\gamma_0)} \\
+T\{M_T(1) - M_T(\gamma_0)\} \log \frac{\Lambda_{T\eta_{2,T,u}}(1) - \Lambda_{T\eta_{2,T,u}}(\gamma_{T,u})}{\Lambda_{T02}(1) - \Lambda_{T02}(\gamma_0)} \\
-T\{M_T(\gamma_{T,u_\gamma}) - M_T(\gamma_0)\} \log \frac{\Lambda_{T\eta_{2,T,u}}(1) - \Lambda_{T\eta_{2,T,u}}(\gamma_{T,u})}{\Lambda_{T\eta_{1,T,u}}(\gamma_{T,u})}
\end{aligned}$$

and by an expansion of $\Lambda_{\eta_{T,u}}$, as T tends to infinity, it has the expansion

$$\begin{aligned}
\widetilde{W}_T(u) = \nu_T(\gamma_0) \frac{u_{\eta_1}^T \dot{\Lambda}_{\eta_{01}}(\gamma_0)}{\Lambda_{01}(\gamma_0)} \\
+\{\nu_T(1) - \nu_T(\gamma_0)\} \frac{u_{\eta_2}^T \{\dot{\Lambda}_{\eta_{02}}(1) - \dot{\Lambda}_{\eta_{02}}(\gamma_0)\}}{\Lambda_{02}(1) - \Lambda_{02}(\gamma_0)} \\
-T\{M_T(\gamma_{T,u_\gamma}) - M_T(\gamma_0)\} \log \frac{\Lambda_{02}(1) - \Lambda_{02}(\gamma_0)}{\Lambda_{01}(\gamma_0)} + o_p(1).
\end{aligned}$$

The empirical process ν_T converges weakly to the transformed Brownian motion $B \circ \Lambda_0$ and the process $T\{M_T(\gamma_{T,u_\gamma}) - M_T(\gamma_0)\}$ satisfies Billingsley's tightness criterion (15.21), it is centered and its variance is

$$TE_0\{\nu_T(\gamma_{T,u_\gamma}) - \nu_T(\gamma_0)\}^2 = T\{\Lambda_{T0}(\gamma_{T,u_\gamma}) - \Lambda_{T0}(\gamma_0)\} = u_\gamma \lambda_0(\gamma_0) + o(1),$$

therefore $T\{M_T(\gamma_{T,u_\gamma}) - M_T(\gamma_0)\}$ converges weakly to a transformed Brownian motion with the variance function $u_\gamma \lambda_{01}(\gamma_0)$. It follows that the process $\widetilde{W}_T$ converges weakly on $\mathcal{U}_T^A$ to a centered Gaussian process. By the independence of the increments of the process ν_T, the covariance of $\nu_T(\gamma_0)$ and $\nu_T(\gamma_{T,u}) - \nu_T(\gamma_0)$ is zero if $\gamma_{T,u} > \gamma_0$, otherwise it is the variance of $\nu_T(\gamma_{T,u}) - \nu_T(\gamma_0)$ which converges to zero. In the same way, the covariance of $\nu_T(1) - \nu_T(\gamma_0)$ and $\nu_T(\gamma_{T,u}) - \nu_T(\gamma_0)$ is zero if $\gamma_{T,u} < \gamma_0$, otherwise it converges to zero. Then $\nu_T(\gamma_0)$, $\nu_T(\gamma_{T,u}) - \nu_T(\gamma_0)$ and $\nu_T(\gamma_{T,u}) - \nu_T(\gamma_0)$ are asymptotically independent.

The weak convergence of $\widehat{u}_{T,\eta} = T^{\frac{1}{2}}(\widehat{\eta}_T - \eta_0)$ to a centered Gaussian variable and the convergence rate of $\gamma_{T,u}$ to γ_0 imply the weak convergence of the process

$$\begin{aligned}\widetilde{W}_T(\widehat{u}_T) &= \nu_T(\gamma_0)\frac{\widehat{u}_{T,\eta_1}^T \dot{\Lambda}_{\eta_{01}}(\gamma_0)}{\Lambda_{01}(\gamma_0)} \\ &+\{\nu_T(1) - \nu_T(\gamma_0)\}\frac{\widehat{u}_{T,\eta_2}^T \{\dot{\Lambda}_{\eta_{02}}(1) - \dot{\Lambda}_{\eta_{02}}(\gamma_0)\}}{\Lambda_{02}(1) - \Lambda_{02}(\gamma_0)} \\ &-T^{\frac{1}{2}}\{\nu_T(\widehat{\gamma}_T) - \nu_T(\gamma_0)\}\log\frac{\Lambda_{02}(1) - \Lambda_{02}(\gamma_0)}{\Lambda_{01}(\gamma_0)} + o_p(1),\end{aligned}$$

the variance of the process $T^{\frac{1}{2}}\{\nu_T(\widehat{\gamma}_T) - \nu_T(\gamma_0)\}$ is $\lambda_{01}(\gamma_0)E_0\widehat{u}_T$ and it is bounded on $\mathcal{U}_T^A$.

Theorem 5.6. *The variable $T(\widehat{\gamma}_T - \gamma_0)$ is asymptotically independent of $\widehat{\eta}_T$ and it converges weakly to the location of the location of the maximum of an uncentered Gaussian process.*

Proof. Let u in $\mathcal{U}_T^A$, its estimator $\widehat{u}_T = T(\widehat{\gamma}_T - \gamma_0)$ maximizes the process $X_T(\theta_{T,u})$ with respect to u_γ and it is bounded in probability under P_0, by Theorem 5.5. and Gaussian. At $\gamma_{T,u}$, with u in $\mathcal{U}_T^A$, the process X_T has an approximation based on the expansions

$$\begin{aligned}N_T(\gamma_0)\log\frac{\Lambda_{T\eta_1}(\gamma)}{\Lambda_{T01}(\gamma_0)} &- \{\Lambda_{T\eta_1}(\gamma) - \Lambda_{T01}(\gamma_0)\} \qquad (5.16)\\ &= T^{-\frac{1}{2}}M_T(\gamma_0)\frac{u_{\eta_1}^T \dot{\Lambda}_{T\eta_{01}}(\gamma_0)}{\Lambda_{T01}(\gamma_0)} - \frac{\{u_{\eta_1}^T \dot{\Lambda}_{T\eta_{01}}(\gamma_0)\}^2}{2T\Lambda_{T01}(\gamma_0)}\{1 + o_p(1)\},\end{aligned}$$

$$\begin{aligned}\{N_T(1) - N_T(\gamma_0)\}\log &\frac{\Lambda_{T\eta_2}(1) - \Lambda_{T\eta_2}(\gamma)}{\Lambda_{T02}(1) - \Lambda_{T02}(\gamma)} \\ &-(\Lambda_{T\eta_2} - \Lambda_{T02})(1) + (\Lambda_{T\eta_2} - \Lambda_{T02})(\gamma_0) \qquad (5.17)\\ &= T^{-\frac{1}{2}}\{M_T(1) - M_T(\gamma_0)\}\frac{u_{\eta_2}^T\{\dot{\Lambda}_{T\eta_{02}}(1) - \dot{\Lambda}_{T\eta_{02}}(\gamma_0)\}}{\Lambda_{T02}(1) - \Lambda_{T02}(\gamma_0)} \\ &- \frac{u_{\eta_2}^{T2}\{\dot{\Lambda}_{T\eta_{02}}(1) - \dot{\Lambda}_{T\eta_{02}}(\gamma_0)\}^2}{2T\{\Lambda_{T02}(1) - \Lambda_{T02}(\gamma_0)\}}\{1 + o_p(1)\},\end{aligned}$$

the first term of these expansions are centered and their variances are $O_p(T^{-1})$. For $k = 1, 2$, the function $\Lambda_{T\eta_{k,T,u}} - \Lambda_{T0k}$ is approximated by $T^{-\frac{1}{2}}u_{\eta_k}^T\dot{\Lambda}_{0k} + o(T^{-\frac{1}{2}})$, then the mean and the variance of the left-hand terms of (5.16) and (5.17) are $O(T^{-1}\|u\|_\eta^2)$. The remainder term of $X_T(\theta)$

is

$$
\begin{aligned}
X_{3T}(\theta) &= \{N_T(\gamma_0) - N_T(\gamma)\} \log \frac{\Lambda_{T\eta_2}(1) - \Lambda_{T\eta_2}(\gamma)}{\Lambda_{T\eta_1}(\gamma)} \\
&\quad -\{\Lambda_{T\eta_1}(\gamma) + \Lambda_{T\eta_2}(\gamma) - \Lambda_{T\eta_1}(\gamma_0) - \Lambda_{T\eta_2}(\gamma_0)\} \qquad (5.18) \\
&= \{M_T(\gamma_0) - M_T(\gamma)\} \log \frac{\Lambda_{T02}(1) - \Lambda_{T02}(\gamma_0)}{\Lambda_{T01}(\gamma_0)} \\
&\quad -\{\Lambda_{T01}(\gamma) + \Lambda_{T02}(\gamma) - \Lambda_{T01}(\gamma_0) - \Lambda_{T02}(\gamma_0)\}\{1 + o(1)\} \\
&\quad + o_p(T^{-\frac{1}{2}})
\end{aligned}
$$

its mean and its variance are $O_p(T^{-1}u_\gamma)$. The process TX_T is therefore approximated by the sum of (5.16) and (5.17) which do not depend on the parameter γ and are asymptotically independent of (5.18). As $\widehat{u}_{T,\gamma}$ maximizes X_{3T}, it is asymptotically independent of $\widehat{u}_{T,\eta} = T^{\frac{1}{2}}(\widehat{\eta}_T - \eta_0)$ which maximizes the sum of (5.16) and (5.17) in $\mathcal{U}_T^A$.

The process $TX_{3T}(\theta_{T,u_\gamma})$ converges weakly to an uncentered Gaussian process with a finite variance on $\mathcal{U}_T^A$ and the result follows. □

5.4 Likelihood ratio test for parametric processes

Let us consider the hypothesis H_0 of an intensity $\lambda_0 = \lambda_{\eta_0}$ depending on a single parameter against the alternative of an intensity with a change of parameter at an unknown time γ according to the model (5.9). Under H_0, $\gamma_0 = 1$ and the estimators of the intensities converge to λ_0, from the estimating equations. The parameters under the alternative are denoted $\gamma_{Tu} = 1 - T^{-1}u$, $\eta_{Tk} = \eta_0 + T^{-\frac{1}{2}}h_{Tk}$ and $\lambda_{Tk}(t) = \lambda_{\eta_{Tk}}(t)$, $k = 1, 2$. The logarithm of the likelihood ratio for the processes $N_T(\gamma_T)$ and $N_T(1) - N_T(\gamma_T)$ is

$$
\begin{aligned}
l_T(\theta) - l_{T0} = N_T(\gamma_{T,u}) \log \frac{\Lambda_{T1}(\gamma_{T,u})}{\Lambda_{T0}(1)} \\
+\{N_T(1) - N_T(\gamma_{T,u})\} \log \frac{\Lambda_{T2}(1) - \Lambda_{T2}(\gamma_{T,u})}{\Lambda_{T0}(1)} \\
+\{\Lambda_{T1}(1) - \Lambda_{T0}(\gamma_{T,u}) - \Lambda_{T2}(1) + \Lambda_{T2}(\gamma_{T,u})\},
\end{aligned}
$$

a second order expansion of the logarithm and the expansions $T^{\frac{1}{2}}\{\Lambda_{T1}(1) - \Lambda_{T0}(\gamma_{T,u})\} = (h_{T1} - h_{T0})^T \dot{\Lambda}_0(1) + o(1)$ and

$$
T\{\Lambda_{T0}(1) - \Lambda_{T1}(\gamma_{T,u})\} = T\{\Lambda_{T0}(1) - \Lambda_{T1}(1)\} - u\lambda_0(1) + o(1)
$$

imply

$$TN_T(\gamma_{T,u})\log\frac{\Lambda_{T1}(\gamma_{T,u})}{\Lambda_{T0}(1)} - T\{\Lambda_{T0}(\gamma_{T,u} - \Lambda_{T1}(1))\}$$
$$= -\frac{\{(h_1-h_0)^T\dot\Lambda_0(1)\}^2}{2\Lambda_{T0}(1)} + o_p(1).$$

The process

$$\{N_T(1)-N_T(\gamma_{T,u})\}\log\{\Lambda_{T2}(1)-\Lambda_{T2}(\gamma_{T,u})\}+\Lambda_{T2}(1)-\Lambda_{T2}(\gamma_{T,u})$$

is asymptotically equivalent to $\{\Lambda_{T2}(1)-\Lambda_T(\gamma_{T,u})\}\log\{\Lambda_{T2}(1)-\Lambda_{T2}(\gamma_{T,u})\}$ and it diverges as γ_T converges to 1 so a test of the hypothesis H_0 cannot be performed with a restriction of the process to the observations of $N_T(\gamma_T)$ and $N_T(1)-N_T(\gamma_T)$. The complete log-likelihood ratio of the process is

$$X_T(\theta,\eta_0) = l_T(\theta) - l_{T0}(\eta_0) = \int_0^\gamma \log\frac{\lambda_{T\eta_1}}{\lambda_{T\eta_0}}\,dN_T + \int_\gamma^1 \log\frac{\lambda_{T\eta_2}}{\lambda_{T\eta_0}}\,dN_T$$
$$+\{\Lambda_{T0}(1)-\Lambda_{T1}(\gamma_{T,u})+\Lambda_{T2}(1)-\Lambda_{T2}(\gamma_{T,u})\},$$

up to a constant term, and the test statistic is

$$S_T = T2\{l_T(\widehat\theta_T) - l_{T0}(\widehat\eta_{T0})\}.$$

The process X_T converges uniformly in probability under H_0 to the function

$$X(\theta,\eta_0) = \int_0^\gamma \log\frac{\lambda_{T\eta_1}}{\lambda_{T\eta_0}}\,d\lambda_0 + \int_\gamma^1 \log\frac{\lambda_{T\eta_2}}{\lambda_{T\eta_0}}\,d\lambda_0$$

which is concave with respect to $\lambda_{T\eta_1}$ and $\lambda_{T\eta_2}$ and by the estimators $\widehat\eta_{Tk}$ maximizing the likelihood of the process N are consistent. Their convergence rate is still $T^{-\frac12}$ due to the weak convergence of the variable $T^{\frac12}\dot X_{T,\eta}(\theta_0)$ to a centered Gaussian variable with variance $-\ddot X_\eta(\theta_0)$. The properties of Section 5.3 are still satisfied. Let $\widehat h_{Tk} = T^{\frac12}(\widehat\eta_{Tk}-\eta_k)$, for $k=0,1,2$.

Proposition 5.4. *The statistic S_T converges weakly under H_0 to a χ^2_d variable.*

Proof. The statistic is the value at $\widehat\theta_T$ and $\widehat\eta_{T0}$ of the process $2s_T(\theta,\eta_0)$ with

$$s_T(\theta_T,\eta_{T0}) = T\int_0^{\gamma_T} \log\frac{\lambda_{T1}}{\lambda_{T0}}\,dM_T + T\int_{\gamma_T}^1 \log\frac{\lambda_{T2}}{\lambda_{T0}}\,dM_T$$
$$+T\int_0^{\gamma_T}\Big[\lambda_0\log\frac{\lambda_{T1}}{\lambda_{T0}} - \{\lambda_{T1}-\lambda_{T0}\}\Big]$$
$$+T\int_{\gamma_T}^1\Big[\lambda_0\log\frac{\lambda_{T2}}{\lambda_{T0}} - \{\lambda_{T2}-\lambda_{T0}\}\Big] \quad (5.19)$$

where $T^{\frac{1}{2}}(\eta_{T1} - \eta_{T0}) = h_{T1} - h_{T0}$ converges to a limit $h_1 - h_0$ under H_0 and $T^{\frac{1}{2}}\{\lambda_{T1}(1) - \lambda_{T0}(1)\} = (h_{T1} - h_{T0})^T \dot{\lambda}_{T\eta_1,0}(1) + o(1)$. The first term of the expression of $s_T(\theta_T, \eta_{T0})$ is asymptotically equivalent to

$$T^{\frac{1}{2}} \int_0^1 (h_{T1} - h_{T0})^T \dot{\lambda}_{T\eta_1,0} \lambda_0^{-1} \, dM_T,$$

at the estimator values, $\widehat{h}_{T0} - \widehat{h}_{T1} = T^{\frac{1}{2}} \dot{X}_{T,\eta_1}(\theta_0)^T I_{01}^{-1} + o_p(1)$, where

$$T^{\frac{1}{2}} \dot{X}_{T,\eta_1}(\theta_0) = T^{\frac{1}{2}} \int_0^1 \frac{\dot{\lambda}_{T\eta_1,0}}{\lambda_{T\eta_0}} \, dM_T,$$

$$I_{01} = \int_0^1 \dot{\lambda}^{\otimes 2}_{\eta_1,0}(s) \lambda_0^{-1}(s) \, ds.$$

The variable $T^{\frac{1}{2}} \dot{X}_{T,\eta_1}(\theta_0)$ converges weakly to a centered Gaussian variable with variance I_{01} hence the first term is asymptotically equivalent to

$$T \dot{X}_{T,\eta_1}(\theta_0)^T I_{01}^{-1} \dot{X}_{T,\eta_1}(\theta_0)$$

and it converges weakly to a χ^2_d variable.

By the weak convergence of the process $T\{M_T(\gamma_{T,u_\gamma}) - M_T(\gamma_0)\}$ and the convergence rate of $\widehat{\eta}_{T0} - \widehat{\eta}_{T1}$, the second term of the expression of $s_T(\theta_T, \eta_{T0})$ converges in probability to zero. By a second order expansion of the logarithm, the third term is asymptotically equivalent to $\frac{1}{2}\{(\widehat{h}_{T0} - \widehat{h}_{T1})^T \dot{\lambda}_0(1)\}^2 \lambda_0^{-1}(1)$ and the last term converges in probability to zero.

Finally, the statistic S_T is asymptotically equivalent to

$$T^{\frac{1}{2}} \dot{X}_{\eta_1}(\theta_0)^T I_{01}^{-1} T^{\frac{1}{2}} \int_0^1 \dot{\lambda}_{\eta_1} \lambda_0^{-1} \, dM_T$$

and its limit follows. □

Under a fixed alternative with γ in $]0,1[$ and distinct parameters λ_1 and λ_2, one of them is different from λ_0 and the test statistic S_T tends to infinity.

Proposition 5.5. *Under local alternatives with parameters θ_T converging to θ_0 and such that the limit of $T^{\frac{1}{2}} \rho(\theta_T, \theta_0)$ converges to a non-zero limit, the statistic S_T converges weakly to an uncentered χ^2_d.*

Proof. Under the local alternatives, the process $2s_T$ is the sum of a process defined at the estimators centered at θ_T, converging to a χ^2_d, and a process at the parameter θ_T, asymptotically equivalent to $\{(h_0 - h_1)^T \dot{\lambda}_0(1)\}^2 \lambda_0^{-1}(1)$. □

5.5 Counting process with right censoring

On a probability space $(\Omega, \mathcal{A}, P)$, let X and C be two independent positive random variables such that $P(X < C)$ is strictly positive, and let

$$T = X \wedge C, \quad \delta = 1\{X \leq C\}$$

denote the observed variables when X is right-censored by C. Let F be the distribution function of the variable X, let $\bar{F}^{-}(x) = P(X \geq x)$ be its survival function and let $\bar{F}(x) = 1 - F(x)$. For the distribution function G of the censoring variable C, the survival function is denoted $\bar{G}$. The survival function $\bar{H} = \bar{G}\bar{F}$ of the variable T has the point is $\tau_H = \sup\{t : \bar{H}(t) > 0\} = \tau_F \wedge \tau_G$ where $\tau_F = \sup\{t : \bar{F}(t) > 0\}$ and $\tau_G = \sup\{t : \bar{G}(t) > 0\}$. The cumulated hazard function related to F is defined for every $t < \tau$ by

$$\Lambda(t) = \int_0^T \frac{dF}{\bar{F}^{-}},$$

conversely

$$\bar{F}(t) = \exp\{-\Lambda^c(t)\} \prod_{s \leq t} \{1 - \Delta\Lambda(s)\}.$$

Let us consider an independent and identically distributed sample of right-censored variables and censoring indicators, $(T_i, \delta_i)_{i \leq n}$. Let $\mathbb{F} = (\mathbb{F}_t)_{t \in \mathbb{R}_+}$ denote the history generated by the observations before t, $\mathbb{F}_t$ is generated by the events $\{\delta_i 1\{T_i \leq s\}, 1\{T_i \leq s\}; 0 < s \leq t, i = 1, \ldots, n\}$. Let

$$N_n(t) = \sum_{1 \leq i \leq n} \delta_i 1\{T_i \leq t\}$$

the number of observations before t and

$$Y_n(t) = \sum_{1 \leq i \leq n} 1\{T_i \geq t\}$$

the number of individuals at risk at t. The estimation relies on a martingale property.

Proposition 5.6. *On a space $(\Omega, \mathcal{A}, P, \mathbb{F})$, let M be a square integrable martingale, there exists an unique increasing and predictable process $< M >$ such that $M^2 - < M >$ is a martingale.*

The process $< M >$ is the predictable variation process of M^2. From proposition 5.6, for a square integrable martingale, $EM_t^2 = E < M >_t$ and for every $0 < s < t$, $E(M_t - M_s)^2 = EM_t^2 - EM_s^2 = E(< M >_t - < M >_s)$. In the following, when the variables are not supposed to have a density, tho centered martingale $M = N - \Lambda$ has a predictable variation process $< M >= \Lambda$.

The predictable compensator of N_n is $\widetilde{N}_n(t) = \int_0^T Y_n(s)\, d\Lambda(s)$ and the function Λ is consistently estimated by the integral of Y_n^{-1} with respect to N_n for every t such that $Y_n(t) > 0$. If $t < \tau$, the cumulated hazard function is written as

$$\Lambda(t) = \int_0^T \frac{\bar{G}\, dF}{\bar{H}^-}$$

and Nelson's estimator of Λ is

$$\begin{aligned}\widehat{\Lambda}_n(t) &= \int_0^T \frac{1\{Y_n(s) > 0\}}{Y_n(s)}\, dN_n(s) \\ &= \sum_{1 \le i \le n} \delta_i \frac{1\{Y_n(X_i) > 0\}}{Y_n(X_i)} 1\{X_i \le t\},\end{aligned} \tag{5.20}$$

with the convention 0/0. The consistency of the estimator $\widehat{\Lambda}_n$ is a consequence of the convergence of the empirical processes $n^{-1}N_n$ and $n^{-1}Y_n$. Its weak convergence is proved by Rebolledo's theorem (1978) for the weak convergence of the L^2 local martingales related to jump processes.

Theorem 5.7. *On every interval* $[0, a]$ *such that* $a < T_{n:n}$, *the process* A_n *defined by* $n^{\frac{1}{2}}(\widehat{\Lambda}_n - \Lambda)$ *converges weakly to a centred Gaussian process* B, *with independent increments and having the finite covariance function* C.

Proof. Let $\widetilde{N}_n(t) = \int_0^T 1\{Y_n > 0\} Y_n(s)\, d\Lambda(s)$ and $\bar{M}_n = N_n - \widetilde{N}_n$. For every $t < T_{n:n}$, we have $Y_n(t) > 0$ and

$$A_n(t) = n^{\frac{1}{2}}(\widehat{\Lambda}_n - \Lambda)(t) = n^{\frac{1}{2}} \int_0^T \frac{1}{Y_n}\, d\bar{M}_n,$$

A_n is a $(P, \mathbb{F})$-martingale on $[0, a]$ with predictable compensator

$$< A_n > (t) = n \int_0^T \frac{1}{Y_n} (1 - \Delta\Lambda)\, d\Lambda$$

which converges in probability to $\int_0^t (\bar{H}^-)^{-1}(1 - \Delta\Lambda)\, d\Lambda$. The size the jumps of A_n^c at t converges to 0 since $n^{-1}Y_n(t)$ is bounded. The weak convergence of the process A_n is a consequence of Rebolledo's convergence theorem. □

Let $0 < T_{n:1} \le \ldots \le T_{n:n}$ be the ordered sequence of the observation times T_i and $\delta_{n:k}$ be the censoring indicator for the time $T_{n:k}$. The distribution function H of the sample $(T_i)_{i\leq n}$ has the empirical estimator $H_n(t) = \frac{1}{n}\sum_{i=1}^n 1\{T_i \leq t\}$, this estimator is modified for the estimation of F under censoring. The classical estimator of the survival function $\bar{F} = 1 - F$ is the product-limit Kaplan–Meier estimator, it is right-continuous with left-hand limits step function, constant between two observations times such that $\delta_i = 1$,

$$\widehat{\bar{F}}_n(T_i) = \prod_{j:T_j\leq T_i} \frac{Y_n(T_j)-1}{Y_n(T_j)}.$$

The estimator $\widehat{\bar{F}}_n$ only depends on the censoring variables through the size of of its jumps and it equals $1 - H_n$ when there is no censoring. At every time t, it is also expressed with the jumps $\Delta\widehat{\Lambda}_n$ of $\widehat{\Lambda}_n$

$$\begin{aligned}
\widehat{\bar{F}}_n(t) &= \prod_{T_{n:k}\leq t} \{1 - \Delta\widehat{\Lambda}_n(T_{n:k})\} \\
&= \widehat{\bar{F}}_n(t^-)\{1 - \Delta\widehat{\Lambda}_n(t)\} \\
&= \prod_{T_{n:k}\leq t} \left\{1 - \frac{1}{Y_n(T_{n:k})}\right\}^{\delta_{n:k}} \\
&= \prod_{T_{n:k}\leq t} \left\{1 - \frac{\delta_{n:k}}{n-k+1}\right\}.
\end{aligned}$$

For every $t \geq T_{n:n}$, $\widehat{\bar{F}}_n(t) = 0$ if $\delta_{n:n} = 1$ is and $\widehat{\bar{F}}_n(t)$ is strictly positive if $T_{n:n}$ is a censoring variable. The consistency at t such that $Y_n(t) > 0$ is a consequence of the martingale properties. For s and t in $[0, a]$, let

$$C(s,t) = \int_0^{s\wedge t} (\bar{H}^-)^{-1}(1 - \Delta\Lambda)\, d\Lambda.$$

The Gaussian limiting process B jumps at the discontinuities of F. On the sub-intervals $[0, a]$ of $[0, T_{n:n}]$, the weak convergence of $n^{\frac{1}{2}}(\widehat{F}_n - F)$ to a Gaussian process has been proved by Breslow and Crowley (1974) who established a bound for the difference between $\widehat{\Lambda}_n$ and $-\log\{\widehat{\bar{F}}_n\}$.

Lemma 5.4. *If* $t < T_{n:n}$,

$$\lim_{n\to\infty} n^{\frac{1}{2}} \sup_{[0,a]} |-\log\{\widehat{\bar{F}}_n(t)\} - \widehat{\Lambda}_n(t)| = 0$$

in probability.

Proof. Tho proof are adapted from Breslow and Crowley (1974). For every $x > 0$

$$0 < -\log\left\{1 - \frac{1}{1+x}\right\} - \frac{1}{1+x} < \frac{1}{x(1+x)}.$$

Let $y = \frac{1}{1+x}$, this assertion is obtained from the derivative of $-\log(1-y)$ written $\frac{1}{1-y} = 1 + \frac{y}{1-y}$ therefore

$$-\log(1-y) = y + \int_0^y \frac{u}{(1-u)^2}\,du < y + \frac{y^2}{(1-y)^2}.$$

If $t < T_{n:n}$ and $Y_n(t) > 1$, this bound at $x = Y_n(t) - 1$ implies

$$\begin{aligned}
0 &< \sum_{k\le n} 1\{T_{n:k} \le t\}\Big[-\log\Big\{1 - \frac{\delta_{n:k}}{Y_n(T_{n:k})}\Big\} - \frac{\delta_{n:k}}{Y_n(T_{n:k})}\Big] \\
&< \sum_{k\le n} 1\{T_{n:k} \le t\}\frac{\delta_{n:k}}{(n-k)(n-k+1)}, \text{ since } Y_n(T_{n:k}) = n-k+1, \\
&= \sum_{k=1}^{N_n(t)} \frac{1}{(n-k)(n-k+1)} < \sum_{k=1}^{N_n(t)} \frac{1}{(n-k)^2} \\
&< \int_0^{N_n(t)} \frac{1}{(n-y)^2}\,dy = \frac{N_n(t)}{n(n-N_n(t))}.
\end{aligned}$$

It follows that

$$0 < -\log\{\widehat{\bar{F}}_n(t)\} - \widehat{\Lambda}_n(t) < \frac{N_n(t)}{n(n-N_n(t))}.$$

□

Let $B_n = n^{\frac{1}{2}}(\widehat{F}_n - F)$.

Theorem 5.8. *On every interval $[0,a]$ such that $a < T_{n:n}$ the process B_n converges weakly to a centered process B.*

Proof. On any interval $[0,a]$ such that $a < T_{n:n}$

$$\begin{aligned}
B_n(t) &= -n^{\frac{1}{2}}[e^{-\widehat{\Lambda}_n(t)} - e^{-\Lambda(t)}] - n^{\frac{1}{2}}[\exp\{\log \widehat{F}_n(t)\} - e^{-\widehat{\Lambda}_n(t)}] \\
&= -e^{-\Lambda(t)}\, n^{\frac{1}{2}}\{\widehat{\Lambda}_n(t) - \Lambda(t)\} - e^{-\Lambda_n^*(t)}\, n^{\frac{1}{2}}\{\Lambda_n^*(t) - \Lambda(t)\}^2 \\
&\quad + n^{\frac{1}{2}} e^{-\Lambda_n^{**}(t)}\{-\log(\widehat{\bar{F}}_n(t)\} - \widehat{\Lambda}_n(t)\},
\end{aligned}$$

where $\Lambda_n^*(t)$ is between $\Lambda(t)$ and $\widehat{\Lambda}_n(t)$ and $\Lambda_n^{**}(t)$ is between $\log \widehat{\bar{F}}_n(t)$ and $\widehat{\Lambda}_n(t)$. As $n^{\frac{1}{2}} \sup_{t\in[0,a]} |\widehat{\Lambda}_n(t) - \Lambda(t)|$ converges weakly, the first two terms converge to zero in probability, the last term converges to zero in probability by Lemma 5.4. The equality $e^{-\Lambda(t)} = \bar{F}(t)$ ends the proof. □

The asymptotic covariance of the process $\bar{F}^{-1}B_n$ is $C(s \wedge t)$ at s and t. The asymptotic distribution B of the process B_n depends on the functions F, $\bar{F}$ and $\bar{G}$ through the covariances $C(s \wedge t)$. Estimating the variance $C(t)$ by

$$\widehat{C}_n(t) = n \int_0^T \frac{1}{Y_n(Y_n - 1)} \, dN_n,$$

an uniform confidence interval for F over an interval $[0, a]$ is deduced from the quantiles of $\sup_{t \in [0, \widehat{C}_n(a)]} |W(t)|$, where W is the centred Gaussian process with covariance $s \wedge t$ at s and t. The weak convergence of B_n has been extended by Gill (1983) to the interval $[0, T_{n:n}]$, it relies on the following proposition which expresses B_n as a martingale up to the $(P, \mathbb{F})$-stopping time $T_{n:n}$.

Theorem 5.9. *For $t < \tau_F$, if $\int_0^{\tau_F} \bar{H}^{-1} \, d\Lambda$ is finite*

$$\widehat{\Lambda}_n(t \wedge T_{n:n}) = \int_0^{t \wedge T_{n:n}} \frac{d\widehat{F}_n(s)}{\widehat{\bar{F}}_n^-(s)}, \quad \widehat{F}_n(t) = \int_0^T \widehat{\bar{F}}_n^-(s) \, d\widehat{\Lambda}_n(s),$$

$$\frac{F - \widehat{F}_n}{1 - F}(t) = \int_0^{t \wedge T_{n:n}} \frac{1 - \widehat{F}_n(s^-)}{1 - F(s)} \{d\widehat{\Lambda}_n(s) - d\Lambda(s)\}$$

and $n^{\frac{1}{2}}(F - \widehat{F}_n)\bar{F}^{-1}$ converges weakly to B_F, a centred Gaussian process on $[0, \tau_F[$, with covariances

$$K(s,t) = \int_0^{s \wedge t} (\bar{F}^{-1}\bar{F}^-)^2 \, dC.$$

This is a consequence of the uniqueness of a locally bounded solution of the equation

$$Z(t) = 1 - \int_0^T \frac{Z^-}{1 - \Delta B}(dA - dB), \; \Delta A \leq 1, \Delta B < 1$$

in the form

$$Z(t) = \frac{\prod_{s \leq t}(1 - \Delta A(s)) \exp\{-A^c(t)\}}{\prod_{s \leq t}(1 - \Delta B(s)) \exp\{-B^c(t)\}},$$

with $A = \widehat{\Lambda}_n$, $B = \Lambda$ and $Z = (1 - \widehat{F})/(1 - F)$, (Doléans–Dade, 1970).

5.6 Change-point in a hazard function

On a probability space $(\Omega, \mathcal{F}, P_0)$, let T^0 be a time variable under an independent and non-informative right-censoring at a random time C. The observed variables are the censored time $T = T^0 \wedge C$ and the censoring indicator $\delta = 1\{T^0 \leq C\}$. The hazard function λ_0 of T^0 is also the hazard function of T and the distribution function of T^0 is defined as $F_0 = \exp\{-\Lambda_0\}$ with the cumulative hazard function $\Lambda_0(t) = \int_0^T \lambda_0(s), ds$. Equivalently, the hazard function of T^0 is defined from its density function f_0 as $\lambda_0 = \bar{F}_0^{-1} f_0$, with the survival function $\bar{F}_0 = 1 - F_0$. The survival function of the censoring time is denoted $\bar{G}$.

We consider a model for hazard function with a discontinuity at an unknown change-point $\gamma > 0$ under a probability distribution P_γ

$$\lambda(t, \gamma) = \delta_\gamma(t)\lambda_1(t) + \{1 - \delta_\gamma(t)\}\lambda_2(t), \tag{5.21}$$

the parameter $\gamma > 0$ belongs to an open and bounded interval Γ of $\mathbb{R}_+$. Under the probability distribution $P_0 = P_{\lambda_0,\gamma_0}$ of the observations, the intensity has a change-point at γ_0 such that $P(X \leq \gamma_0)$ is different from 0 and 1 and λ is discontinuous at γ_0, with $\lambda_{01}(\gamma_0) \neq \lambda_{02}(\gamma_0)$. The functions λ_1 and λ_2 are supposed to be continuous and regular. In parametric models, they are indexed by distinct parameters θ_1 and θ_2 of open and bounded parameter sets Θ_1 and Θ_2, and regular with respect to their parameter. Regular nonparametric hazard functions will also be considered.

The log-likelihood for a n-sample $(T_i, \delta_i)_{i=1,\dots,n}$ of (T, δ) is expressed with the counting processes of the uncensored observations and of the individual at risk before and after the change-point. Under a probability $P_{\lambda,\gamma}$ and with an interval of observations $[0, \tau]$, $\tau < \tau_H$, let

$$N_{1,\gamma,n}(t) = N_n(t \wedge \gamma) = \sum_{i=1}^n \delta_i 1_{\{T_i \leq t \wedge \gamma\}},$$

$$N_{2,\gamma,n}(t) = N_n(t) - N_n(\gamma) = \sum_{i=1}^n \delta_i 1_{\{\gamma < T_i \leq t\}}$$

and let

$$Y_{1,\gamma,n}(t) = Y_n(t) - Y_n(\gamma) = \sum_{i=1}^n 1_{\{t \leq T_i < \gamma\}},\ t < \gamma,$$

$$Y_{2,\gamma,n}(t) = Y_n(t) = \sum_{i=1}^n 1_{\{t \leq T_i\}},\ t \geq \gamma$$

be the restrictions of N_n and Y_n to $[0,\gamma]$ and, respectively, $]\gamma,\tau]$. Let $\mathcal{F}_t$ is the σ-algebra generated by N_n and Y_n up to t and let $\mathbb{F} = (\mathcal{F}_t)_{t\geq 0}$. The cumulative hazard functions

$$\Lambda_{1,\gamma}(t) = \int_0^{t\wedge\gamma} \lambda_1(s)\,ds, \quad \Lambda_{2,\gamma}(t) = \int_\gamma^T \lambda_2(s)\,ds$$

are equal at γ and $\Lambda(t) = \int_0^t \lambda(s)\,ds$ is written as

$$\Lambda(t) = \Lambda_1(t\wedge\gamma) + \Lambda_2(t\vee\gamma) - \Lambda_2(\gamma).$$

The survival function of T^0 under $P_{\lambda,\gamma}$ is deduced as

$$\bar{F}_\gamma(t) = \frac{\bar{F}_1(t\wedge\gamma)\bar{F}_2(t\vee\gamma)}{\bar{F}_2(\gamma)}$$

hence $\bar{F}_\gamma(t) = \{1-\delta_\gamma(t)\}\bar{F}_1(t)+\delta_\gamma(t)\bar{F}_2(t)$ where $\bar{F}_k = \exp(-\Lambda_k)$, $k=1,2$. The functions Λ_1 and Λ_2 estimated separately from the sub-samples on the observation intervals $[0,\gamma]$ and, respectively, $]\gamma,\tau]$, as

$$\begin{aligned}\widehat{\Lambda}_{1,n,\gamma}(t) &= \int_0^{t\wedge\gamma} \frac{1_{\{Y_{1,\gamma,n}>0\}}\,dN_{1,\gamma,n}}{Y_{1,\gamma,n}},\\ \widehat{\Lambda}_{2,n,\gamma}(t) &= \int_\gamma^T \frac{1_{\{Y_{2,\gamma,n}>0\}}\,dN_{2,\gamma,n}}{Y_{2,\gamma,n}}.\end{aligned} \tag{5.22}$$

For nonparametric hazard functions, at a fixed value γ, $\lambda(t)$ is estimated by smoothing the cumulative hazard function with a symmetric kernel K with a bandwidth h_n converging to zero

$$\begin{aligned}\widehat{\lambda}_n(t,\gamma) &= \delta_\gamma(t)\sum_{i=1}^n K_h(T_i-t)\delta_i 1_{\{T_i\leq\gamma\}}\frac{1_{\{Y_{1,\gamma,n}(T_i)>0\}}}{Y_{1,\gamma,n}(T_i)}\\ &\quad+\{1-\delta_\gamma(t)\}\sum_{i=1}^n K_h(T_i-t)\delta_i 1_{\{T_i>\gamma\}}\frac{1_{\{Y_{2,\gamma,n}(T_i)>0\}}}{Y_{2,\gamma,n}(T_i)},\end{aligned}$$

In the parametric case, the intensities under a probability distribution $P_{\eta,\gamma}$ are defined in a model with parameters γ and η as $\lambda_{k,\gamma} = \lambda_{\eta_k,\gamma}$ with a change-point at γ in $]0,1[$, for $k=1,2$, where η_1 and η_2 are distinct parameters of an open subspace $\mathcal{H}$ of $\mathbb{R}^d$ and the functions $\eta_k \mapsto \lambda_{\eta_k,\gamma}(t)$ belongs to $L^2(\mathcal{H})$ uniformly on $[0,\tau]$. Let θ be the parameter with components γ, η_1 and η_2, and let θ_0 be its value under the probability distribution P_0 of the observations.

Under P_θ, the logarithm of the likelihood of the sample is

$$l_n(\eta,\gamma) = \int_0^T \log\lambda_{\eta,\gamma}(t)\,dN_n(t) - \int_0^T Y_n(t)\lambda_{\eta,\gamma}(t)\,dt,$$

let $\widehat{\eta}_{k,n,\gamma}$ be the maximum likelihood estimator of the parameter η_k, for $k = 1, 2$, the parametric hazard function is estimated by

$$\widehat{\lambda}_{n,\gamma} = \delta_\gamma \lambda_{\widehat{\eta}_{1,n,\gamma}} + \{1 - \delta_\gamma\} \lambda_{\widehat{\eta}_{2,n,\gamma}}.$$

The parameter γ is estimated by

$$\widehat{\gamma}_n = \inf\{\gamma; \max\{\widehat{l}_n(\gamma^-), \widehat{l}_n(\gamma)\} = \sup_{s \in]0,\tau[} \widehat{l}_n(s)$$

and the parameters are estimated by $\widehat{\eta}_n = \widehat{\eta}_{n,\widehat{\gamma}_n}$ and $\widehat{\lambda}_n = \widehat{\lambda}_{\widehat{\theta}_n}$.

5.7 Parametric estimators under right-censoring

The predictable compensator of N_n with respect to the filtration $\mathbb{F}$ under P_θ is $\widetilde{N}_n(t) = \int_0^T Y_n(s)\lambda_\gamma(s)$. The log-likelihood ratio of the observations on the interval $[0, \tau]$, under P_θ with respect to P_0, is

$$X_n(\theta) = n^{-1}\{l_n(\theta) - l_n(\theta_0)\}.$$

It splits like $l_n = l_n^- 1_{\{\gamma < \gamma_0\}} + l_n^+ 1_{\{\gamma > \gamma_0\}}$ according to the sign of $\gamma - \gamma_0$

$$\begin{aligned} X_n^-(\theta) &= n^{-1} \int_0^\gamma \log \frac{\lambda_1}{\lambda_{01}} \, dN_n - n^{-1} \int_0^\gamma (\lambda_1 - \lambda_{01}) Y_n \, ds \\ &+ n^{-1} \int_\gamma^{\gamma_0} \log \frac{\lambda_2}{\lambda_{01}} \, dN_n - n^{-1} \int_\gamma^{\gamma_0} (\lambda_2 - \lambda_{01}) Y_n \, ds \\ &+ n^{-1} \int_{\gamma_0}^\tau \log \frac{\lambda_2}{\lambda_{02}} \, dN_n - n^{-1} \int_{\gamma_0}^\tau (\lambda_2 - \lambda_{02}) Y_n \, ds \end{aligned}$$

and the expression of X_n^+ is similar with an inversion of λ_1 and λ_2 between γ and γ_0.

Under P_0, the processes X_n^- and X_n^- are expressed with the local martingales $M_{1n} = N_{1n} - \int_0^{\cdot\wedge\gamma_0} Y_{1n} \, d\Lambda_{01}$ and $M_{2n} = N_{2n} - \int_0^{\cdot\wedge\gamma_0} Y_{2n} \, d\Lambda_{02}$ which are centered and independent

$$\begin{aligned} X_n^-(\theta) &= n^{-1} \int_0^{\gamma_0} \log \frac{\lambda_1}{\lambda_{01}} \, dM_{1n} + n^{-1} \int_0^{\gamma_0} (\lambda_{01} \log \frac{\lambda_1}{\lambda_{01}} - \lambda_1 + \lambda_{01}) Y_n \, ds \\ &+ n^{-1} \int_\gamma^{\gamma_0} \log \frac{\lambda_2}{\lambda_1} \, dM_{1n} + n^{-1} \int_\gamma^{\gamma_0} (\lambda_{01} \log \frac{\lambda_2}{\lambda_1} - \lambda_2 + \lambda_1) Y_n \, ds \\ &+ n^{-1} \int_{\gamma_0}^\tau \log \frac{\lambda_2}{\lambda_{02}} \, dM_{2n} + n^{-1} \int_{\gamma_0}^\tau (\lambda_{02} \log \frac{\lambda_2}{\lambda_{02}} - \lambda_2 + \lambda_{02}) Y_n \, ds \end{aligned}$$

if $\gamma < \gamma_0$, and X_n^+ is similar with $\gamma > \gamma_0$

$$X_n^+(\theta) = n^{-1}\int_0^{\gamma_0} \log\frac{\lambda_1}{\lambda_{01}}\,dM_{1n} + n^{-1}\int_0^{\gamma_0}\Big(\lambda_{01}\log\frac{\lambda_1}{\lambda_{01}} - \lambda_1 + \lambda_{01}\Big)Y_n\,ds$$
$$+ n^{-1}\int_{\gamma_0}^{\gamma} \log\frac{\lambda_1}{\lambda_2}\,dM_{2n} + n^{-1}\int_{\gamma_0}^{\gamma}\Big(\lambda_{02}\log\frac{\lambda_1}{\lambda_2} - \lambda_1 + \lambda_2\Big)Y_n\,ds$$
$$+ n^{-1}\int_{\gamma_0}^{\tau} \log\frac{\lambda_2}{\lambda_{02}}\,dM_{2n} + n^{-1}\int_{\gamma_0}^{\tau}\Big(\lambda_{02}\log\frac{\lambda_2}{\lambda_{02}} - \lambda_2 + \lambda_{02}\Big)Y_n\,ds.$$

The processes Y_n and N_n converge a.s. uniformly over R_+ under P_θ and their limits are expressed by the distribution functions F and G

$$\lim_{n\to\infty}\|n^{-1}Y_n - \bar{G}\bar{F}_\theta\| = 0,$$
$$\lim_{n\to\infty}\|n^{-1}N_n - \textstyle\int_0^{\cdot}\bar{G}\,dF_\theta\| = 0.$$

Assuming that the time variables have a parametric distribution F_θ, we denote $y_\theta = \bar{G}\bar{F}_\theta$ the limit of $n^{-1}Y_n$ under P_θ so that $\int_0^{\cdot}\bar{G}\,dF_\theta = \int_0^{\cdot} y_\theta\,d\Lambda_\theta$. The process $n^{-1}X_n$ converge a.s. uniformly under P_0 to the function

$$X(\theta) = \int_0^{\gamma_0}\Big\{\lambda_{01}\log\frac{\lambda_1}{\lambda_{01}} - (\lambda_1 - \lambda_{01})\Big\}y_{\theta_0}\,ds$$
$$+\int_{\gamma}^{\gamma_0}\Big[\{\lambda_{01}\delta_{\gamma_0}(\gamma) - \lambda_{02}\delta_{\gamma}(\gamma_0)\}\log\frac{\lambda_2}{\lambda_1} - (\lambda_2 - \lambda_1)\Big]y_{\theta_0}\,ds$$
$$+\int_{\gamma_0}^{\infty}\Big\{\lambda_{02}\log\frac{\lambda_2}{\lambda_{02}} - (\lambda_2 - \lambda_{02})\Big\}y_{\theta_0}\,ds.$$

For $\varepsilon > 0$, let $V_\varepsilon(\theta_0)$ be an ε-neighborhood of θ_0 in the parameter space $[0,\tau]\times\mathcal{H}^{\otimes 2}$ endowed with the semi-norm $\rho(\theta,\theta_0) = (|\gamma-\gamma_0| + \|\eta-\eta_0\|^2)^{\frac{1}{2}}$.

Lemma 5.5. *For ε sufficiently small, there exists a constant $\kappa_0 > 0$ such that for every θ in $V_\varepsilon(\theta_0)$, $X(\theta) \le -\kappa_0\rho^2(\theta,\theta_0)$.*

Proof. The first and third terms of X in $V_\varepsilon(\theta_0)$ are $O((\lambda_k - \lambda_{0k})^2)$ since

$$\psi(\lambda_{0k}) = \lambda_{0k}\log\frac{\lambda_k}{\lambda_{0k}} - (\lambda_k - \lambda_{0k}) = -\frac{(\lambda_k - \lambda_{0k})^2}{2\lambda_{0k}} + o((\lambda_k - \lambda_{0k})^2) \quad (5.23)$$

and its second term is a $O(|\gamma - \gamma_0|)$ and it is negative. □

Let $W_n = n^{\frac{1}{2}}(X_n - X)$, we assume that the condition

$$E_0\Big\{\sup_{\theta\in V_\varepsilon(\theta_0)} l_n^2(\theta)\Big\} < \infty. \quad (5.24)$$

Lemma 5.6. *For every $\varepsilon > 0$, there exists $\kappa_1 > 0$ such that for n sufficiently large*

$$E_0 \sup_{\theta\in V_\varepsilon(\theta_0)} |W_n(\theta)| \le \kappa_1\varepsilon.$$

This is a consequence of the boundedness of the functions of the hazard functions in $V_\varepsilon(\theta_0)$ and of the weak convergence of the processes $n^{-\frac{1}{2}} M_{kn}$ and $n^{\frac{1}{2}}(n^{-1}Y_n - y_\gamma)$, uniformly in $V_\varepsilon(\theta_0)$.

The log-likelihood is maximal as all terms have the same order which implies that the the maximum likelihood estimator of γ_0 is such that $\widehat{\gamma}_n - \gamma_0$ has the same order as $\|\widehat{\eta}_n - \eta_0\|^2$.

The parameters η_1 and η_2 are first estimated by maximization of $X_n(\theta)$ as γ is fixed. The first derivative of X_n with respect to the parameters are expressed according to the local martingales under P_0

$$\begin{aligned}
\dot{X}_{n,\eta_1}(\theta) &= n^{-1}\int_0^{\gamma\wedge\gamma_0} \frac{\dot{\lambda}_1}{\lambda_1}\Big\{dM_{1n} + (\lambda_{01} - \lambda_1)Y_n\,ds\Big\} \\
&\quad + 1_{\{\gamma>\gamma_0\}} n^{-1}\Big\{\int_{\gamma_0}^{\gamma} \frac{\dot{\lambda}_1}{\lambda_1}\,dM_{2n} + \int_{\gamma_0}^{\gamma} \frac{\dot{\lambda}_1}{\lambda_1}(\lambda_{02} - \lambda_1)Y_n\,ds\Big\}, \\
\dot{X}_{n,\eta_2}(\theta) &= n^{-1}\int_{\gamma\vee\gamma_0}^{\infty} \Big\{\frac{\dot{\lambda}_2}{\lambda_2}\,dM_{2n} + (\lambda_{02} - \lambda_2)Y_n\,ds\Big\} \\
&\quad + 1_{\{\gamma\le\gamma_0\}} n^{-1}\Big\{\int_{\gamma}^{\gamma_0} \frac{\dot{\lambda}_2}{\lambda_2}\,dM_{1n} + \int_{\gamma}^{\gamma_0} \frac{\dot{\lambda}_2}{\lambda_2}(\lambda_{01} - \lambda_2)Y_n\,d\Lambda_0\Big\},
\end{aligned}$$

and

$$\begin{aligned}
\ddot{X}_{n,\eta_1}(\theta) &= n^{-1}\int_0^{\gamma\wedge\gamma_0} \Big(\frac{\ddot{\lambda}_1}{\lambda_1} - \frac{\dot{\lambda}_1^{\otimes 2}}{\lambda_1^2}\Big)\,dN_{1n} - n^{-1}\int_0^{\gamma\wedge\gamma_0} \ddot{\lambda}_1 Y_n\,ds \\
&\quad + 1_{\{\gamma>\gamma_0\}} n^{-1}\Big\{\int_{\gamma_0}^{\gamma} \Big(\frac{\ddot{\lambda}_1}{\lambda_1} - \frac{\dot{\lambda}_1^{\otimes 2}}{\lambda_1^2}\Big)\,dN_{2n} - \int_{\gamma_0}^{\gamma} \ddot{\lambda}_1 Y_n\,ds\Big\}, \\
\ddot{X}_{n,\eta_2}(\theta) &= n^{-1}\int_{\gamma\vee\gamma_0}^{\infty} \Big(\frac{\ddot{\lambda}_2}{\lambda_2} - \frac{\dot{\lambda}_2^{\otimes 2}}{\lambda_2^2}\Big)\,dN_{2n} - n^{-1}\int_{\gamma\vee\gamma_0}^{\infty} \ddot{\lambda}_2 Y_n\,ds \\
&\quad + 1_{\{\gamma\le\gamma_0\}}\Big\{\int_{\gamma}^{\gamma_0} n^{-1}\Big(\frac{\ddot{\lambda}_2}{\lambda_2} - \frac{\dot{\lambda}_2^{\otimes 2}}{\lambda_2^2}\Big)\,dN_{1n} - \int_{\gamma}^{\gamma_0} \ddot{\lambda}_2 Y_n\,ds\Big\},
\end{aligned}$$

and the second derivative of $X_n(\theta)$ with respect to η_1 and η_2 is zero.

Let $U_n(\theta) = n^{\frac{1}{2}}\dot{X}_n(\theta)$ and let $I_k(\theta)$ be the limit of $\dot{X}_{n,\eta_k}(\theta)$ as n tends to infinity, for $k = 1, 2$. Under P_0 the variable $U_n(\theta_0)$ is a centered and it converges weakly to a Gaussian variable U_0 with a $2d$-dimensional variance matrix I_0 with block diagonals $I_1(\theta_0)$ and $I_2(\theta_0)$, and it is zero elsewhere. The matrices $I_{01} = I_1(\theta_0)$ and $I_{02} = I_2(\theta_0)$ are positive definite and such that $I_{0k} = \int_0^{\gamma_0} \dot\lambda_{0k}^{\otimes 2} \lambda_{0k}^{-1} y_{\gamma_0}\, ds$. At θ they are

$$I_1(\theta) = -\int_0^{\gamma\wedge\gamma_0} \Big(\frac{\ddot\lambda_1}{\lambda_1} - \frac{\dot\lambda_1^{\otimes 2}}{\lambda_1^2}\Big)\lambda_{01} y_\gamma\, ds + \int_0^{\gamma\wedge\gamma_0} \ddot\lambda_1 y_\gamma\, ds$$

$$- 1_{\{\gamma>\gamma_0\}}\Big\{\int_{\gamma_0}^{\gamma} \Big(\frac{\ddot\lambda_1}{\lambda_1} - \frac{\dot\lambda_1^{\otimes 2}}{\lambda_1^2}\Big)\lambda_{02} y_\gamma\, ds - \int_{\gamma_0}^{\gamma} \ddot\lambda_1 y_\gamma\, ds\Big\},$$

$$I_2(\theta) = -\int_{\gamma\vee\gamma_0}^{\infty} \Big(\frac{\ddot\lambda_2}{\lambda_2} - \frac{\dot\lambda_2^{\otimes 2}}{\lambda_2^2}\Big)\lambda_{02} y_\gamma\, ds + \int_{\gamma\vee\gamma_0}^{\infty} \ddot\lambda_2 y_\gamma\, ds$$

$$- 1_{\{\gamma\le\gamma_0\}}\Big\{\int_{\gamma_0}^{\gamma} \Big(\frac{\ddot\lambda_2}{\lambda_2} - \frac{\dot\lambda_2^{\otimes 2}}{\lambda_2^2}\Big)\lambda_{01} y_\gamma\, ds - \int_{\gamma_0}^{\gamma} \ddot\lambda_2 y_\gamma\, ds\Big\}.$$

Theorem 5.10. *Under P_0, the maximum likelihood estimator $\widehat\theta_n$ is a.s. consistent.*

Proof. The log-likelihood ratio is maximum at $\widehat\theta_n$ and by its uniform convergence $0 \le X_n(\widehat\theta_n) \le \sup_\theta |X_n(\theta) - X(\theta)| + X(\widehat\theta_n)$ where $X(\widehat\theta_n) \ge 0$ therefore $X(\widehat\theta_n)$ converges a.s. to zero. As its limit X is zero at θ_0 where it reaches its maximum, $\widehat\theta_n$ converges a.s. to θ_0. □

Let $\mathcal{U}_n = \{u = (u_1, u_2^T)^T; u_1 = n(\gamma - \gamma_0), u_2 = n^{\frac12}(\eta - \eta_0)\colon\ \gamma \in]0,\tau[, \eta \in \mathcal{H}^{\otimes 2}\}$. For every u in $\mathcal{U}_n$, let $\gamma_{n,u} = \gamma_0 + n^{-1}u_1$ and $\eta_{n,u} = \eta_0 + n^{-\frac12}u_2$, and let $\theta_{n,u}$ be the vector with components $\gamma_{n,u}$ and $\eta_{n,u}$. For $\varepsilon > 0$, let $\mathcal{U}_{n,\varepsilon} = \{u \in \mathcal{U}_n : \rho(\theta_{n,u}, \theta_0) \le \varepsilon\}$. The convergence rates of the estimators are deduced from Lemmas 5.5 and 5.6, with the same proof as for the other models.

Theorem 5.11. *For every $\varepsilon > 0$*

$$\overline{\lim}_{n,A\to\infty} P_0\{\sup_{u\in\mathcal{U}_{n,\varepsilon}, \|u\|>A} n^{-1} l_n(\theta_{n,u}) \ge 0\} = 0,$$

$$\overline{\lim}_{n,A\to\infty} P_0\{n|\widehat\gamma_n - \gamma_0| > A) + P_0(n^{\frac12}\|\widehat\eta_n - \eta_0\| > A\} = 0.$$

Proposition 5.7. *Under P_0, the variables $n^{\frac12}(\widehat\eta_{1n} - \eta_1)$ and $n^{\frac12}(\widehat\eta_{2n} - \eta_2)$ are asymptotically independent and they converge weakly to centered Gaussian variables with respective variances I_{0k}^{-1}, for $k = 1, 2$.*

Proof. From the consistency of the estimators and by Rebolledo's weak convergence theorem for the martingales related to counting processes, for every sequence $(\gamma_n)_n$ converging to γ_0, $U_n(\gamma_n, \eta_0)$ converges weakly to a centered Gaussian process with independent increments and variance I_0^{-1}. By a first order expansion of $U_n(\widehat\gamma_n, \widehat\eta_n)$, $n^{\frac12}(\widehat\eta_n - \eta_0) = I_0^{-1} U_n(\widehat\gamma_n, \eta_0) + o(1)$ and the weak convergence of the estimators $\widehat\eta_n$ follows. □

Corollary 5.1. *Under P_0, the processes $\widehat{\lambda}_{1n} = \lambda_{\widehat{\eta}_{1n}}$ and $\widehat{\lambda}_{2n} = \lambda_{\widehat{\eta}_{2n}}$ are asymptotically independent and for $k = 1,2$, $n^{\frac{1}{2}}(\widehat{\lambda}_{kn} - \lambda_{0k})$ converges weakly, to a centered Gaussian process with covariance function $v_k(s,s') = \dot{\lambda}_{0k}^T(s) I_{0k}^{-1} \dot{\lambda}_{0k}(s')$.*

The limiting distributions of the log-likelihood and of the estimators are determined by an asymptotic expansion of $l_n(\widehat{\theta}_n)$ in a neighborhood of $\theta_0 = (\gamma_0, \eta_0)$, according to the convergence rates of the estimators. An uniform approximation of the process defined on $\mathcal{U}_n$ by the map $u \mapsto nX_n(\theta_{n,u})$ splits the log-likelihood ratio into two terms according to the regular parameter η and the change-point parameter γ. The process $X_n(\theta_{n,u})$ is approximated by the sum of a term which do not depend on γ

$$\widetilde{X}_n(u) = n^{-\frac{1}{2}}\Big\{\int_0^{\gamma_0} \frac{u_{\eta,1}^T \dot{\lambda}_{01}}{\lambda_{01}}\, dM_{1n} + \int_{\gamma_0}^{\tau} \frac{u_{\eta,2}^T \dot{\lambda}_{02}}{\lambda_{02}}\, dM_{2n}\Big\}$$

$$-n^{-1}\Big\{\int_0^{\gamma_0} \frac{u_{\eta,1}^T \ddot{\lambda}_{01} u_{\eta,1}}{2\lambda_{01}} Y_{1n}\, ds + \int_{\gamma_0}^{\tau} \frac{u_{\eta,2}^T \ddot{\lambda}_{02} u_{\eta,2}}{2\lambda_{02}} Y_{2n}\, ds\Big\}$$

and a process $Q_n = Q_n^- 1_{\{\gamma_{n,u}<\gamma_0\}} + Q_n^+ 1_{\{\gamma_{n,u}>\gamma_0\}})$ sum of integrals between γ_0 and $\gamma_{n,u}$

$$Q_n^-(u) = \int_{\gamma_{n,u}}^{\gamma_0} \log\frac{\lambda_2}{\lambda_1}\, dM_{1n} + \int_{\gamma_{n,u}}^{\gamma_0} \varphi_1(\lambda, s) Y_{1n}\, ds,$$

$$Q_n^+(u) = \int_{\gamma_0}^{\gamma_{n,u}} \log\frac{\lambda_1}{\lambda_2}\, dM_{2n} + \int_{\gamma_0}^{\gamma_{n,u}} \varphi_2(\lambda, s) Y_{2n}\, ds, \tag{5.25}$$

where $\varphi_1(\lambda) = \lambda_{01}\log\lambda_2\lambda_1^{-1} - (\lambda_2 - \lambda_1)$ and $\varphi_2(\lambda) = \lambda_{02}\log\lambda_1\lambda_2^{-1} - (\lambda_1 - \lambda_2)$.

Theorem 5.12. *For every $A > 0$, the process $nX_n(\theta_{n,u})$ is uniformly approximated on $\mathcal{U}_n^A$ as*

$$nX_n(\theta_{n,u}) = Q_n(u_\gamma) + \widetilde{X}_n(u_\eta) + o_p(1),$$

as n tends to infinity. The variable $\widetilde{X}_n(u_{\eta_0})$ converges weakly to a Gaussian variable with variance $\widetilde{\sigma}_\eta^2 = u_\eta^T I_0 u_\eta$ and mean

$$\widetilde{\mu}_{\eta_0} = -\frac{1}{2}\Big(\int_0^{\gamma_0} \frac{u_{\eta,1}^T \ddot{\lambda}_{01} u_{\eta,1}}{\lambda_{01}} y_{\gamma_0}\, ds + \int_{\gamma_0}^{\tau} \frac{u_{\eta,2}^T \ddot{\lambda}_{02} u_{\eta,2}}{\lambda_{02}} y_{\gamma_0}\, ds\Big).$$

This is a consequence of the approximation of $\log(1+x) - x$ in a neighborhood of zero for $\widetilde{X}_n$.

Theorem 5.13. *The variable $n(\widehat{\gamma}_n - \gamma_0)$ is asymptotically independent of $\widehat{\eta}_n$ and it converges weakly to the location of the maximum of an uncentered Gaussian processes with independent increments.*

Proof. The process Q_n^- and Q_n^+ have the approximations

$$Q_n^-(u) = 1_{\{u<0\}}\Big\{\int_{\gamma_{n,u}}^{\gamma_0} \log\frac{\lambda_{02}}{\lambda_{01}}\,dM_{1n} + \int_{\gamma_{n,u}}^{\gamma_0} \varphi_1(\lambda_0,s)Y_{1n}(s)\,ds\Big\} + o_p(1),$$

$$Q_n^+(u) = 1_{\{u>0\}}\Big\{\int_{\gamma_0}^{\gamma_{n,u}} \log\frac{\lambda_1}{\lambda_2}\,dM_{2n} + \int_{\gamma_0}^{\gamma_{n,u}} \varphi_2(\lambda_0,s)Y_{2n}(s)\,ds\Big\} + o_p(1)$$

they do not depend on the parameter η. As the martingales are centered, their asymptotic means are $\mu^-(u) = 1_{\{u<0\}}|u|\varphi_1(\lambda_0,\gamma_0)y_1(\gamma_0)$ and, respectively $\mu_n^+(u) = 1_{\{u>0\}}u\varphi_2(\lambda_0,\gamma_0)y_2(\gamma_0)$. In the expansion of the process X_n given in Theorem 5.12, $\widetilde{X}_n$ does not depend on γ. Theorem 5.11 implies $\widehat{u}_{n\gamma} = n(\widehat{\gamma}_n - \gamma_0)$ is bounded in probability under P_0 so there exists $A > 0$ such that $\widehat{u}_{n\gamma}$ maximizes the process $n^{-\frac{1}{2}}Q_n(u)$ on $\mathcal{U}_n^A$.

The integrals with respect to the martingales and the integrals of the functions $n^{\frac{1}{2}}\varphi_k y_k$ on the intervals $]\gamma_{n,u},\gamma_0]$ and $]\gamma_0,\gamma_{n,u}]$ converge weakly to zero as the length of the intervals tends to zero. The process $n^{-\frac{1}{2}}Q_n(u)$ is then asymptotically equivalent to the integrals of the functions φ_k with respect to $n^{\frac{1}{2}}(n^{-1}Y_{kn} - y_k)(s)\,ds$ where the empirical processes $n^{\frac{1}{2}}(n^{-1}Y_{kn} - y_k)$ converge weakly to independent transformed Brownian bridges B_k with variance functions v_k, then the integrals converge weakly to independent centered Gaussian processes with independent increments and with variances $\varphi_k(\lambda_0,\gamma_0)v_k(\gamma_0)$. Finally, $\widehat{u}_n$ converges weakly to the maximum of the sum of these processes which is bounded in probability. The asymptotic independence of the estimators $\widehat{\gamma}_n$ and $\widehat{\eta}_n$ is a consequence of the asymptotic independence of the asymptotically Gaussian processes $\widetilde{X}_n$ and Q_n. □

If the parametric family $\lambda_\eta = \lambda \circ \varphi_\eta$ depends on an unknown baseline intensity λ, the log-likelihood process X_n is estimated using a kernel estimator $\widehat{\lambda}_{n,h}$ smoothing the cumulative intensity Λ, like in Section 5.6. Under similar conditions, the convergence rate of the change-point estimator is still n^{-1}.

5.8 Estimation by minimum of variance

In this section, the change-point γ_0 is estimated by a simple criterion which consists in the minimization of the least variance of the nonparametric estimated cumulative intensity. At the true change-point value γ_0, the process

$$W_{\Lambda,n}(t) = \{1 - \delta_{\gamma_0}(t)\}n^{\frac{1}{2}}\{\widehat{\Lambda}_{1n}(t) - \Lambda_{01}(t)\} + \delta_{\gamma_0}(t)n^{\frac{1}{2}}\{\widehat{\Lambda}_{2n}(t) - \Lambda_{02}(t)\}$$

converges weakly to a centred Gaussian process W_Λ. With a change-point at γ belonging to the interval $]0,\tau[$, the intensity (5.21) determines the cumulative intensity Λ_θ. Let

$$J_{1n,\gamma}(t) = 1_{\{Y_{1n,\gamma}(t)>0\}},\ t \le \gamma,$$
$$J_{2n,\gamma}(t) = 1_{\{Y_{2n,\gamma}(t)>0\}},\ t > \gamma.$$

The quadratic estimation error for the cumulative hazard function Λ with a change-point at γ in $]0,\tau[$ is

$$V_n(\gamma) = E\int_0^\gamma J_{1n,\gamma}Y_{1n,\gamma}^{-1}\,d\Lambda_{\theta,1} + E\int_\gamma^\tau J_{2n,\gamma}Y_{2n,\gamma}^{-1}\,d\Lambda_{\theta,2},$$

it is consistently estimated by

$$\widehat{V}_n(\gamma) = \int_0^\gamma \frac{J_{1,\gamma,n}}{Y_{1,\gamma,n}^2}\,dN_{1,\gamma,n} + \int_\gamma^\tau \frac{J_{2,\gamma,n}}{Y_{2,\gamma,n}^2}\,dN_{2,\gamma,n}.$$

Since the minimal variance is reached in the true model under P_0, $V_n(\gamma)$ is minimum at the function $\Lambda_0(t,\gamma_0)$, by continuity of V_n with respect to λ_{01}, λ_{02} and γ. Its estimator $\widehat{V}_n$ is therefore minimum in a neighborhood of γ_0 which is estimated by

$$\widehat{\gamma}_n = \arg\inf_\gamma \widehat{V}_n(\gamma).$$

The consistency of this estimator is a consequence of the consistency of $\widehat{V}_n$, by the same arguments as for Theorem 5.10.

Proposition 5.8. *The minimum variance estimators of the cumulative hazard function Λ and the parameter γ are consistent.*

Let $Y_{n,\gamma}(t) = Y_{1n,\gamma}(t)\delta_\gamma(t) + Y_{2n,\gamma}(t)\{1-\delta_\gamma(t)\}$ and let

$$y_{0,\gamma}(t) = P_0(t \le T < \gamma)\delta_\gamma(t) + P_0(\gamma < t \le T \le \tau)\{1-\delta_\gamma(t)\}$$

be the a.s. limit of $n^{-1}Y_{n,\gamma}$ under P_0, as n tends to infinity. Under P_0, the counting processes have the partial means $\mu_{01,\gamma}(t) = P_0(T \le t \wedge \gamma)$ and $\mu_{02,\gamma}(t) = P_0(\gamma < T \le t)$ such that $\Lambda_{0k,\gamma}(t) = \int_0^t y_{0k,\gamma}^{-1}\,d\mu_{0k,\gamma}$ for $k = 1,2$. The variable $n\widehat{V}_n$ converges a.s. under P_0 to the function

$$\begin{aligned}
V_0(\gamma) &= \int_0^\gamma 1_{\{y_{01,\gamma}>0\}} y_{01,\gamma}^{-1}\,d\Lambda_{01,\gamma} + \int_\gamma^\tau 1_{\{y_{02,\gamma}>0\}} y_{02,\gamma}^{-1}\,d\Lambda_{02,\gamma},\\
&= \int_0^{\gamma_0} 1_{\{y_{01,\gamma}>0\}} y_{01,\gamma}^{-1}\,d\Lambda_{01,\gamma} + \int_{\gamma_0}^\tau 1_{\{y_{02,\gamma}>0\}} y_{02,\gamma}^{-1}\,d\Lambda_{02,\gamma}\\
&\quad +\delta_{\gamma_0}(\gamma)\Big\{\int_\gamma^{\gamma_0} (1_{\{y_{02,\gamma}>0\}} y_{02,\gamma}^{-1} - 1_{\{y_{01,\gamma}>0\}} y_{01,\gamma}^{-1})\,d\Lambda_{01,\gamma}\Big\}\\
&\quad +\{1-\delta_{\gamma_0}(\gamma)\}\Big\{\int_{\gamma_0}^\gamma (1_{\{y_{01,\gamma}>0\}} y_{01,\gamma}^{-1} - 1_{\{y_{02,\gamma}>0\}} y_{02,\gamma}^{-1})\,d\Lambda_{02,\gamma}\Big\}.
\end{aligned}$$

Let $X(\gamma) = V_0(\gamma) - V_0(\gamma_0)$.

Lemma 5.7. *For ε sufficiently small, there exists a constant $\kappa_0 > 0$ such that*

$$\inf_{\gamma \in V_\varepsilon(\gamma_0)} X(\gamma) \geq \kappa_0 \varepsilon^2.$$

Proof. If $\gamma < t \leq \gamma_0$, $y_{01,\gamma} \equiv 0$ and the integral on $]\gamma, \gamma_0]$ of the process X is positive; if $\gamma_0 < t \leq \gamma$, $y_{02,\gamma} \equiv 0$ and the integral on $]\gamma_0, \gamma]$ is positive, they are larger than a constant times $|\gamma - \gamma_0|$. The differences

$$1_{\{\gamma > \gamma_0\}} \int_0^{\gamma_0} \Big(1_{\{y_{01,\gamma}>0\}} y_{01,\gamma}^{-1} - 1_{\{y_{01}>0\}} y_{01}^{-1} \Big) \, d\Lambda_{01},$$
$$1_{\{\gamma < \gamma_0\}} \int_{\gamma_0}^{\tau} \Big(1_{\{y_{02,\gamma}>0\}} y_{02,\gamma}^{-1} - 1_{\{y_{02}>0\}} y_{02}^{-1} \Big) \, d\Lambda_{02}$$

have the same lower bound. Replacing the first integral by an integral on $[0, \gamma]$ if $\gamma < \gamma_0$, it has the same bound and the same argument applies to the second integral. The integrals on $]\gamma, \gamma_0]$ and $]\gamma_0, \gamma]$ have opposite signs and the result follows. □

Let $X_n(\gamma) = n\{\widehat{V}_n(\gamma) - \widehat{V}_n(\gamma_0)\}$ and let $W_n(\gamma) = n^{\frac{1}{2}}\{X_n(\gamma) - X(\gamma)\}$, the next bound is obtained by similar arguments from the Cauchy–Schwarz inequality.

Lemma 5.8. *For every $\varepsilon > 0$, there exist a constant $\kappa_1 > 0$ such that for n large enough, $0 \leq E_0 \sup_{\gamma \in V_\varepsilon(\gamma_0)} X_n(\gamma) \leq \kappa_1 \varepsilon$.*

By the same proof as Theorem 2.2, we deduce the convergence rate of $\widehat{\gamma}_n$ from Lemmas 5.8 and 5.7, as

$$\overline{\lim}_{n,A\to\infty} P_0(n|\widehat{\gamma}_n - \gamma_0| > A) = 0. \tag{5.26}$$

The asymptotic distribution of $n(\widehat{\gamma}_n - \gamma_0)$ is deduced from the behaviour of the process $X_n(\gamma_{n,u})$ where $\gamma_{n,u} = \gamma_0 + n^{-1} u_n$, where u_n converges to a limit u in a subset $\mathcal{U}_n^A$ of $]0, \tau[$ such that $u \leq A$.

Theorem 5.14. *The variable $n(\widehat{\gamma}_n - \gamma_0)$ converges weakly to the location of the maximum of an uncentered Gaussian process with independent increments on $\mathcal{U}_n^A$, as n and A tend to infinity.*

Proof. We have

$$\begin{aligned} X_n(\gamma) &= n\Big(\int_0^{\gamma_0} \frac{J_{1,\gamma,n}}{Y_{1,\gamma,n}^2}\, dN_{1,\gamma,n} - \int_0^{\gamma_0} \frac{J_{1,n}}{Y_{1,n}^2}\, dN_{1,n}\Big) \\ &+n\Big(\int_{\gamma_0}^{\tau} \frac{J_{2,\gamma,n}}{Y_{2,\gamma,n}^2}\, dN_{2,\gamma,n} - \int_{\gamma_0}^{\tau} \frac{J_{2,n}}{Y_{2,n}^2}\, dN_{2,n}\Big) \\ &+\delta_{\gamma_0}(\gamma) n \int_{\gamma}^{\gamma_0} \Big(\frac{J_{2,\gamma,n}}{Y_{2,\gamma,n}^2} - \frac{J_{1,\gamma,n}}{Y_{1,\gamma,n}^2}\Big)\, dN_{1,\gamma,n} \\ &+\{1-\delta_{\gamma_0}(\gamma)\} n \int_{\gamma_0}^{\gamma} \Big(\frac{J_{1,\gamma,n}}{Y_{1,\gamma,n}^2} - \frac{J_{2,\gamma,n}}{Y_{2,\gamma,n}^2}\Big)\, dN_{2,\gamma,n}, \end{aligned}$$

the mean of the integral on $]\gamma,\gamma_0]$ is the difference of the means of $n\int_\gamma^{\gamma_0} J_{k,\gamma,n} Y_{1,n} Y_{k,\gamma,n}^{-2}\, d\Lambda_1$ for $k = 1,2$ and it is a $O(|\gamma-\gamma_0|)$, by the martingale property of $M_{1,\gamma,n}$ its variance is the mean of $n^2\int_\gamma^{\gamma_0} J_{2,\gamma,n} Y_{2,\gamma,n}^{-4} Y_{1,\gamma,n}\, d\Lambda_1$ which is a $O(n^{-1}|\gamma-\gamma_0|)$, the mean and the variance of the integral on $]\gamma_0,\gamma]$ are similar. The mean of the first difference of integrals is a $O(|\gamma-\gamma_0|)$ and its variance is a $O(n^{-1}|\gamma-\gamma_0|)$, the second difference has the same behaviour. It follows that the process $nX_n(\gamma_{n,u})$ converges weakly on $\mathcal{U}_n^A$ to an uncentered Gaussian process with a finite variance, as n tends to infinity.

The maximum of the process nX_n is achieved at $\widehat{\gamma}_n = \gamma_{\widehat{u}_n}$ and $\widehat{u}_n$ is asymptotically bounded in probability, from (5.26), it converges weakly to the location of the maximum of the Gaussian process limit of $nX_n(\gamma_{n,u})$ which is bounded in probability. □

5.9 Mean squares test of no change

In the model (5.21) with nonparametric hazard functions, the test of the minimal variance for the hypothesis H_0 of a continuous hazard function λ_0 with $\gamma_0 = \tau$ finite, against the alternative of a hazard function with a discontinuity at an unknown change-point is performed with the statistic

$$T_n = n^2\{\widehat{V}(\widehat{\gamma}_n) - \widehat{V}_{0n}\},$$

where $\widehat{V}_{0n}$ is the empirical estimator of the variance of the estimated cumulative hazard function under the hypothesis H_0, $\widehat{V}_{0n} = \int_0^\tau J_n Y_n^{-2}\, dN_n$. Under H_0, $n\{\widehat{V}(\gamma) - \widehat{V}_{0n}\}$ converges in probability to the function $X(\gamma)$ minimum at γ_0 and the least variance estimator $\widehat{\gamma}_n$ is consistent. The expression of the function V_0 reduces to

$$V_0(\gamma) = \int_0^{\gamma_0} 1_{\{y_{01,\gamma}>0\}} y_{01,\gamma}^{-1} \, d\Lambda_{01,\gamma} + \int_\gamma^{\gamma_0} (1_{\{y_{02,\gamma}>0\}} y_{02,\gamma}^{-1} - 1_{\{y_{01,\gamma}>0\}} y_{01,\gamma}^{-1}) \, d\Lambda_{01,\gamma}$$

and Lemma 5.7 is fulfilled under H_0. The process $X_n(\gamma) = n\{\widehat{V}_n(\gamma) - \widehat{V}_{0n}\}$ satisfies Lemma 5.8, then the convergence rate of $\widehat{\gamma}_n$ is still n^{-1} under H_0 and the asymptotic behaviour of $n(\widehat{\gamma}_n - \gamma_0)$ is given by Theorem 5.14.

Proposition 5.9. *The statistic T_n converges weakly under H_0 to the maximum T_0 of an uncentered Gaussian process.*

Proof. The test statistic is written as

$$T_n = n^2 \Big(\int_0^{\widehat{\gamma}_n} \frac{J_{1,n}}{Y_{1,n}^2} \, dN_{1,n} + \int_{\widehat{\gamma}_n}^{\tau} \frac{J_{2,n}}{Y_{2,n}^2} \, dN_{2n} - \int_0^{\tau} \frac{J_n}{Y_n^2} \, dN_n \Big)$$

and it converges weakly to the maximum of an uncentered Gaussian process defined by Theorem 5.14. □

Under a fixed alternative with γ in $]0, \tau[$ and hazard functions λ_1 and λ_2, one of them is different from λ_0 and the test statistic diverges.

Proposition 5.10. *Under local alternatives, the test statistic converges weakly to $T_0 + \mu$ where μ is a non-zero limit.*

Proof. Under local alternatives K_n, the cumulative hazard functions are $\Lambda_{n1} = \Lambda_0 + n^{-\frac{1}{2}} \phi_n$ and $\Lambda_{n2} = \Lambda_{n1} + n^{-\frac{1}{2}} \psi_n$ with sequences of functions $(\phi_n)_n$ and $(\psi_n)_n$ converging to non-zero functions ϕ and ψ, the change occurs at $\gamma_{nu} = \tau - n^{-1} u_n$ with a real sequence $(u_n)_n$ converging to non-zero limit u. They define a sequence of cumulative hazard functions Λ_n depending on γ_{nu}, Λ_{n1} and Λ_{n2}, converging to Λ_0 with the rate $n^{-\frac{1}{2}}$. The time variable T^0 has a sequence of distribution functions F_n converging to the distribution function F_0 under H_0 with the rate $n^{-\frac{1}{2}}$ and the means $y_{kn,\gamma}$ of the processes $n^{-1} Y_{kn,\gamma}$ converge in probability to the functions $y_{k,\gamma}$ with the rate $n^{-\frac{1}{2}}$, the processes $n^{\frac{1}{2}} \{n^{-1} Y_{kn,\gamma} - y_{k,\gamma}\}$ converge to uncentered Gaussian processes with means the non-zero limits v_k of $n^{\frac{1}{2}} (y_{kn,\gamma} - y_{k,\gamma})$. The asymptotic mean of T_n under K_n is the sum of its asymptotic mean under H_0 and an expression depending on u and on integral of expression of the functions v_k, ϕ and ψ with respect to Λ_0. □

5.10 Nonparametric maximum likelihood estimators

The log-likelihood of the sample is expressed with the unknown intensity function λ_γ depending on $\lambda_{1,\gamma}$, $\lambda_{2,\gamma}$ and γ as

$$l_n(\gamma, \lambda) = \int_0^\tau \log \lambda_\gamma(t)\, dN_n(t) - \int_0^\tau Y_n(t)\lambda_\gamma(t)\, dt$$

and the maximum likelihood estimation of the change-point is performed replacing the intensities λ_1 and λ_2 with their kernel estimators. The following conditions imply the weak convergence of the kernel estimators of the hazard functions

C1 K is a symmetric density such that $|x|^2 K(x)$ converges to zero as $|x|$ tends to infinity or K has a compact support with value zero on its frontier;

C2 The hazard functions λ_1 and λ_2 belong to the class $\mathcal{L}$ of hazard functions in $C_s(\mathbb{R}_+)$:

C3 The kernel function satisfies the next integrability conditions: the integrals $m_{jK} = \int u^j K(u) du$ is zero for $j < s$, m_{sK}, $k_\alpha = \int K^\alpha(u) du$, $\alpha \geq 0$, and $\int |K'(u)|^\alpha du$, for $\alpha \leq s$, are finite;

C4 As n tends to infinity, h_n and nh_n^4 converge to zero and nh_n tends to infinity.

Under the above conditions, the processes $(nh_n)^{\frac{1}{2}}\{\widehat{\lambda}_{kn}(t,\gamma) - \lambda_k(t,\gamma)\}$, for $k = 1, 2$, are independent and they converge weakly on every compact subset of $[h, \gamma - h]$ and, respectively $[\gamma + h, \tau - h]$, to Gaussian processes $W_{\lambda_{k,\gamma}}$ under P_γ, their bias converge to zero and their variance is $k_2\, y_\gamma^{-1}\lambda_{k,\gamma}$. The mean integrated squared error for the estimation of λ under P_γ is

$$\begin{aligned}
\sigma^2_{\lambda,\gamma} &= E_\gamma \int_0^\tau \{\widehat{\lambda}_n(t,\gamma) - \lambda(t,\gamma)\}^2\, dt \\
&= (nh)^{-1} k_2 \Big\{ \int_0^\gamma y^{-1}(t)\lambda_1(t)\, dt + \int_\gamma^\tau y^{-1}(t)\lambda_2(t)\, dt \Big\} \\
&\quad + \frac{1}{(s!)^2} m_{sK}^2 h^{2s} \Big\{ \int_0^\gamma \lambda_1^{(s)2}(t)\, dt + \int_\gamma^\tau \lambda_2^{(s)2}(t)\, dt \Big\} + o((nh)^{-1} + h^{2s}).
\end{aligned}$$

The log-likelihood process $X_n = n^{-1}(l_n - l_{0n})$ and its limit under P_0 are defined as in Section 5.7 and we assume that the condition (5.24) is fulfilled. Let

$$N_n(s) = N_{1n}(s)\delta_{\gamma_0}(s) + N_{2n}(s)\{1 - \delta_{\gamma_0}(s)\}$$

and let $Y_n(s) = Y_{1n}(s)\delta_{\gamma_0}(s) + Y_{2n}(s)\{1-\delta_{\gamma_0}(s)\}$. Replacing the intensities λ_1 and λ_2 by their kernel estimators in the expression of X_n, the log-likelihood is estimated by the process

$$\begin{aligned}\widehat{X}_{n,h}(\gamma) &= \int_0^{\gamma_0} \log \frac{\widehat{\lambda}_{1n,h}}{\lambda_{01}}\, dN_{1n} - \int_0^{\gamma_0} (\widehat{\lambda}_{1n,h} - \lambda_{01})Y_{1n} \\ &+ \int_{\gamma_0}^{\tau} \log \frac{\widehat{\lambda}_{2n,h}}{\lambda_{02}}\, dN_{2n} - \int_{\gamma_0}^{\tau} (\widehat{\lambda}_{2n,h} - \lambda_{02})Y_{2n} \\ &+ \int_{\gamma}^{\gamma_0} \log \frac{\widehat{\lambda}_{2n,h}}{\widehat{\lambda}_{1n,h}}\, dN_n - \int_{\gamma}^{\gamma_0} (\widehat{\lambda}_{2n,h} - \widehat{\lambda}_{1n,h})Y_n.\end{aligned}$$

Under P_0, $\widehat{X}_{n,h}$ converges a.s. uniformly to the function X of Section 5.7 and it is maximum at $\widehat{\gamma}_{n,h}$.

Theorem 5.15. *Under P_0, the maximum likelihood estimator $\widehat{\gamma}_{n,h}$ of γ_0 is a.s. consistent.*

Proof. The function X is concave with respect to λ and it is zero at λ_0 where it reaches its maximum, furthermore the process $\widehat{X}_{n,h} - X_n$ a.s. converges uniformly to zero on $]0,\tau[$ by the consistency of the estimated intensities, therefore $X_{n,h}$ converges a.s. uniformly to X on $]0,\tau[$. We have

$$0 \le \widehat{X}_{n,h}(\widehat{\gamma}_{n,h}) \le \sup_\gamma |\widehat{X}_{n,h}(\gamma) - X(\gamma)| + X(\widehat{\gamma}_{n,h})$$

where $0 = X(\gamma_0) \ge X(\widehat{\gamma}_{n,h})$ hence $\lim_{n\to\infty} X(\widehat{\gamma}_{n,h}) = 0$ and by concavity of X, $\widehat{\gamma}_n$ converges a.s. to γ_0. □

Let $\widehat{W}_{n,h} = n^{\frac{1}{2}}(\widehat{X}_{n,h} - X)$.

Lemma 5.9. *For every $\varepsilon > 0$, there exists $\kappa_1 > 0$ such that for n sufficiently large*

$$\mathrm{E}_0 \sup_{|\gamma-\gamma_0|^{\frac{1}{2}} \le \varepsilon} |W_n(\gamma)| \le \kappa_1 \varepsilon.$$

Proof. The process $\widehat{X}_{n,h}$ is written as integrals with respect to the local martingales as

$$\begin{aligned}\widehat{X}_{n,h}(\gamma) &= n^{-1} \int_0^{\gamma_0} \log \frac{\widehat{\lambda}_{1n,h}}{\lambda_{01}}\, dM_{1n} + n^{-1} \int_0^{\gamma_0} \psi(\widehat{\lambda}_{1n,h}, \lambda_{01})Y_{1n} \\ &+n^{-1} \int_{\gamma_0}^{\tau} \log \frac{\widehat{\lambda}_{2n,h}}{\lambda_{02}}\, dM_{2n} + n^{-1} \int_{\gamma_0}^{\tau} \psi(\widehat{\lambda}_{2n,h}, \lambda_{02})Y_{2n} \\ &+n^{-1} \int_{\gamma}^{\gamma_0} \log \frac{\widehat{\lambda}_{2n,h}}{\widehat{\lambda}_{1n,h}}\, dM_n + n^{-1} \int_{\gamma}^{\gamma_0} \psi(\widehat{\lambda}_{2n,h}, \widehat{\lambda}_{1n,h})Y_n,\end{aligned}$$

the integrals of $\widehat{W}_{n,h}(\gamma)$ with respect to the martingales converge in probability to zero by the consistency of the estimators, the differences

$$n^{\frac{1}{2}} \int_0^{\gamma_0} \psi(\widehat{\lambda}_{1n,h}, \lambda_{01})(n^{-1}Y_{1n} - y_1),\ n^{\frac{1}{2}} \int_{\gamma_0}^{T} \psi(\widehat{\lambda}_{2n,h}, \lambda_{02})(n^{-1}Y_{2n} - y_2)$$

are $o_p(1)$ by the weak convergence of the processes $n^{\frac{1}{2}}(n^{-1}Y_n - y_\gamma)$, uniformly in $V_\varepsilon(\gamma_0)$, as a consequence of the uniform weak convergence of empirical process. The integrals $n^{-\frac{1}{2}} \int_0^{\gamma_0} \psi(\widehat{\lambda}_{1n,h}, \lambda_{01}) y_1$, $n^{-\frac{1}{2}} \int_{\gamma_0}^{T} \psi(\widehat{\lambda}_{2n,h}, \lambda_{02}) y_2$ and the integrals of the functions ψ with respect to the functions y_k are $O_p((nh_n^4)^{\frac{1}{2}})$ by the weak convergence of the intensity estimators, they converge to zero under the conditions. Finally, the integrals

$$n^{\frac{1}{2}} \int_\gamma^{\gamma_0} \psi(\widehat{\lambda}_{2n,h}, \widehat{\lambda}_{1n,h})(n^{-1}Y_n - y),\ n^{-\frac{1}{2}} \int_\gamma^{\gamma_0} \log\{\widehat{\lambda}_{2n,h}\widehat{\lambda}_{1n,h}^{-1}\}\, dM_n$$

converge to centered Gaussian process with variances $O(|\gamma - \gamma_0|)$ in $V_\varepsilon(\gamma_0)$ which yields the result. □

Lemmas 5.5 and 5.9 provide the convergence rate of $\widehat{\gamma}_{n,h}$

$$\overline{\lim}_{n,A\to\infty} P_0\{n|\widehat{\gamma}_{n,h} - \gamma_0| > A) = 0. \tag{5.27}$$

Expansion of the logarithms in the expression of $\widehat{W}_{n,h}$ prove the approximation

$$\widehat{W}_{n,h} = O_p(|\gamma - \gamma_0|^{\frac{1}{2}}) + o_p(1).$$

Theorem 5.16. *The variable $n(\widehat{\gamma}_n - \gamma_0)$ converges weakly to the location of the maximum of an uncentered Gaussian process with independent increments as n and A tends to infinity.*

Proof. As $\widehat{u}_{n,h} = n(\widehat{\gamma}_{n,h} - \gamma_0)$ is bounded in probability under P_0 from (5.27), there exists $A > 0$ such that $\widehat{u}_{n,h}$ maximizes the process $\widehat{X}_{n,h}$ on $\mathcal{U}_n^A$. Let u in $\mathcal{U}_n^A$ and let $\gamma_{n,u} = \gamma_0 + n^{-1}u$, the process $\widehat{W}_{n,h}(\gamma_{n,u})$ converges weakly on $\mathcal{U}_n^A$ to a functional of an uncentered Gaussian process with independent increments and $\widehat{u}_n$ converges weakly to the maximum of this Gaussian process in $\mathcal{U}_n^A$. □

Theorem 5.17. *The nonparametric estimator $\widehat{\lambda}_{n,h} = \widehat{\lambda}_{n,\widehat{\gamma}_n}$ is asymptotically independent of $\widehat{\gamma}_{n,h}$ and it has the same limiting distribution as $\widehat{\lambda}_{n,h,\gamma_0}$ where γ_0 is known.*

Proof. The difference between the estimated intensities is

$$\begin{aligned}\widehat{\lambda}_{n,h}(t)-\widehat{\lambda}_{0n,h}(t) &= \{\widehat{\lambda}_{1n,h}(t)-\widehat{\lambda}_{01n,h}(t)\}\delta_{\gamma_0}(t)\\ &\quad+\{\widehat{\lambda}_{2n,h}(t)-\widehat{\lambda}_{02n,h}(t)\}\{1-\delta_{\gamma_0}(t)\}\\ &\quad+1_{\{\gamma_0<t\leq\widehat{\gamma}_{n,h}}\{\widehat{\lambda}_{1n,h}(t)-\widehat{\lambda}_{02n,h}(t)\}\\ &\quad+1_{\{\widehat{\gamma}_{n,h}<t\leq\gamma_0\}}\{\widehat{\lambda}_{2n,h}(t)-\widehat{\lambda}_{01n,h}(t)\},\end{aligned}$$

the convergence rates of the first two terms is $(nh)^{\frac{1}{2}}$ and the last two terms converge to $(\lambda_1-\lambda_2)1_{(\gamma_0,\widehat{\gamma}_{n,h})}$ where the length of the interval has the order $O_p(n^{-1})$ and they converge in probability to zero. The difference of the estimators for $t\leq\gamma_0$ is

$$\sum_{i=1}^{n}\delta_i K_h(T_i-t)\Big\{1_{\{T_i\leq\widehat{\gamma}_{n,h}\}}\frac{J_{1n,\widehat{\gamma}_{n,h}}}{Y_{1n,\widehat{\gamma}_{n,h}}(T_i)}-1_{\{T_i\leq\gamma_0\}}\frac{J_{1n,\gamma_0}}{Y_{1n,\gamma_0}(T_i)}\Big\},$$

it converges in probability to zero, the difference $Y_{1n,\gamma_0}(T_i)-Y_{1n,\widehat{\gamma}_{n,h}}(T_i)$ being the cardinal of the variables T_i between $\widehat{\gamma}_{n,h}$ and γ_0, it converges in probability to $f_0(\gamma_0)$. By the same arguments, the difference of the estimators for $t>\gamma_0$ converges in probability to zero. □

5.11 Likelihood ratio test of models without change

In the model with nonparametric hazard functions, the log-likelihood ratio test for the hypothesis H_0 of a continuous function λ_0 against the alternative of an intensity with a change-point is performed with the statistic

$$T_n=2h_n\{\widehat{l}_{n,h}(\widehat{\gamma}_{n,h})-\widehat{l}_{0n,h}\},$$

where $\widehat{l}_{0n}=\int_0^\tau\log\widehat{\lambda}_{n,h}\,dN_n-\int_0^\tau\widehat{\lambda}_{n,h}Y_n\,ds$ is the estimated log-likelihood under H_0 when the intensity λ_0 has the kernel estimator $\widehat{\lambda}_{n,h}$.

We assume that the conditions of Section 5.10 are satisfied. Under H_0, the variables $\tau-\widehat{\gamma}_{n,h}$ and $Q_n(\widehat{\gamma}_{n,h})$ are $o_p(1)$, the estimator $\widehat{\lambda}_{1n,h}$ belongs to a $n^{-\frac{1}{2}}$-neighborhood of λ_0 and $\widehat{\lambda}_{2n,h}$ converges in probability to zero on the interval $]\widehat{\gamma}_{n,h},\tau]$. The estimator of the intensity (5.21) is such that $(nh_n)^{\frac{1}{2}}(\widehat{\lambda}_n-\lambda_0)$ converges weakly to a centered Gaussian process W_0.

Theorem 5.18. *Under the null hypothesis and the conditions 1-4, the statistic T_n converges weakly to $T_0=-\int_0^\tau W_0^2(s)\lambda_0^{-1}(s)y(s)\,ds$.*

Proof. In the nonparametric model and under H_0

$$\widehat{l}_{n,h}(\widehat{\gamma}_{n,h}) - \widehat{l}_{0n,h} = \int_0^{\widehat{\gamma}_{n,h}} \log \frac{\widehat{\lambda}_{1n,h}}{\widehat{\lambda}_{0n,h}}\, dM_{1n} + \int_0^{\widehat{\gamma}_{n,h}} \psi(\widehat{\lambda}_{1n,h}, \widehat{\lambda}_{0n,h}) Y_{1n}$$
$$+ \int_{\widehat{\gamma}_{n,h}}^{\tau} \log \frac{\widehat{\lambda}_{2n,h}}{\widehat{\lambda}_{0n,h}}\, dM_{2n} + \int_{\widehat{\gamma}_{n,h}}^{\tau} \psi(\widehat{\lambda}_{2n,h}, \widehat{\lambda}_{0n,h}) Y_{2n},$$

by (5.23), $\psi(\widehat{\lambda}_{kn,h}, \widehat{\lambda}_{0n,h}) = O_p((nh)^{-1})$ for $k = 1,2$, therefore the last integral of $h\{\widehat{l}_{n,h}(\widehat{\gamma}_{n,h}) - \widehat{l}_{0n,h}\}$ is a $o_p(1)$. An expansion of the logarithm of $\widehat{\lambda}_{kn,h}\widehat{\lambda}_{0n,h}^{-1}$ as $(\widehat{\lambda}_{kn,h} - \widehat{\lambda}_{0n,h})\widehat{\lambda}_{0n,h}^{-1} - \frac{1}{2}(\widehat{\lambda}_{kn,h} - \widehat{\lambda}_{0n,h})^2\widehat{\lambda}_{0n,h}^{-2} + o_p(1)$ and the weak convergence of the martingale $n^{-\frac{1}{2}}M_{kn}$ to a centered Gaussian process on $[0,\tau]$ imply that the first and the third integrals of $h\{\widehat{l}_{n,h}(\widehat{\gamma}_{n,h}) - \widehat{l}_{0n,h}\}$ are $o_p(1)$. The second integral is asymptotically equivalent to $-nh\int_0^{\tau}(\widehat{\lambda}_{kn,h} - \widehat{\lambda}_{0n,h})^2\widehat{\lambda}_{0n,h}^{-1} y_1\, ds$, where $(nh_n)^{\frac{1}{2}}(\widehat{\lambda}_n - \lambda_0)$ converges weakly to W_0 and the test statistic converges weakly to the variable T_0. □

Under alternatives of a change-point at γ in $]0,\tau[$, the test statistic diverges. Under local alternatives H_n, the intensity has a change-point at $\gamma_n = \tau - n^{-1}u_n$ with a sequence $(u_n)_n$ converging to a limit u and for $k = 1,2$, the intensities are $\lambda_{kn} = \lambda_0(1 + a_n^{-\frac{1}{2}} b_{kn})$ where $a_n = nh_n$ and $(b_{kn})_n$, are sequences of functions such that b_{1n} converges uniformly to a function b_1 on $[0,\gamma_n]$ and b_{2n} converges uniformly to a function b_2 on $]\gamma_n,\tau]$. This implies the existence of sequences of functions η and ξ and of uniformly convergent sequences of functions $(\eta_n)_n$ and $(\xi_n)_n$ such that $\lambda_{2n} = \lambda_{0n}(1 + a_n^{-\frac{1}{2}}\eta_n)$ and $\lambda_{1n} = \lambda_{2n}(1 - a_n^{-\frac{1}{2}}\xi_n)$, with $\eta_n - \eta$ and $\xi_n - \xi$ converging uniformly to zero.

Theorem 5.19. *Under the local alternatives H_n, T_n converges weakly to*

$$\int_0^{\tau} (W_0 + b_1)^2 \lambda_0^{-1} y\, ds.$$

Proof. Under the alternative, the second term in the expression of l_n^- depends on $\varphi_{1n} = \lambda_{1n}\log\{\lambda_{1n}\lambda_{01}^{-1}\} - \lambda_{1n} + \lambda_{01}$ and it is expanded as

$$\varphi_{1n} = \frac{1}{a_n\lambda_0}\Big(\frac{b_{1n}^2}{2} + o(1)\Big)$$

and the other terms have the same expansion as under H_0, they are $o_p(1)$. It follows that T_n is asymptotically equivalent to $n^{-1}\int_0^{\tau}\widehat{b}_{1n}^2 Y_n\, ds$ where the function b_{1n} is estimated by the process $\widehat{b}_{1n} = (nh)^{\frac{1}{2}}(\widehat{\lambda}_{1n} - \lambda_0)$ and $(nh)^{\frac{1}{2}}(\widehat{\lambda}_{1n} - \lambda_{1n})$ converges weakly under H_n to the centered Gaussian process W_0, therefore $\widehat{b}_{1n}$ converges weakly to $W_0 + b_1$, the limit of T_n is deduced. □

5.12 Counting process with a multiplicative intensity

On a filtered probability space $(\Omega, \mathcal{A}, \mathbb{F}, P_0)$, let N be an adapted counting process observed on the increasing interval $[0, n]$ and let

$$N_n(t) = N(nt) = \sum_{i \geq 1} 1\{T_i \leq nt\}$$

denote the normalized process on $[0,1]$. We assume that N_n has a multiplicative predictable process $\widetilde{N}_{0n}(t) = \int_0^t Y_n(s)\lambda_0(s)\,ds$ with respect to $(\mathbb{F}, P_0)$, where the process $Y_n(t) = Y(nt)$ is predictable. Under a probability measure P, the predictable process of N_n is $\widetilde{N}_n(t) = \int_0^t Y_n(s)\lambda(s)\,ds$. Examples of such processes are the renewal processes and the semi-markovian jump processes, $N_n(t)$ being the number of jumps between two consecutive transient states, when the variables T_i are the duration times between two jumps.

Under a probability measure P, the function λ is defined by the model (5.21), with a change-point at an unknown point $\gamma > 0$ under P_γ. Then the process

$$\begin{aligned} M_n(t) &= n^{-\frac{1}{2}}\Big\{N_n(t) - \int_0^t Y_n(s)\lambda(s)\,ds\Big\} \\ &= n^{-\frac{1}{2}}\Big\{N_n(t) - \int_0^{t\wedge\gamma} Y_n(s)\lambda_1(s)\,ds - \int_{t\wedge\gamma}^t Y_n(s)\lambda_2(s)\,ds\Big\}, \end{aligned}$$

is the martingale of the compensated jumps of N_n on $[0,1]$. At fixed γ, the estimators of the cumulative intensities in each phase, $\widehat{\Lambda}_{1n,\gamma}$ and $\widehat{\Lambda}_{1n,\gamma}$, are defined by (5.22) and the change-point is estimated by the same methods as in the previous sections.

The log-likelihood ratio of the distribution of N_n under P_γ with respect to P_0 is written like for a censored sample, up to a variable which does not depend on the parameters, as the difference of the log-likelihood processes

$$\begin{aligned} l_n(\gamma) - l_{0n} &= \int_0^{\gamma_0} \log\frac{\lambda_1}{\lambda_{01}}\,dM_{1n} + \int_0^{\gamma_0} \psi(\lambda_1, \lambda_{01})Y_{1n}\,ds \\ &+ \int_{\gamma_0}^1 \log\frac{\lambda_2}{\lambda_{02}}\,dM_{2n} + \int_{\gamma_0}^1 \psi(\lambda_2, \lambda_{02})Y_{2n}\,ds \\ &+ \int_\gamma^{\gamma_0} \log\frac{\lambda_2}{\lambda_1}\,dN_n - \int_\gamma^{\gamma_0} (\lambda_2 - \lambda_1)Y_n\,ds. \end{aligned}$$

We assume that the functions λ_1 and λ_2 belong to $C_b^2(]0,\gamma])$ and, respectively $C_b^2(]\gamma,1[)$, and there exists of a function y on $[0,1]$ such that

$$\lim_n \sup_{t\in[0,1]} |n^{-1}Y_n(t) - y(t)| = 0, \text{ a.s.}$$

The converge of the process $n^{-1}Y_n$ implies $X_n = n^{-1}\{l_n(\gamma) - l_{0n}\}$ converges a.s. to the function X defined in Section 5.7, it has the bound given by Lemma 5.5.

In models with known or parametric functions λ_1 and λ_2, the estimators of the parameters are a.s. consistent. Under a weak law of large numbers for the process $n^{\frac{1}{2}}(n^{-1}Y_n - y)$, Lemma 5.6 or a similar bound for the nonparametric cases are still satisfied which yield the convergence rate of the estimators and the weak convergence of $n(\widehat{\gamma}_n - \gamma_0)$ to the location of the maximum of a Gaussian process is proved under the same conditions.

If the intensity functions are unknown, the cumulative intensity functions Λ_1 and Λ_2 are estimated by their empirical estimators (5.20) and the minimization of the estimated variance of the process $W_{\Lambda,n}$ provides an estimator of the change-point by the same procedure as in Section 5.8. The maximum likelihood estimation of the change-point require the nonparametric estimation of the intensity functions and the asymptotic results of Sections 5.10 and 5.11 are still valid.

Chapter 6

Change-points in proportional hazards model

Abstract. This chapter studies regression models for hazard functions, with covariate processes and change-points at unknown time or covariate thresholds. We consider the nonparametric estimator of the baseline cumulative hazard function and the estimators that maximize the partial likelihood for the regression and the change-point parameters, their convergence rates are established. Tests for the hypothesis of models without changes are defined by the same methods. The weak convergences of the estimators and the test statistics for the hypothesis of models without change are proved.

6.1 Proportional hazards with changes

The proportional hazards regression model introduced by Cox (1972) assumes that conditionally on a vector of covariates Z, the hazard function of a random time T^0 observed under right-censoring on an interval $[0, a]$ is

$$\lambda(t \mid Z) = \lambda(t) \exp\{\beta^T Z\},$$

where β is a vector of unknown regression parameters and λ is an unknown and unspecified baseline hazard function. Inference on the regression parameters is based on a partial likelihood and the asymptotic properties of the estimators of β and of the cumulative hazard function gave raise to many papers, among them Cox (1975), Tsiatis (1981) for time-independent covariates, Andersen and Gill (1982) and Prentice and Self (1983) in a more general set-up. Several authors also considered a non-regular model with a two-phase regression on time-dependent covariates defined by a change-point at an unknown time.

Let $Z = (Z_1^T, Z_2^T, Z_3)^T$ be a vector of time-dependent covariates, where Z_1 and Z_2 are respectively p and q-dimensional left-continuous processes with right hand limits and Z_3 is a real valued process. First, we assume that conditionally on Z the hazard function of a survival time T^0 has the form

$$\lambda_\theta(t \mid Z) = \lambda(t) \exp\{r_\theta(Z(t))\} \tag{6.1}$$

with a change of regression on Z_2 according to an unknown threshold of the covariate Z_3

$$r_\theta(Z(t)) = \alpha^T Z_1(t) + \beta^T Z_2(t) 1_{\{Z_3(t) \leq \zeta\}} + \gamma^T Z_2(t) 1_{\{Z_3(t) > \zeta\}}, \tag{6.2}$$

where $\theta = (\zeta, \xi^T)^T$, with $\xi = (\alpha^T, \beta^T, \gamma^T)^T$ the vector of the regression parameters, and λ is an unknown baseline hazard function. The regression parameters α, and respectively β and γ, belong respectively to bounded subsets of $\mathbb{R}^p$, and respectively $\mathbb{R}^q$, the threshold ζ is a parameter lying in a bounded interval $]\zeta_1, \zeta_2[$ strictly included in the support of Z_3. The true parameter values θ_0 and λ_0 are supposed to be identifiable, that is, θ_0 is such that $\beta_0 \neq \gamma_0$ and a change-point actually occurs at ζ_0. In the same framework, a simpler model was defined by adding a constant to the regression on a covariate Z_1 after a change-point according to another variable Z_2, $r_\theta(Z(t)) = \alpha^T Z_1(t) + \beta 1_{\{Z_2 \leq \zeta\}}$.

Secondly, we consider a model (6.1) where the hazard rate of T_0 conditionally on $Z = (Z_1^T, Z_2^T)^T$ has a change of the regression parameter on Z_2 according to an unknown time τ

$$r_\theta(Z(t)) = \alpha^T Z_1(t) + (\beta + \gamma 1_{\{t > \tau\}})^T Z_2(t), \tag{6.3}$$

the parameter is now $\theta = (\tau, \xi^T)^T$ with $\xi = (\alpha^T, \beta^T, \gamma^T)^T$ with a threshold τ belonging to an open sub-interval strictly included in the observation interval $[0, a]$ of the time variable T^0, and λ is an unknown baseline hazard function bounded in $[0, a]$. Models (6.2) and (6.3) generalize the models previously studied with random variables Z.

We suppose that the time variable T^0 with hazard function (6.1) may be right-censored at a non-informative censoring time C such that C is independent of T^0 conditionally on the process Z. Let $(\Omega, \mathcal{F}, P_{\theta,\lambda})_{\theta,\lambda}$ be a family of complete probability spaces provided with a history $\mathbb{F} = (\mathcal{F}_t)_t$, where $\mathcal{F}_t \subseteq \mathcal{F}$ is an increasing and right-continuous filtration. We assume that under $P_{\theta,\lambda}$, T^0 satisfies (6.1), C and Z having the same distribution under all probabilities $P_{\theta,\lambda}$. Under the true parameter values, let $P_0 = P_{\theta_0,\lambda_0}$ and let E_0 be the expectation of the random variables. The observations are a sequence of censored times $T_i = T_i^0 \wedge C_i$ and the censoring indicators $\delta_i = 1_{\{T_i^0 \leq C_i\}}$ such that the processes $N(t) = \sum_{i \geq 1} \delta_i 1_{\{T_i \leq t\}}$ and Z are $\mathbb{F}$-adapted.

The inference will be based on a sample $(T_i, \delta_i, Z_i)_{1\leq i\leq n}$ of n independent and identically distributed observations or it satisfies an uniform law of large numbers. As in the classical Cox model for i.i.d. individuals, we assume that the variables T_i are observed on a time interval $[0, a]$ such that $Pr(T \geq a) > 0$. In the model (6.1), θ_0 is estimated by the value $\widehat{\theta}_n$ that maximizes the partial likelihood

$$L_n(\theta) = \prod_{i=1}^{n}\Big\{\frac{\exp\{r_\theta(Z_i(T_i))\}}{\sum_j Y_j(T_i)\exp\{r_\theta(Z_j(T_i))\}}\Big\}^{\delta_i} \tag{6.4}$$

where $Y_i(t) = 1_{\{T_i\geq t\}}$ indicates whether individual i is still under observation at t. Let

$$S_n^{(0)}(t;\theta) = \sum_{i=1}^{n} Y_i(t)\exp\{r_\theta(Z_i(t))\},$$

the logarithm of the partial likelihood is

$$l_n(\theta) = \log L_n(\theta) = \sum_{i=1}^{n} \delta_i\{r_\theta(Z_i(T_i)) - \log S_n^{(0)}(T_i;\theta)\} \tag{6.5}$$

and the estimator of the cumulative baseline hazard function Λ_0 is

$$\widehat{\Lambda}_n(t) = \sum_{i=1}^{n}\int_0^t \frac{dN_i(s)}{S_n^{(0)}(s;\widehat{\theta}_n)} \tag{6.6}$$

where $N_i(t) = \delta_i 1_{\{T_i\leq t\}}$. In models (6.1) and (6.2), the estimator $\widehat{\theta}_n$ is obtained in two steps procedure by maximization of the logarithm of the partial likelihood l_n with respect to ξ at a fixed value of the change-point parameters, then the estimated partial likelihood, where the parameter ξ is replaced by this estimator, is maximized with respect to the change-point parameter.

The maximum likelihood in models (6.1) differs from the inference in linear regression models due to the partial likelihood L_n which cannot be simply related to random walks because all individual contributions depend on the process $S_n^{(0)}$, they are therefore all dependent and it is not possible to split (6.5) into terms for individuals with a covariate before or after the threshold.

The results are similar in a multiplicative regression model where the exponential linear model of the intensity is replaced by a multiplicative model $\lambda(t)r(\theta, Z(t))$ with a change-point in the form

$$r(\theta, Z(t)) = r(\theta_1, Z(t)1_{Z(t)\leq\gamma})r(\theta_2, Z(t)1_{Z(t)>\gamma})$$

for a model with change-point at a threshold of the covariate and

$$r(\theta, Z(t)) = r(\theta_1, Z(t)1_{t\le\gamma})r(\theta_2, Z(t)1_{t>\gamma}),$$

with a time change-point. When the indicator is zero, the function r reduces to an unidentifiable constant and the other terms of the parametric regression model are identifiable up to this multiplicative constant, therefore the convention is $r(\theta, 0) \equiv 1$. The parameters of the other multiplicative terms with other covariates are then identifiable.

6.2 Change-point at a covariate threshold

In model (6.1), for a fixed value of the change-point parameter ζ, the regression parameter ξ belonging to a space Ξ is estimated by

$$\widehat{\xi}_n(\zeta) = \arg\max_{\xi\in\Xi} l_n(\zeta, \xi),$$

and the estimated partial likelihood $\widehat{l}_n(\zeta) = l_n(\zeta, \widehat{\xi}_n(\zeta))$ is maximum at the estimator $\widehat{\zeta}_n$ of ζ_0 which satisfies the relationship

$$\widehat{\zeta}_n = \inf\Big\{\zeta \in [\zeta_1, \zeta_2] : \max\{\widehat{l}_n(\zeta^-), \widehat{l}_n(\zeta)\} = \sup_{\zeta\in[\zeta_1,\zeta_2]} \widehat{l}_n(\zeta)\Big\}, \qquad (6.7)$$

where $\widehat{l}_n(\zeta^-)$ denotes the left-hand limit of $\widehat{l}_n$ at ζ. This defines the maximum likelihood estimator of ξ_0 as $\widehat{\xi}_n = \widehat{\xi}_n(\widehat{\zeta}_n)$ and $\widehat{\theta}_n = (\widehat{\zeta}_n, \widehat{\xi}_n^T)^T$.

Assumptions and notation for the asymptotic properties of the estimators for i.i.d. individuals are given in the following. The processes Z_1 and Z_2 have left-continuous sample paths with right-hand limits, with values in sets $\mathcal{Z}_1 \subset \mathbb{R}^p$ and $\mathcal{Z}_2 \subset \mathbb{R}^q$. The process Z_3 has its values in a subset $\mathcal{Z}_3$ of $\mathbb{R}$. For t in $[0, a]$, $\theta = (\zeta, \xi^T)^T$ and $k = 0, 1, 2$, we denote the covariate vector in the model (6.2) with a threshold at ζ and the derivatives of $S_n^{(0)}$ with respect to the regression parameters by

$$\widetilde{Z}(t;\zeta) = \Big(Z_1^T(t), Z_2^T(t)1_{\{Z_3(t)\le\zeta\}}, Z_2^T(t)1_{\{Z_3(t)>\zeta\}}\Big)^T,$$

$$S_n^{(k)}(t;\theta) = \sum_i Y_i(t)\widetilde{Z}_i^{\otimes k}(t;\zeta)\exp\{r_\theta(Z_i(t)\},\ k = 0, 1, 2,$$

where $x^{\otimes 0} = 1$, $x^{\otimes 1} = x$ and $x^{\otimes 2} = xx^T$, for x in $\mathbb{R}^{p+2q}$. For $1 \le i \le n$, let $N_i(t) = \delta_i 1_{\{T_i\le t\}}$ be the counting process for individual i and let

$$M_i(t) = N_i(t) - \int_0^t Y_i(s)\exp\{r_{\theta_0}(Z_i(s))\}\,d\Lambda_0(s)$$

be the martingale of the compensated jumps of N_i on $[0, a]$. We also denote their normalized sums as $\bar{N}_n = \sum_{i \le n} N_i$ and

$$\begin{aligned}
\mathcal{M}_n^{(0)}(t) &= n^{-\frac{1}{2}} \left\{ \bar{N}_n(t) - \int_0^t S_n^{(0)}(\theta_0)\, d\Lambda_0 \right\}, \\
\mathcal{M}_n^{(1)}(t) &= n^{-\frac{1}{2}} \left\{ \sum_i \int_0^t \widetilde{Z}_i(\zeta_0)\, dN_i - \int_0^t S_n^{(1)}(\theta_0)\, d\Lambda_0 \right\} \\
&= n^{-\frac{1}{2}} \sum_i \int_0^t \widetilde{Z}_i(\zeta_0)\, dM_i. \qquad (6.8)
\end{aligned}$$

Adapting the classical notation, we define

$$\begin{aligned}
s^{(k)}(t;\theta) &= E_0[Y_i(t)\widetilde{Z}_i^{\otimes k}(t;\zeta) \exp\{r_\theta(Z_i(t)\}], \\
V_n(t;\theta) &= \{S_n^{(2)} S_n^{(0)-1} - [S_n^{(1)} S_n^{(0)-1}]^{\otimes 2}\}(t;\theta), \\
v(t;\theta) &= \{s^{(2)} s^{(0)-1} - [s^{(1)} s^{(0)-1}]^{\otimes 2}\}(t;\theta), \\
I(\theta) &= \int_0^a v(s;\theta) s^{(0)}(s;\theta_0) \lambda_0(s)\, ds.
\end{aligned}$$

We denote the first p components of $s^{(1)}$ by

$$s_1^{(1)}(t;\theta) = E_0[Y_i(t) Z_{1i}(t) \exp\{r_\theta(Z_i(t)\}].$$

Let also $s_2^{(1)-}(\theta)$ and $s_2^{(1)+}(\theta)$ be the q-dimensional components of $s^{(1)}$ related to the component Z_2 of Z under restrictions on the location of Z_3 with respect to the parameter ζ,

$$\begin{aligned}
s_2^{(1)-}(t;\theta) &= E_0[Y_i(t) Z_{2i}(t) 1_{\{Z_{3i}(t) \le \zeta\}} \exp\{\alpha^T Z_{1i}(t) + \beta^T Z_{2i}(t)\}], \\
s_2^{(1)+}(t;\theta) &= E_0[Y_i(t) Z_{2i}(t) 1_{\{Z_{3i}(t) > \zeta\}} \exp\{\alpha^T Z_{1i}(t) + \gamma^T Z_{2i}(t)\}].
\end{aligned}$$

For $\zeta < \zeta'$, let

$$\begin{aligned}
s_2^{(1)}(]\zeta,\zeta'],\alpha,\beta) &= s_2^{(1)-}(\zeta',\alpha,\beta) - s_2^{(1)-}(\zeta,\alpha,\beta), \\
s_2^{(1)}(]\zeta,\zeta'],\alpha,\gamma) &= s_2^{(1)+}(\zeta,\alpha,\gamma) - s_2^{(1)+}(\zeta',\alpha,\gamma).
\end{aligned}$$

Similar notation is used for the processes $S_n^{(k)}$,

$$\begin{aligned}
S_n^{(k)-}(t;\theta) &= \sum_i Y_i(t) \widetilde{Z}_i^{\otimes k}(t;\zeta) 1_{\{Z_{3i}(t) \le \zeta\}} \exp\{\alpha^T Z_{1i}(t) + \beta^T Z_{2i}(t)\}, \\
S_n^{(k)+}(t;\theta) &= \sum_i Y_i(t) \widetilde{Z}_i^{\otimes k}(t;\zeta) 1_{\{Z_{3i}(t) > \zeta\}} \exp\{\alpha^T Z_{1i}(t) + \gamma^T Z_{2i}(t)\}, \\
S_{1n}^{(1)}(t;\theta) &= \sum_i Y_i(t) Z_{1i}(t) \exp\{r_\theta(Z_i(t)\}\},
\end{aligned}$$

$$S_{2n}^{(1)-}(t;\theta) = \sum_i Y_i(t)Z_{2i}(t)1_{\{Z_{3i}(t)\le\zeta\}} \exp\{\alpha^T Z_{1i}(t) + \beta^T Z_{2i}(t)\},$$

$$S_{2n}^{(1)+}(t;\theta) = \sum_i Y_i(t)Z_{2i}(t)1_{\{Z_{3i}(t)>\zeta\}} \exp\{\alpha^T Z_{1i}(t) + \gamma^T Z_{2i}(t)\}, \text{ etc.}$$

$S_n^{(k)\pm}(t;]\zeta,\zeta'],\xi)$ is similar to $S_n^{(k)\pm}(t;\theta)$ with a restriction of the covariates Z_{3i} to the interval $]\zeta,\zeta']$ and $S_n^{(k)-1}$ denotes the inverse of $S_n^{(k)}$.

Using the logarithm of the partial likelihood (6.5), the estimator $\widehat{\theta}_n$ maximizes the process

$$\begin{aligned} X_n(\theta) &= n^{-1}\{l_n(\theta) - l_n(\theta_0)\} \\ &= n^{-1}\sum_{i\le n}\Big\{(r_\theta - r_{\theta_0})(Z_i(T_i)) - \log\frac{S_n^{(0)}(T_i;\theta)}{S_n^{(0)}(T_i;\theta_0)}\Big\} \qquad (6.9) \\ &= n^{-1}\sum_{i\le n}\int_0^a \Big\{(r_\theta - r_{\theta_0})(Z_i(t)) - \log\frac{S_n^{(0)}(t;\theta)}{S_n^{(0)}(t;\theta_0)}\Big\}\, dN_i(t) \end{aligned}$$

and we define the function

$$\begin{aligned} X(\theta) = \int_0^a \Big\{&(\alpha-\alpha_0)^T s_1^{(1)}(\theta_0) + (\beta-\beta_0)^T s_2^{(1)-}(\zeta\wedge\zeta_0,\alpha_0,\beta_0) \\ &+(\gamma-\gamma_0)^T s_2^{(1)+}(\zeta\vee\zeta_0,\alpha_0,\gamma_0) + (\beta-\gamma_0)^T s_2^{(1)}(]\zeta_0,\zeta],\alpha_0,\gamma_0) \\ &+(\gamma-\beta_0)^T s_2^{(1)}(]\zeta,\zeta_0],\alpha_0,\beta_0) - s^{(0)}(\theta_0)\log\frac{s^{(0)}(\theta)}{s^{(0)}(\theta_0)}\Big\}\, d\Lambda_0. \qquad (6.10) \end{aligned}$$

The quadratic norms of vectors in $\mathbb{R}^{p+2q}$ and matrices in $(R^{p+2q})^{\otimes 2}$ are denoted $\|\cdot\|$. The asymptotic properties of the estimators will be established under the following conditions:

C1. The variable $Z_3(t)$ has a density $h_3(t,\cdot)$ which is strictly positive, bounded and continuous in a neighborhood of ζ_0, $\sup_{t\in[0,a]}\lambda_0(t)$ is finite and $P_0(T\ge a)>0$.

C2. The parameter space Ξ is bounded and there exists a convex and bounded parameter space Θ including θ_0 such that for $k=0,1,2$, the means

$$E_0 \sup_{t\in[0,1]}\sup_{\theta\in\Theta}\{(\|Z_1(t)\|^k + \|Z_2(t)\|^k)e^{r_\theta(Z(t))}\}^2, \qquad (6.11)$$

$$\sup_{z\in[\zeta_1,\zeta_2]} E_0\Big[\sup_{t\in[0,1]}\sup_{\theta\in\Theta}\{(\|Z_1(t)\|^k + \|Z_2(t)\|^k)e^{r_\theta(Z(t))}\}^j \mid Z_3(t)=z\Big]$$

are finite, for $j = 1, 2$, and

$$\sup_{z,z' \in [\zeta_1,\zeta_2]} \sup_{t\in[0,1]} \sup_{\theta\in\Theta} |E_0\{e^{r_\theta(Z(t))} \mid Z_3(t) = z\} - E_0\{e^{r_\theta(Z(t))} \mid Z_3(t) = z'\}|$$

converges to zero as $|z - z'|$ converges to zero.

C3. The variables $\sup_{t\in[0,1]} \sup_{\theta\in\Theta} \|n^{-1}S_n^{(k)}(t;\theta) - s^{(k)}(t;\theta)\|$ converge a.s. to zero under P_0, $k = 0, 1, 2$.

Under Condition C2, the variance

$$\int_0^a E_0 \inf_\beta [Y(t)\{(\beta_0 - \gamma_0)^T Z_2(t)\}^{\otimes 2} e^{\alpha_0^T Z_1(t) + \beta^T Z_2(t)} \mid Z_3(t) = \zeta_0] \, d\Lambda_0, \tag{6.12}$$

is positive definite, where the infimum is over β between β_0 and γ_0. If Z is a random variable, Condition C3 is satisfied by the Glivenko–Cantelli theorem. If Z_1 or Z_2 are processes, it may be proved by the arguments of Theorem 4.1, of Andersen and Gill (1982).

6.3 Convergence of the estimators

The proof of the consistency of the estimators $\widehat{\zeta}_n$ and $\widehat{\xi}_n$ is based on the a.s. convergence of X_n to X, uniformly in Θ, and on properties of X in the neighborhood of θ_0. They rely on the following lemmas.

Lemma 6.1. *Under Conditions* C1–C3, *the process $n^{-1}\bar{N}_n$ converges a.s. under P_0 to the function $\widetilde{N_0}(t) = \int_0^t s^{(0)}(\theta_0)\, d\Lambda_0$ and the process $n^{-1}X_n(\theta)$ converges a.s. under P_0 to the function X, uniformly on Θ.*

This is a simple consequence of the conditions and of the Glivenko–Cantelli theorem. The local variations of the function X in Ξ are deduced from the next lemma. The functions $s^{(k)}$, $k = 0, 1, 2$, are sums of mean integrals for $Z_3(t) \le \zeta$ and $Z_3 > \zeta$ so they have a left and right first order derivatives and expansions with respect to ζ.

Lemma 6.2. *Under Conditions* C1–C2, *$s^{(0)}$ is bounded away from zero on $[0, a] \times \Theta$, $s^{(1)}(t; \zeta, \xi)$ and $s^{(2)}(t; \zeta, \xi)$ are the first two partial derivatives of $s^{(0)}(t; \zeta, \xi)$ with respect to ξ, and the functions $s^{(k)}$ are continuous on Θ, uniformly in $t \in [0, a]$, for $k = 0, 1, 2$. As $\|\theta - \theta'\|$ converges to zero,*

$s^{(k)}(t;\theta') - s^{(k)}(t;\theta) = O(|\zeta - \zeta'| + \|\xi - \xi'\|)$ *and*

$$s^{(0)}(\theta') - s^{(0)}(\theta) = (\xi' - \xi)^T s^{(1)}(\theta) + \frac{1}{2}(\xi' - \xi)^T s^{(2)}(\theta)(\xi' - \xi) + (\zeta' - \zeta)\dot{s}^{(0)}_{\zeta}(\theta) + o(|\zeta - \zeta'| + \|\xi - \xi'\|^2)$$

uniformly on $[0, a] \times \Theta$, $\dot{s}^{(0)}_{\zeta}(\theta) = h_3(\zeta)\, E_0\{e^{\alpha^T Z_1}(e^{\beta^T Z_2} - e^{\gamma^T Z_2}) \mid Z_3 = \zeta\}$.

Theorem 6.1. *Under conditions* C1–C3, *there exists a neighborhood* $\mathcal{B}_0$ *of* θ_0 *such that if* $\widehat{\theta}_n$ *lies in* $\mathcal{B}_0$, *then it converges in probability to* θ_0 *as* n *tend to infinity.*

Proof. For every θ in Θ, the first derivatives of the function X with respect to α, β and γ are zero at θ_0 and the second derivative of the function $X(\theta)$ with respect to ξ, at fixed ζ, is the matrix $-I(\theta)$. The assumptions that λ_0 is bounded and $Var\,\widetilde{Z}(t;\zeta)$ is positive definite imply that $I(\theta)$ is positive definite in a neighborhood of θ_0, and the function $\xi \mapsto X(\zeta, \xi)$ is concave for every θ in a neighborhood of θ_0.

In a neighborhood of θ_0, X has partial derivatives with respect to ζ, at fixed ξ, $\dot{X}^-_{\zeta}(\zeta, \xi)$ for $\zeta < \zeta_0$ and $\dot{X}^+_{\zeta}(\zeta, \xi)$ for $\zeta > \zeta_0$

$$\dot{X}^-_{\zeta}(\theta) = \int_0^a E_0\Big[Y(t)\Big\{(\beta - \gamma)^T Z_2(t) e^{\alpha_0^T Z_1(t) + \beta_0^T Z_2(t)} - e^{\alpha^T Z_1(t)}(e^{\beta^T Z_2(t)} - e^{\gamma^T Z_2(t)}) \frac{s^{(0)}(t;\theta_0)}{s^{(0)}(t;\theta)}\Big\}\Big]\, d\Lambda_0(t),$$

$$\dot{X}^+_{\zeta}(\theta) = \int_0^a E_0\Big[Y(t)\Big\{(\beta - \gamma)^T Z_2(t) e^{\alpha_0^T Z_1(t) + \gamma_0^T Z_2}(t) - e^{\alpha^T Z_1(t)}(e^{\beta^T Z_2(t)} - e^{\gamma^T Z_2(t)}) \frac{s^{(0)}(t;\theta_0)}{s^{(0)}(t;\theta)}\Big\}\Big]\, d\Lambda_0(t).$$

If θ tends to θ_0 with $\zeta < \zeta_0$, the continuity of $s^{(0)}(t;\theta)$ with respect to θ (Lemma 6.2) implies that $\dot{X}^-_{\zeta}(\theta)$ tends to

$$\dot{X}^-_{\zeta}(\theta_0) = \int_0^a h_3(t;\zeta_0) E_0\Big[Y e^{\alpha_0^T Z_1(t)}\Big\{(\beta_0 - \gamma_0)^T Z_2(t) e^{\beta_0^T Z_2(t)} + e^{\gamma_0^T Z_2(t)} - e^{\beta_0^T Z_2(t)}\Big\} \mid Z_3(t) = \zeta_0\Big] h_3(t;\zeta)\, d\Lambda_0(t) \tag{6.13}$$

and there exist β_* between β_0 and γ_0, and $\widetilde{\beta}$ betweenβ_0 and β_* such that

$$\begin{aligned}&(\beta_0 - \gamma_0)^T Z_2(t) e^{\beta_0^T Z_2(t)} + e^{\gamma_0^T Z_2(t)} - e^{\beta_0^T Z_2(t)}\\ &= (\beta_0 - \gamma_0)^T Z_2(t) e^{\beta_0^T Z_2(t)} - (\beta_0 - \gamma_0)^T Z_2(t) e^{\beta_*^T Z_2(t)}\\ &= (\beta_0 - \gamma_0)^T Z_2(t) (\beta_0 - \beta_*)^T Z_2(t) e^{\widetilde{\beta}^T Z_2(t)}\end{aligned}$$

where $(\beta_0 - \gamma_0)^T Z_2$ and $(\beta_0 - \beta_*)^T Z_2$ have the same sign. By condition (6.12), $\dot{X}_\zeta^-(\theta_0)$ is strictly positive therefore $\dot{X}_\zeta^-(\theta)$ is strictly positive in a neighborhood of θ_0. Similarly, if θ tends to θ_0 with $\zeta < \zeta_0$, $\dot{X}_\zeta^+(\theta)$ tends to

$$\dot{X}_\zeta^+(\theta_0) = \int_0^a h_3(t;\zeta_0) E_0\Big[Y(t) e^{\alpha_0^T Z_1(t)} \Big\{(\beta_0 - \gamma_0)^T Z_2(t) e^{\gamma_0^T Z_2(t)} \qquad (6.14)$$
$$+ e^{\gamma_0^T Z_2(t)} - e^{\beta_0^T Z_2(t)}\Big\} \mid Z_3(t) = \zeta_0^+\Big]\, d\Lambda_0(t),$$

where

$$(\beta_0 - \gamma_0)^T Z_2(t) e^{\gamma_0^T Z_2(t)} = (\beta_0 - \gamma_0)^T Z_2(t) (\gamma_0 - \beta_*)^T Z_2(t) e^{\widetilde{\beta}^T Z_2(t)}$$

and it is strictly negative. This implies the existence of a neighborhood $\mathcal{B}_0$ of θ_0 where X attains a strict maximum at θ_0 and where X is concave. As X_n converges uniformly to X (Lemma 6.1), the consistency of $\widehat{\theta}_n$ follows, by the same arguments as in the previous Chapters. □

Let W_n be the partial log-likelihood process defined by

$$W_n(\theta) = n^{\frac{1}{2}}(X_n - X)(\theta), \qquad (6.15)$$

with X_n and X given by (6.9) and (6.10). The rates of convergence of $\widehat{\zeta}_n$ and $\widehat{\xi}_n$ are deduced from the limiting behaviour of the process W_n following the arguments of Theorem 3.2 and using the next lemmas. Let

$$\mathcal{U}_n = \{u_n = (u_{1n}, u_{2n}^T)^T : u_{1n} = n(\zeta_n - \zeta_0), u_{2n} = n^{\frac{1}{2}}(\xi_n - \xi_0),$$
$$\zeta_n \in [\zeta_1, \zeta_2], \xi_n \in \Xi\}.$$

For $x = (x_1, x_2^T)^T$ with x_1 in $\mathbb{R}$ and x_2 in $\mathbb{R}^{p+2q}$, let $\rho(x) = (|x_1| + \|x_2\|^2)^{\frac{1}{2}}$ and let $V_\varepsilon(\theta_0)$ an ε-neighborhood of θ_0 with respect to the semi-norm ρ. Reversely, for $u = (u_1, u_2^T)^T$ in $\mathcal{U}_n$, let $\zeta_{n,u} = \zeta_0 + n^{-1}u_1$, let $\xi_{n,u} = \xi_0 + n^{-\frac{1}{2}}u_2$ and $\theta_{n,u} = (\zeta_{n,u}, \xi_{n,u}^T)^T$ in Θ, and let $\mathcal{U}_{n,\varepsilon} = \{u \in \mathcal{U}_n : \rho(u) \leq n^{\frac{1}{2}}\varepsilon\}$.

Lemma 6.3. *Under Conditions* C1–C3, *for ε sufficiently small, there exists a constant $\kappa_0 > 0$ such that for every θ in $V_\varepsilon(\theta_0)$, $X(\theta) \geq -\kappa_0\{\rho(\theta, \theta_0)\}^2$.*

Proof. Since $X(\theta_0)$ and $\dot{X}_\xi(\theta_0)$ are zero, a Taylor expansion of the function X, for ε sufficiently small and for θ in $V_\varepsilon(\theta_0)$, implies

$$X(\theta) = -|\zeta - \zeta_0| \dot{X}_\zeta^-(\theta_0) - \frac{1}{2}(\xi - \xi_0)^T I(\theta^*)(\xi - \xi_0) + o(|\zeta - \zeta_0|), \; \zeta < \zeta_0,$$
$$= -\frac{1}{2}(\xi - \xi_0)^T I(\theta^*)(\xi - \xi_0) + o(|\zeta - \zeta_0|), \; \zeta < \zeta_0,$$
$$X(\theta) = -\frac{1}{2}(\xi - \xi_0)^T I(\theta^*)(\xi - \xi_0) + o(|\zeta - \zeta_0|), \; \zeta > \zeta_0,$$

whore θ^* is between θ and θ_0 and with (6.13) and (6.14). The matrix $I(\theta^*)$ is positive definite for all θ^* in a neighborhood of θ_0 and by Lemma 6.1, $\|I(\theta) - I(\theta_0)\|$ tends to zero with $\rho(\theta, \theta_0)$, moreover, $\dot{X}_{\zeta}^{-}(\theta_0)$ is strictly positive if $\zeta < \zeta_0$ and strictly negative if $\zeta > \zeta_0$ then the result follows. □

Lemma 6.4. *Under Conditions* C1–C3, *for every* $\varepsilon > 0$, *there exists a constant* $\kappa_1 > 0$ *such that for* n *large enough* $E_0 \sup_{\theta \in V_\varepsilon(\theta_0)} |W_n(\theta)| \leq \kappa_1 \varepsilon$, *as* n *tends to infinity.*

Proof. The process W_n is the difference $W_{1n} - W_{2n}$ of the processes

$$W_{2n}(\theta) = n^{-\frac{1}{2}} \sum_i \Big[\log \frac{S_n^{(0)}(T_i;\theta)}{S_n^{(0)}(T_i;\theta_0)} - \int_0^a \log\Big\{\frac{s^{(0)}(\theta)}{s^{(0)}(\theta_0)}\Big\} s^{(0)}(\theta_0)\, d\Lambda_0\Big]$$

and

$$\begin{aligned}
W_{1n}(\theta) &= n^{\frac{1}{2}} \Big[n^{-1} \sum_i \int_0^a \{r_\theta(Z_i(t)) - r_{\theta_0}(Z_i(t))\}\, dN_i(t) \\
&\quad - E_0 \int_0^a \{r_\theta(Z(t)) - r_{\theta_0}(Z(t))\} e^{r_{\theta_0}(Z(t))}\, d\Lambda_0(t)\Big], \\
&= n^{-\frac{1}{2}} \sum_{i \leq n} [(\alpha - \alpha_0)^T \int_0^a \{Z_{1i}\, dN_i - s_1^{(1)}(\theta_0)\, d\Lambda_0\} \\
&\quad + (\beta - \beta_0)^T \int_0^a \{Z_{2i} 1_{\{Z_{3i} \leq \zeta_0\}}\, dN_i - s_2^{(1)-}(\zeta_0, \alpha_0, \beta_0)\, d\Lambda_0\} \\
&\quad + (\gamma - \gamma_0)^T \int_0^a \{Z_{2i} 1_{\{Z_{3i} > \zeta_0\}}\, dN_i - s_2^{(1)+}(\zeta_0, \alpha_0, \gamma_0)\, d\Lambda_0\} \\
&\quad + (\beta - \gamma)^T \int_0^a \{Z_{2i} 1_{\{\zeta_0 < Z_{3i} \leq \zeta\}}\, dN_i - s_2^{(1)}(]\zeta_0, \zeta], \alpha_0, \gamma_0)\, d\Lambda_0\} \\
&\quad + (\gamma - \beta)^T \int_0^a \{Z_{2i} 1_{\{\zeta < Z_{3i} \leq \zeta_0\}}\}\, dN_i - s_2^{(1)}(]\zeta, \zeta_0], \alpha_0, \beta_0)\, d\Lambda_0\}].
\end{aligned}$$

Let $\mathcal{G}_n^{(k)}(\theta) = n^{\frac{1}{2}}(n^{-1} S_n^{(k)} - s^{(k)})(\theta)$ be the empirical process associated to $S_n^{(1)}$ and let $\mathcal{G}_n^{(1)}(\theta, \theta_0)$ be the empirical process defined in the same way but with the parameter θ_0 in the exponential, the process W_{1n} is the sum

$$\begin{aligned}
W_{1n}(\theta) &= n^{-\frac{1}{2}} \sum_i \int_0^a \{r_\theta(Z_i(t)) - r_{\theta_0}(Z_i(t))\}\, dM_i(t) \\
&\quad + n^{\frac{1}{2}} \int_0^a \Big[n^{-1} \sum_i \{r_\theta(Z_i(t)) - r_{\theta_0}(Z_i(t))\} e^{r_{\theta_0}(Z_i(t))} \\
&\quad - E_0\{r_\theta(Z(t)) - r_{\theta_0}(Z(t))\} e^{r_{\theta_0}(Z(t))}\Big]\, d\Lambda_0(t),
\end{aligned}$$

its norm $E_0 \sup_{\theta\in V_\varepsilon(\theta_0)} \|W_{1n}(\theta)\|_2$ is bounded by the sum of the norms

$$n^{-1}\sum_i\Big\{\int_0^a E_0 \sup_{\theta\in V_\varepsilon(\theta_0)} |r_\theta(Z_i(t)) - r_{\theta_0}(Z_i(t))|^2 e^{r_{\theta_0}(Z_i(t))}\, d\Lambda_0(t)\Big\}^{\frac{1}{2}}$$
$$+n^{-1}\sum_i\Big\{\int_0^a E_0 \sup_{\theta\in V_\varepsilon(\theta_0)} |\xi^T\mathcal{G}_n^{(1)}(\theta) - \xi_0^T\mathcal{G}_n^{(1)}(\theta_0)|^2\, d\Lambda_0(t)\Big\}^{\frac{1}{2}}$$

and it is a $O(\rho(\theta,\theta_0))$ in a neighborhood of θ_0. For the second term, we have

$$W_{2n}(\theta) = n^{-\frac{1}{2}}\int_0^a \log\frac{S_n^{(0)}(\theta)}{S_n^{(0)}(\theta_0)}\, d\bar{M}_n + \int_0^a \mathcal{G}_n^{(0)}(\theta_0)\log\frac{s^{(0)}(\theta)}{s^{(0)}(\theta_0)}\, d\Lambda_0$$
$$+n^{-\frac{1}{2}}\int_0^a\Big\{\log\frac{S_n^{(0)}(\theta)}{s^{(0)}(\theta)} - \log\frac{S_n^{(0)}(\theta_0)}{s^{(0)}(\theta_0)}\Big\}S_n^{(0)}(\theta_0)\, d\Lambda_0$$

where the last term is expanded in a neighborhood of θ_0 as

$$\int_0^a\Big\{\frac{\mathcal{G}_n^{(0)}(\theta)}{s^{(0)}(\theta)} - \frac{\mathcal{G}_n^{(0)}(\theta_0)}{s^{(0)}(\theta_0)}\Big\}\{1+o_p(1)\}S_n^{(0)}(\theta_0)\, d\Lambda_0.$$

The covariance of $\mathcal{G}_n^{(0)}(s;\theta_0)$ and $\mathcal{G}_n^{(0)}(t;\theta_0)$ is $C^{(0)}(s,t) = s^{(0)}(s\wedge t;\zeta_0, 2\xi_0) - s^{(0)}(s;\theta_0)s^{(0)}(t;\theta_0)$ and by the uniform law of large numbers, the square of the norm $E_0 \sup_{\theta\in V_\varepsilon(\theta_0)} W_{2n}^2(\theta)$ is bounded by

$$\int_0^a E_0 \sup_{V_\varepsilon(\theta_0)}\Big\{\log\frac{S_n^{(0)}(t;\theta)}{S_n^{(0)}(t;\theta_0)}\Big\}^2 n^{-1}S_n^{(0)}(t;\theta_0)\, d\Lambda_0$$
$$+\int_0^a\int_0^a C^{(0)}(s,t) \sup_{V_\varepsilon(\theta_0)}\sup_{x\in[0,a]} \frac{\{s^{(0)}(x;\theta) - s^{(0)}(x;\theta_0)\}^2}{s^{(0)2}(x;\theta_0)}\, d\Lambda_0(s)\, d\Lambda_0(t)$$
$$+\int_0^a\int_0^a E_0\Big[\sup_{V_\varepsilon(\theta_0)}\Big\{\frac{\mathcal{G}_n^{(0)}(s;\theta)}{s^{(0)}(s;\theta)} - \frac{\mathcal{G}_n^{(0)}(s;\theta_0)}{s^{(0)}(s;\theta_0)}\Big\}\Big\{\frac{\mathcal{G}_n^{(0)}(t;\theta)}{s^{(0)}(t;\theta)} - \frac{\mathcal{G}_n^{(0)}(t;\theta_0)}{s^{(0)}(t;\theta_0)}\Big\}$$
$$.S_n^{(0)}(s;\theta_0)S_n^{(0)}(t;\theta_0)\Big]\, d\Lambda_0(s)\, d\Lambda_0(t)\{1+o(1)\},$$

by Conditions C3, it is a $O(\rho^2(\theta,\theta_0))$ and the inequality is a consequence of the Cauchy–Schwarz inequality. □

Theorem 6.2. *Under conditions* C1–C3, *for* $\varepsilon > 0$ *sufficiently small*

$$\overline{\lim}_{n,A\to\infty} P_0(\sup_{u\in\mathcal{U}_{n,\varepsilon},\rho(u)>A} X_n(\theta_{n,u}) \geq 0) = 0,$$
$$\overline{\lim}_{n,A\to\infty} P_0(n^{\frac{1}{2}}\rho(\widehat{\theta}_n,\theta_0) > A) = 0.$$

Let $A > 0$ and $\mathcal{U}_n^A = \{u \in \mathcal{U}_n; |u_1| + \|u_2\|^2 \leq A\}$. The limiting distribution of $(n(\widehat{\zeta}_n - \zeta_0), n^{\frac{1}{2}}(\widehat{\xi}_n - \xi_0))$ will be deduced from Theorem 6.2 and from the behaviour of the restriction of the log-likelihood ratio process (6.5) by the map $u \mapsto l_n(\theta_{n,u}) - l_n(\theta_0)$ to the compact set $\mathcal{U}_n^A$, for A sufficiently large. We define a process Q_n on $\mathbb{R}$ and a variable $\widetilde{l}_n$ by

$$Q_n(u_1) = \sum_i \delta_i \Big\{(\gamma_0 - \beta_0)^T Z_{2i}(T_i)(1_{\{\zeta_{nu} < Z_{3i}(T_i) \leq \zeta_0\}} - 1_{\{\zeta_0 < Z_{3i}(T_i) \leq \zeta_{nu}\}}) - \frac{S_n^{(0)}(T_i; \zeta_{nu}, \xi_0) - S_n^{(0)}(T_i; \theta_0)}{S_n^{(0)}(T_i; \theta_0)}\Big\},$$

$$\widetilde{l}_n = n^{-\frac{1}{2}} \sum_i \int_0^a \Big\{\widetilde{Z}_i(\zeta_0) - \frac{S_n^{(1)}(\theta_0)}{S_n^{(0)}(\theta_0)}\Big\}\, dM_i, \tag{6.16}$$

Theorem 6.3. *Under Conditions* C1–C3, *the process* l_n *has the following asymptotic approximation uniformly on* $\mathcal{U}_n^A$, *for every* $A > 0$

$$l_n(\theta_{n,u}) - l_n(\theta_0) = Q_n(u_1) + u_2^T \widetilde{l}_n - \frac{1}{2} u_2^T I(\theta_0) u_2 + o_p(1).$$

Proof. For $A > 0$, let $u = (u_1, u_2) \in \mathcal{U}_n^A$ with u_1 in $\mathbb{R}$ and u_2 in $\mathbb{R}^{p+2q}$, and let $\theta_{n,u} = (\zeta_{n,u}, \xi_{n,u}^T)^T$ with $\zeta_{n,u} = \zeta_0 + n^{-1} u_1$ and $\xi_{n,u} = \xi_0 + n^{-\frac{1}{2}} u_2$. For $1 \leq i \leq n$, we have $(r_{\theta_{n,u}} - r_{\theta_0})(Z_i(T_i)) = n^{-\frac{1}{2}} u_2^T \widetilde{Z}_i(T_i; \zeta_{n,u}) + (\gamma - \beta_0)^T Z_{2i}(T_i) 1_{\{\zeta_{n,u} < Z_{3i}(T_i) \leq \zeta_0\}} - (\gamma_0 - \beta)^T Z_{2i}(T_i) 1_{\{\zeta_0 < Z_{3i}(T_i) \leq \zeta_{n,u}\}})\}$. Using C3 and the continuity of the functions $s^{(k)}$, a Taylor expansion for $\xi_{n,u}$ close to ξ_0 implies

$$S_n^{(0)}(\theta_{n,u}) = S_n^{(0)}(\zeta_{n,u}, \xi_0) + n^{-\frac{1}{2}} u_2^T S_n^{(1)}(\zeta_{n,u}, \xi_0) + \frac{1}{2} n^{-1} u_2^T S_n^{(2)}(\zeta_{n,u}, \xi_0) u_2 + o_p(1),$$

and a second order asymptotic expansion of the logarithm, uniformly on $\mathcal{U}_n^A$, yields

$$\begin{aligned}
\log \frac{S_n^{(0)}(\theta_{n,u})}{S_n^{(0)}(\theta_0)} &= \log\Big\{1 + \frac{S_n^{(0)}(\theta_{n,u}) - S_n^{(0)}(\theta_0)}{S_n^{(0)}(\theta_0)}\Big\} \\
&= \frac{S_n^{(0)}(\theta_{n,u}) - S_n^{(0)}(\theta_0)}{S_n^{(0)}(\theta_0)} - \frac{1}{2}\Big\{\frac{S_n^{(0)}(\theta_{n,u}) - S_n^{(0)}(\theta_0)}{S_n^{(0)}(\theta_0)}\Big\}^2 \\
&\quad + o_p(1) \\
&= \frac{S_n^{(0)}(\zeta_{n,u}, \xi_0) - S_n^{(0)}(\theta_0)}{S_n^{(0)}(\theta_0)} + n^{-\frac{1}{2}} u_2^T \frac{S_n^{(1)}(\zeta_{n,u}, \xi_0)}{S_n^{(0)}(\theta_0)} \\
&\quad - \frac{n^{-1}}{2} u_2^T V_n(\zeta_{n,u}, \xi_0) u_2 + o_p(1)
\end{aligned}$$

By the uniform convergence of $n^{-1}\bar{N}_n$, the process $l_n(\theta_{n,u}) - l_n(\theta_0)$ has the expansion $u_2^T C_n(u_2) - \frac{1}{2}u_2^T I(\theta_0)u_2 + Q_n(u_1) + o_p(1)$ uniformly on $\mathcal{U}_n^A$, where

$$\begin{aligned} C_n(u) &= n^{-\frac{1}{2}} \sum_i \int_0^a \Big\{ \widetilde{Z}_i(\zeta_{n,u}) - \frac{S_n^{(1)}(\zeta_{n,u}, \xi_0)}{S_n^{(0)}(\theta_0)} \Big\} \, dN_i \\ &= n^{-\frac{1}{2}} \sum_i \int_0^a \Big\{ \widetilde{Z}_i(\zeta_{n,u}) - \frac{S_n^{(1)}(\zeta_{n,u}, \xi_0)}{S_n^{(0)}(\theta_0)} \Big\} \, dM_i \\ &\quad + n^{-\frac{1}{2}} \int_0^a \Big\{ S_n^{(1)}(]\zeta_0, \zeta_{n,u}], \alpha_0, \gamma_0) - S_n^{(1)}(]\zeta_0, \zeta_{n,u}], \alpha_0, \beta_0) \\ &\qquad + S_n^{(1)}(]\zeta_{n,u}, \zeta_0], \alpha_0, \beta_0) - S_n^{(1)}(]\zeta_{n,u}, \zeta_0], \alpha_0, \gamma_0) \Big\} \, d\Lambda_0. \end{aligned}$$

Let $a_{1i}(u_1) = \int_0^a \{\widetilde{Z}_i(\zeta_{n,u}) - \widetilde{Z}_i(\zeta_0)\} \, dM_i$ and $a_{2i}(u_1) = \int_0^a \{S_n^{(1)}(\zeta_{n,u}, \xi_0) - S_n^{(1)}(\theta_0)\} S_n^{(0)-1}(\theta_0) \, dM_i$, $1 \le i \le n$. The variables $a_{1i}(u_1)$ and $\sum_i a_{2i}(u_1)$ have the mean zero as integrals of predictable processes with respect to M_i and $\sum_i M_i$, respectively. Let 0_q be the q-dimensional vector with value zero, the increments of $\widetilde{Z}_i(t)$ and $S_n^{(1)}(t; \xi_0)$ for ζ between ζ_0 and $\zeta_{n,u}$ are written as the vectors $\widetilde{Z}_i(t; \zeta_{n,u}) - \widetilde{Z}_i(t; \zeta_0) = (0_p, Z_{2i}^T(t), -Z_{2i}^T(t))^T (1_{\{\zeta_0 < Z_{3i}(t) \le \zeta_{n,u}\}} - 1_{\{\zeta_{n,u} < Z_{3i}(t) \le \zeta_0\}})$ and

$$\begin{aligned} & S_n^{(1)}(t; \zeta_{n,u}, \xi_0) - S_n^{(1)}(t; \theta_0) \\ &\quad = \sum_i Y_i(t) e^{\alpha_0^T Z_{1i}(t)} (1_{\{\zeta_0 < Z_{3i}(t) \le \zeta_{n,u}\}} - 1_{\{\zeta_{n,u} < Z_{3i}(t) \le \zeta_0\}}) \\ &\quad\quad . \Big\{ (Z_{1i}^T(t), Z_{2i}^T(t), 0_q^T)^T \Big\} e^{\beta_0^T Z_{2i}(t)} - (Z_{1i}^T, 0_q^T, Z_{2i}^T)^T e^{\gamma_0^T Z_{2i}(t)}. \end{aligned}$$

Taking the supremum over the intervals $]\zeta_0, \zeta_{n,u}]$ and $]\zeta_{n,u}, \zeta_0]$, we have $E_0 \sup_{|u_1| \le A} \|a_{\ell i}(u_1)\|^2 = O(n^{-1})$ therefore $E_0 \sup_{|u_1| \le A} n^{-\frac{1}{2}} \sum_i a_{\ell i}(u_1)$ is a $o(1)$, for $\ell = 1, 2$, and the result follows. □

Let $\zeta_{n,v} = \zeta_0 + n^{-1} v_n$, with a real sequence $(v_n)_n$ converging to a non-zero limit v. The process Q_n is written as a sequence of partial sums according to $Z_{3i} \le \zeta_{n,v}$ or $Z_{3i} > \zeta_{n,v}$. By Theorem 6.2, the sample paths of Q_n belong to the space D_A of right-continuous functions with left-hand limits on $[-A, A]$, for any $A > 0$.

Theorem 6.4. *Under conditions* C1–C3, *the variable* $\widetilde{l}_n$ *converges weakly to a centered Gaussian variable with variance* $I(\theta_0)$, *the process* Q_n *converges weakly to an uncentered Gaussian process* Q *in* D_A, *for every* $A > 0$, *and they are asymptotically independent.*

Proof. The weak convergence of the variable $\widetilde{l}_n$ in (6.16) to a Gaussian variable $\mathcal{N}(0, I(\theta_0))$ is the same as in Theorem 4.1 of Andersen and Gill (1982). The process Q_n is the difference $Q_n = Q_n^+ - Q_n^-$ of processes Q_n^+ and Q_n^- defined by $Q_n^+ = 0$ on $\mathbb{R}_-$, $Q_n^- = 0$ on $\mathbb{R}_+$ and

$$Q_n^+(v) = \sum_i \delta_i \Big\{ (\beta_0 - \gamma_0)^T Z_{2i}(T_i) 1_{\{\zeta_0 < Z_{3i}(T_i) \le \zeta_{n,v}\}} - S_n^{(0)-1}(T_i;\theta_0)$$
$$\cdot \sum_j Y_j(T_i) e^{\alpha_0^T Z_{1j}(T_i)} (e^{\beta_0^T Z_{2j}(T_i)} - e^{\gamma_0^T Z_{2j}(T_i)}) 1_{\{\zeta_0 < Z_{3j(T_j)} \le \zeta_{n,v}\}} \Big\},$$

$$Q_n^-(v) = \sum_i \delta_i \Big\{ (\gamma_0 - \beta_0)^T Z_{2i}(T_i) 1_{\{\zeta_{n,v} < Z_{3i}(T_i) \le \zeta_0\}} - S_n^{(0)-1}(T_i;\theta_0)$$
$$\cdot \sum_j Y_j(T_i) e^{\alpha_0^T Z_{1j}(T_i)} (e^{\beta_0^T Z_{2j}(T_i)} - e^{\gamma_0^T Z_{2j}(T_i)}) 1_{\{\zeta_{n,v} < Z_{3j}(T_j) \le \zeta_0\}} \Big\}.$$

The process Q_n is a sum of integrals of functionals of the partial sum processes $S_n^{(k)}$ with respect to the processes M_i or the function Λ_0. Using the notations of Section 6.2, we have

$$Q_n^+(v) = (\beta_0 - \gamma_0)^T \sum_i \int_0^a Z_{2i}(t) 1_{\{\zeta_0 < Z_{3i}(t) \le \zeta_{n,v}\}} \, dM_i(t)$$
$$- \sum_i \int_0^a \frac{S_n^{(0)-}(t;]\zeta_0, \zeta_{n,v}], \xi_0) - S_n^{(0)+}(t;]\zeta_0, \zeta_{n,v}], \xi_0)}{S_n^{(0)}(t;\theta_0)} \, dM_i(t)$$
$$+ (\beta_0 - \gamma_0)^T \int_0^a S_{2n}^{(1)-}(t;]\zeta_0, \zeta_{n,v}], \xi_0) \, d\Lambda_0(t)$$
$$+ \int_0^a \{S_n^{(0)-}(t;]\zeta_0, \zeta_{n,v}], \xi_0) - S_n^{(0)+}(t;]\zeta_0, \zeta_{n,v}], \xi_0)\} \, d\Lambda_0(t),$$

its mean under P_0 is

$$E_0 Q_n^+(v) = n(\beta_0 - \gamma_0)^T \int_0^a s_2^{(1)-}(t;]\zeta_0, \zeta_{n,v}], \xi_0) \, d\Lambda_0(t)$$
$$+ n \int_0^a \{s^{(0)-}(t;]\zeta_0, \zeta_{n,v}], \xi_0) - s^{(0)+}(t;]\zeta_0, \zeta_{n,v}], \xi_0)\} \, d\Lambda_0(t),$$

$E_0 Q_n^+(v)$ is asymptotically equivalent, as n tends to infinity, to

$$\mu_0^+(v) = v(\beta_0 - \gamma_0)^T \int_0^a h_3(t,\zeta_0) s_2^{(1)-}(t;\theta_0) \, d\Lambda_0(t)$$
$$+ v \int_0^a h_3(t,\zeta_0) \{s^{(0)-}(t;\theta_0) - s^{(0)+}(t;\theta_0)\} \, d\Lambda_0(t).$$

The variance of $Q_n^+(v)$ under P_0 is

$$V_0^+(v) = \int_0^a (\beta_0 - \gamma_0)^T s_2^{(2)}(]\zeta_0, \zeta_{n,v}], \xi_0)(\beta_0 - \gamma_0)\, d\Lambda_0$$
$$+ \int_0^a E_0 \frac{\{S_n^{(0)-}(]\zeta_0, \zeta_{n,v}], \xi_0) - S_n^{(0)+}(]\zeta_0, \zeta_{n,v}], \xi_0)\}^2}{S_n^{(0)}(\theta_0)}\, d\Lambda_0$$
$$-2\int_0^a (\beta_0 - \gamma_0)^T E_0\Big\{S_{2n}^{(1)}(]\zeta_0, \zeta_{n,v}], \xi_0)$$
$$\cdot \frac{S_n^{(0)-}(]\zeta_0, \zeta_{n,v}], \xi_0) - S_n^{(0)+}(]\zeta_0, \zeta_{n,v}], \xi_0)}{S_n^{(0)}(\theta_0)}\Big\}\, d\Lambda_0$$
$$+Var_0\Big\{\int_0^a (\beta_0 - \gamma_0)^T S_{2n}^{(1)-}(t;]\zeta_0, \zeta_{n,v}], \xi_0)\, d\Lambda_0(t)$$
$$+ \int_0^a \{S_n^{(0)-}(t;]\zeta_0, \zeta_{n,v}], \xi_0) - S_n^{(0)+}(t;]\zeta_0, \zeta_{n,v}], \xi_0)\}\, d\Lambda_0(t)\Big\},$$

and $Var_0 Q_n^+(v)$ is asymptotically equivalent to $V_0^+(v)$ defined as v times the mean of the above processes at θ_0. It follows that the process Q_n^+ converges weakly to a Gaussian process with mean μ_0^+ and a finite covariance function, as an empirical process of partial sums. In the same way, the process Q_n^- is written as

$$Q_n^-(v) = (\beta_0 - \gamma_0)^T \sum_i \int_0^a Z_{2i}(t) 1_{\{\zeta_{n,v} < Z_{3i}(t) \le \zeta_0\}}\, dM_i(t)$$
$$-\sum_i \int_0^a \frac{S_n^{(0)-}(t;]\zeta_{n,v}, \zeta_0], \xi_0) - S_n^{(0)+}(t;]\zeta_{n,v}, \zeta_0], \xi_0)}{S_n^{(0)}(t; \theta_0)}\, dM_i(t)$$
$$+(\beta_0 - \gamma_0)^T \int_0^a S_{2n}^{(1)-}(t;]\zeta_{n,v}, \zeta_0], \xi_0)\, d\Lambda_0(t)$$
$$+ \int_0^a \{S_n^{(0)-}(t;]\zeta_{n,v}, \zeta_0], \xi_0) - S_n^{(0)+}(t;]\zeta_{n,v}, \zeta_0], \xi_0)\}\, d\Lambda_0(t),$$

$E_0 Q_n^-(v)$ is asymptotically equivalent, as n tends to infinity, to

$$\mu_0^-(v) = |v|(\beta_0 - \gamma_0)^T \int_0^a h_3(t, \zeta_0) s_2^{(1)-}(t; \theta_0)\, d\Lambda_0(t)$$
$$+v \int_0^a h_3(t, \zeta_0)\{s^{(0)-}(t; \theta_0) - s^{(0)+}(t; \theta_0)\}\, d\Lambda_0(t),$$

and the variance of $Q_n^-(v)$ is finite and proportional to $|v|$. Then the process Q_n^- converges weakly to a Gaussian process with mean μ_0^- and a finite covariance function. Finally the process Q_n converges weakly to the difference Q of the processes Q_n+ and Q_n^-. □

Theorem 6.5. *Under P_0 and Conditions* C1–C3, *the variables $\widehat{\zeta}_n$ and $\widehat{\xi}_n$ are asymptotically independent, $n(\widehat{\zeta}_n - \zeta_0)$ converges weakly to the location of the supremum of the process Q, $n^{\frac{1}{2}}(\widehat{\xi}_n - \xi_0) = I^{-1}(\theta_0)\widetilde{l}_n + o_p(1)$ and it converges weakly to a centered Gaussian variable with variance $I^{-1}(\theta_0)$.*

Proof. Theorem 6.3 implies that $\widehat{v}_n = n(\widehat{\zeta}_n - \zeta_0)$ maximizes the process Q_n, its asymptotic distribution is deduced from Theorem 6.4. The variable $\widehat{u}_n = n^{\frac{1}{2}}(\widehat{\xi}_n - \xi_0)$ maximizes $u^T\widetilde{l}_n - \frac{1}{2}u^T I(\theta_0)u + o_p(1)$ so it has the expansion $\widehat{u}_n = I^{-1}(\theta_0)\widetilde{l}_n + o_p(1)$, its limiting distribution is obtained from Theorem 6.5. They are bounded in probability, as a consequence of Theorem 6.2. Their asymptotic independence is a consequence of the asymptotic independence of $\widetilde{l}_n$ and Q_n, they have Gaussian limiting distributions and they satisfy $E_0\widetilde{l}_n = 0$ and $E_0\{\widetilde{l}_n Q_n(t)\} = o_p(1)$ for every t in $[0, a]$. □

If ζ_0 is known, Theorem 6.3 reduced to a second order expansion of the partial log-likelihood, the estimator of ξ_0 is such that $n^{\frac{1}{2}}(\widehat{\xi}_n - \xi_0)$ has the same asymptotic distribution as in Theorem 6.5 and it is an efficient estimator of ξ_0. With ζ_0 unknown, $\widehat{\xi}_n$ is then an adaptive estimator of ξ_0.

The weak convergence of $n^{\frac{1}{2}}(\widehat{\Lambda}_n - \Lambda_0)$ may be established using the approach of Andersen and Gill (1982). Its asymptotic behaviour follows from Theorem 6.5 and from the next result, which does not depend on the knowledge of ζ_0. For s and t in $[0, a]$, let

$$C(s,t) = \int_0^{s\wedge t} s^{(0)-1}(\theta_0)\, d\Lambda_0.$$

Theorem 6.6. *Under* C1–C3, *the process defined for t in $[0, a]$ by*

$$n^{\frac{1}{2}}(\widehat{\Lambda}_n - \Lambda_0)(t) + n^{\frac{1}{2}}(\widehat{\xi}_n - \xi_0)^T \int_0^t \frac{s^{(1)}}{s^{(0)}}(\theta_0)\, d\Lambda_0 \qquad (6.17)$$

converges weakly to a centered Gaussian process with the covariance function C in $[0, a]^2$, and it is asymptotically independent of the variable $n^{\frac{1}{2}}(\widehat{\xi}_n - \xi_0)$.

Proof. By definition of the predictable compensator of $\bar{N}_n$,

$$n^{\frac{1}{2}}(\widehat{\Lambda}_n - \Lambda_0)(t) = \int_0^T \frac{d\mathcal{M}_n^{(0)}}{n^{-1}S_n^{(0)}(\widehat{\theta}_n)} - \int_0^T \frac{n^{-\frac{1}{2}}\{S_n^{(0)}(\widehat{\theta}_n) - S_n^{(0)}(\theta_0)\}}{n^{-1}S_n^{(0)}(\widehat{\theta}_n)}\, d\Lambda_0.$$

The first term in the right member is the integral of the left-continuous process $nS_n^{(0)-1}(\widehat{\theta}_n)$ with respect to the martingale $\mathcal{M}_n^{(0)}$ and it converges weakly to a centered Gaussian process with covariance $\int_0^{s\wedge t} s^{(0)-1}(\theta_0)\,d\Lambda_0$, by Rebolledo's convergence theorem. The asymptotic equivalence of the second term and $n^{\frac{1}{2}}(\widehat{\xi}_n - \xi_0)^T \int_0^t s^{(1)}(\theta_0)s^{(0)-1}(\theta_0)\,d\Lambda_0$ is obtained from the expansion

$$n^{-\frac{1}{2}}\Big\{S_n^{(0)}(\widehat{\theta}_n) - S_n^{(0)}(\theta_0)\Big\} = n^{\frac{1}{2}}(\widehat{\xi}_n - \xi_0)^T n^{-1} S_n^{(1)}(\widehat{\zeta}_n, \xi_n^*) + n^{-\frac{1}{2}}\Big\{S_n^{(0)}(\widehat{\zeta}_n, \xi_0) - S_n^{(0)}(\theta_0)\Big\},$$

with ξ_n^* between $\widehat{\xi}_n$ and ξ_0. From condition C3, Lemma 6.1 and the consistency of the parameter estimators, the variables

$$\sup_{t\in[0,a]} \|n^{-1}S_n^{(0)}(t;\widehat{\theta}_n) - s^{(0)}(t;\theta_0)\|,$$
$$\sup_{t\in[0,a]} \sup_{\xi\in]\widehat{\xi}_n,\xi_0[\cup]\xi_0,\widehat{\xi}_n[} \|n^{-1}S_n^{(1)}(t;\widehat{\zeta}_n,\xi) - s^{(1)}(t;\theta_0)\|$$

converge to zero in probability. Moreover, we have

$$\begin{aligned} &n^{-\frac{1}{2}}\Big\{S_n^{(0)}(t;\widehat{\zeta}_n,\xi_0) - S_n^{(0)}(t;\theta_0)\Big\} \\ &= n^{-\frac{1}{2}}\sum_i Y_i e^{\alpha_0^T Z_{1i}(t)}\Big(e^{\beta_0^T Z_{2i}(t)} - e^{\gamma_0^T Z_{2i}(t)}\Big) \\ &\quad \cdot\Big(1_{\{\zeta_0<Z_{3i}(t)\le\widehat{\zeta}_n\}} - 1_{\{\widehat{\zeta}_n<Z_{3i}(t)\le\zeta_0\}}\Big), \end{aligned}$$

it is denoted $n^{-\frac{1}{2}}\sum_i Y_i\phi_i(1_{\{\zeta_0<Z_{3i}\le\widehat{\zeta}_n\}} - 1_{\{\widehat{\zeta}_n<Z_{3i}\le\zeta_0\}})$. From Theorem 6.2, for every $\varepsilon > 0$, there exist A and n_0 such that for every $n \ge n_0$

$$P_0(n|\widehat{\zeta}_n - \zeta_0| > A) \le \frac{\varepsilon}{2}.$$

Let $\Omega_{nA} = \{n|\widehat{\zeta}_n - \zeta_0| \le A\}$, for every $\eta > 0$

$$\begin{aligned} &P_0(\sup_t n^{-\frac{1}{2}}|\sum_i Y_i(t)\phi_i(t)1_{\{\zeta_0<Z_{3i}(t)\le\widehat{\zeta}_n\}}| > \eta) \\ &\le P_0\Big(\sup_t n^{-\frac{1}{2}}\Big|\sum_i Y_i(t)\phi_i(t)1_{\{\zeta_0<Z_{3i}(t)\le\widehat{\zeta}_n\}}1_{\Omega_{nA}}\Big| > \eta\Big) + \frac{\varepsilon}{2} \\ &\le \frac{1}{\eta^2}E_0\Big\{\sup_t Y(t)\phi^2(t)1_{\{\zeta_0<Z_3(t)\le\zeta_0+n^{-1}A\}}\Big\} \\ &+\frac{n-1}{\eta^2}\Big[E_0\Big\{\sup_t Y(t)|\phi|(t)1_{\{\zeta_0<Z_3(t)\le\zeta_0+n^{-1}A\}}\Big\}\Big]^2 + \frac{\varepsilon}{2} \\ &\le \frac{A}{n\eta^2}E_0\Big\{\sup_t h_3(t;\zeta_0)|\phi|^2(t) \mid Z_3(t) = \zeta_0^+\Big\} \\ &+\frac{(n-1)A^2}{n^2\eta^2}\Big[E_0\Big\{\sup_t h_3(t;\zeta_0)|\phi|(t) \mid Z_3(t) = \zeta_0^+\Big\}\Big]^2 + \frac{\varepsilon}{2}, \end{aligned}$$

it is smaller than ε for n large enough and the same result holds for the process $n^{-\frac{1}{2}} \sum_i Y_i \phi_i 1_{\{\widehat{\zeta}_n < Z_{3i} \leq \zeta_0\}}$.

These convergences imply the convergence in probability to zero of the process $n^{-\frac{1}{2}}\{S_n^{(0)}(\widehat{\zeta}_n, \xi_0) - S_n^{(0)}(\theta_0)\}$, uniformly on $[0,a]$, and (6.17) is uniformly approximated by $\int_0^T nS_n^{(0)-1}(\widehat{\theta}_n)\, d\mathcal{M}_n^{(0)}$.

The asymptotic independence of (6.17) and $n^{\frac{1}{2}}(\widehat{\xi}_n - \xi_0)$ is a consequence of the approximation $n^{\frac{1}{2}}(\widehat{\xi}_n - \xi_0) = I(\theta_0)^{-1}\widetilde{l}_n + o_p(1)$ since $\widetilde{l}_n$ and the local martingale $\int_0^{\cdot} nS_n^{(0)-1}(\widehat{\theta}_n)\, d\mathcal{M}_n^{(0)}$ are asymptotically Gaussian with mean zero and their covariance satisfies $E_0\widetilde{l}_n \int_0^t S_n^{(0)-1}(\widehat{\theta}_n)\, d\mathcal{M}_n^{(0)} = 0$, for every t in $[0,a]$. □

As $\widehat{\xi}_n$ and $\widehat{\Lambda}_n$ are adaptive with respect to ζ_0, asymptotic confidence intervals for the components of ξ_0 and for Λ_0 are the same as in the regular Cox model with a change-point at a known time ζ_0. This enables to use the standard softwares for survival data analysis by a maximization of the partial likelihood $L_n(a_k, \xi)$ with respect to the parameter ξ for successive values a_k on a grid in $[\zeta_1, \zeta_2]$, with a path of order $o(n^{-1})$. The maximization of $L_n(a_k, \cdot)$ provides an estimator $\widehat{\xi}_{k,n}$ for ξ_0 and $\widehat{\zeta}_n$ can be approximated by the value $\widetilde{\zeta}_n$ that maximizes the sequence $(L_n(a_k, \widehat{\xi}_{k,n}))_k$. Then $\widehat{\xi}_n$ is approximated by the value of $\widetilde{\xi}_n$ associated with $\widetilde{\zeta}_n$ and $\widehat{\Lambda}_n$ is approximated by the Breslow estimator $\widetilde{\Lambda}_n$ calculated with $S_n^{(0)}(\widetilde{\zeta}_n, \widetilde{\xi}_n)$. Under the conditions, they have the same asymptotic behaviour as $\widehat{\zeta}_n$, $\widehat{\xi}_n$ and $\widehat{\Lambda}_n$, described in Theorems 6.5 and 6.6. This procedure applies to the estimation of consecutive change-points and parameter vectors.

The properties of the estimators are the same in Model (6.1) with a parametric regression function $r_\theta(z)$ of $C^2(\Xi)$, for every z in $\mathcal{Z}$. The notations of the partial derivatives of the processes $S_n^{(0)}(t;\theta)$ and l_n with respect to the components of the vector $\xi = (\alpha^T, \beta^T, \gamma^T)^T$ are modified according to those of the regression function, they become

$$S_n^{(1)}(t;\theta) = \sum_i Y_i(t)\dot{r}_{\theta,\xi}(Z_i(t)) \exp\{r_\theta(Z_i(t))\},$$

$$S_n^{(2)}(t;\theta) = \sum_i Y_i(t)\{\ddot{r}_{\theta,\xi}(Z_i(t)) + \dot{r}^2_{\theta,\xi}(Z_i(t))\} \exp\{r_\theta(Z_i(t))\},$$

$$l'_{n,\xi}(\theta) = \sum_{i=1}^n \int_0^a \Big\{\dot{r}_{\theta,\xi}(Z_i) - \frac{S_n^{(1)}(\theta)}{S_n^{(0)}(\theta_0)}\Big\}\, dN_i,$$

$$l''_{n,\xi}(\theta) = \sum_{i=1}^n \int_0^a \{\ddot{r}_{\theta,\xi}(Z_i) - V_n(\theta)\}\, dN_i.$$

6.4 Maximum likelihood test of models without change

We consider the log-likelihood ratio test for the hypothesis H_0 of a proportional hazards model (6.1) such that the regression function $r_0 = r_{\eta_0}$ has no change of parameter against the alternative of a change at an unknown breakpoint according to model (6.2) with two distinct regression parameter vectors β and γ. The statistic of the partial likelihood ratio test for H_0 is

$$T_n = 2 \sup_{\zeta} \{\widehat{l}_n(\zeta) - \widehat{l}_{0n}\},$$

where $\widehat{l}_{0n}$ is the logarithm of the estimated partial likelihood $l_n(\widehat{\theta}_{0n})$ of the sample under H_0 and $\widehat{l}_n(\zeta) = l_n(\widehat{\xi}_{n,\zeta}, \zeta)$ is the estimated partial likelihood under the alternative. Under the probability P_0 of the hypothesis, the values of the parameter vector reduce to the $p+q$ dimensional vector ξ_0 with components α_0 and β_0, and we consider that ζ_0 is at the end point of the support of the process Z_3 on $[0, a]$, and $\gamma_0 = 0$. From Theorem 6.6, T_n is asymptotically equivalent to the likelihood ratio test statistic.

Under H_0, the function X reduces to

$$\begin{aligned} X(\theta) = \int_0^a \Big\{ & (\xi - \xi_0)^T s^{(1)}(\theta_0) + (\beta - \beta_0)^T s_2^{(1)-}(\zeta, \alpha_0, \beta_0) \\ & + (\gamma - \beta_0)^T s_2^{(1)}(]\zeta, \zeta_0], \alpha_0, \beta_0) - s^{(0)}(\theta_0) \log \frac{s^{(0)}(\theta)}{s^{(0)}(\theta_0)} \Big\} \, d\Lambda_0. \end{aligned}$$

it is negative and maximum at ξ_0 where it is zero. The a.s. uniform convergence of X_n to X and the local concavity of X_n at $\widehat{\theta}_n$, and respectively X at θ_0, where they are maximum imply the a.s. convergence of $\widehat{\theta}_n$ to θ_0, under H_0 and the conditions **C**. The first derivative of the process X_n with respect to ξ is

$$\dot{X}_{n,\xi}(\theta) = n^{-1} \sum_{i=1}^n \int_0^a \Big\{ \widetilde{Z}_i(t) - \frac{S_n^{(1)}(t;\theta)}{S_n^{(1)}(t;\theta)} \Big\} \, dN_i(t),$$

it converges a.s. under H_0 to its expectation function $\dot{X}_\xi$ which is zero at θ_0, and the last q components of the processes $\widetilde{Z}_i$ and $S_n^{(1)}(\theta)$ are zero under H_0. Let U_n be the process of the first derivatives of X_n with respect to the parameters α and β

$$U_n(\theta) = n^{\frac{1}{2}} \dot{X}_{n,\alpha,\beta}(\theta),$$

under P_0, its variance I_{0n} at θ_0 is positive definite as n tends to infinity and it converges a.s. to $I_0 = -\lim_n \ddot{X}_{n,\alpha,\beta}(\theta_0)$. The process $U_n(\theta)$ and its variance $I_n(\theta)$ depend on the parameter γ through the processes $S_n^{(k)}$, and they only depend on α_0 and β_0 at θ_0.

Under H_0, Lemmas 6.1 and 6.2 are still satisfied and the convergence rate of $\widehat{\theta}_n$ is given by Theorem 6.2. The variable $U_n(\widehat{\theta}_n)$ is zero, moreover $U_{0n}(\widehat{\theta}_{0n}) = U_n(\zeta_0, \widehat{\alpha}_{0n}, \widehat{\beta}_{0n}, 0) = 0$ under H_0 and the difference $U_n(\zeta_0, \widehat{\xi}_{0n}) - U_n(\widehat{\zeta}_n, \widehat{\xi}_{0n})$ is a $o_p(1)$ as $\widehat{\zeta}_n$ converges to ζ_0 with the rate n^{-1}. The asymptotic behaviour of the process U_n and the difference of the estimators is similar to those of the parametric densities or regression models studied in Sections 3.2 and 4.6.

Proposition 6.1. *In model (6.2), the test statistic T_n converges weakly under H_0 to a χ^2_q variable T_0.*

Proof. Let ξ_1 be the parameter vector with components α and β, at ζ_0 the process X_n does not depend on the parameter γ and it is denoted $X_{1n}(\xi) = X_n(\xi, \zeta_0) = X_n(\xi_1, \zeta_0)$. The consistency of the maximum likelihood estimators $\widehat{\xi}_{1n} = (\widehat{\alpha}_n^T, \widehat{\beta}_n^T)^T$ and $\widehat{\xi}_{0n}$, and a first order expansion of $U_n(\widehat{\xi}_{1n})$, as n tends to infinity, imply

$$n^{\frac{1}{2}}(\widehat{\xi}_{1n} - \widehat{\xi}_{0n}) = I_0^{-1} U_n(\widehat{\xi}_{0n}, \zeta_0) + o_p(1)$$

with a positive definite matrix I_0^{-1} having the dimension $(p+q) \times (p+q)$, and the variable $n^{\frac{1}{2}}(\widehat{\xi}_n - \widehat{\xi}_{0n})$ converges in probability to zero under P_0. By second order asymptotic expansion of the process X_n, the variable $n\{X_{1n}(\widehat{\xi}_{1n}) - X_{1n}(\widehat{\xi}_{0n})\}$ converges in probability under P_0 to zero, as n tends to infinity.

The process $X_{2n}(\theta) = X_n(\theta) - X_{1n}(\xi_1)$ is written as

$$\begin{aligned} X_{2n}(\theta) &= \sum_{i=1}^n \int_0^a \{(\gamma - \gamma_0)^T \widetilde{Z}_{3i}(t;\zeta) + \gamma_0^T \widetilde{Z}_{2i}(t;\zeta)\}\, dN_i(t) \\ &\quad - \sum_{i=1}^n \int_0^a \log \frac{S_n^{(0)}(t;\theta)}{S_n^{(0)}(t;\xi_1,\zeta_0)}\, dN_i(t), \end{aligned}$$

its first order derivative with respect to γ is

$$U_{2n}(\theta) = \sum_{i=1}^n \int_0^a \widetilde{Z}_{3i}(t;\zeta)\, dN_i(t) - \int_0^a \frac{S_{2n}^{(1)+}(t;\theta)}{S_n^{(0)}(t;\theta)}\, d\bar{N}_n(t)$$

it is zero at $\widehat{\theta}_n$ and

$$\widehat{\gamma}_n - \gamma_0 = U_{2n}(\widehat{\xi}_{1n}, \gamma_0, \widehat{\zeta}_n) I_{2n}^{-1}(\widehat{\xi}_{1n}, \gamma_0, \widehat{\zeta}_n) + o_p(1)$$

where the second order derivative $I_{2n}(\widehat{\xi}_{1n}, \gamma_0, \widehat{\zeta}_n)$ of the process X_{2n} with respect to γ converges in probability to a strictly positive definite matrix $I_{02} = I_2(\theta_0)$.

The expectation of the process $U_{2n}(\theta)$ under P_0 is $n \int_0^a s^{(1)+}(t;\theta)\, d\Lambda_0 - n \int_0^a s^{(1)+}(t;\theta) s^{(0)-1}(t;\theta) s^{(0)}(t;\theta_0)\, d\Lambda_0$ and at $(\widehat{\xi}_{1n}^T, \gamma_0^T, \widehat{\zeta}_n)^T$, the function $ns^{(0)-1}(t;\theta)s^{(0)}(t;\theta_0)$ reduces to $ns_1^{(0)-1}(t;\widehat{\xi}_{1n}, \widehat{\zeta}_n)s_0^{(0)}(t;\xi_0)$ and it converges in probability to zero, so the expectation $U_{2n}(\widehat{\xi}_{1n}, \gamma_0, \widehat{\zeta}_n)$ converges to zero. Its variance converges to I_{02} as n tends to infinity, hence $U_{2n}(\widehat{\xi}_{1n}, \gamma_0, \widehat{\zeta}_n)$ converges weakly to a centered Gaussian variable with the variance I_{02}.

By a second order asymptotic expansion, the process X_{2n} satisfies

$$2nX_{2n}(\widehat{\theta}_n) = U_{2n}^T(\widehat{\xi}_{1n}, \gamma_0, \widehat{\zeta}_n) I_{02}^{-1} U_{2n}^T(\widehat{\xi}_{1n}, \gamma_0, \widehat{\zeta}_n) + o_p(1)$$

and its limit follows. □

Under the alternative K of a model with a change point at ζ strictly inside the support of the observations of the covariate $Z_{3i}(T_i)$, and a regression parameter vector ξ with distinct components β and γ, the process $n^{-1}l_{0n}$ converges a.s. under P_θ to the function $l_0(\theta)$ and the process $X_n(\theta') = n^{-1}\{l_n(\theta') - l_{0n}\}$ converges a.s. under P_θ to the function

$$\begin{aligned} X_\theta(\theta') = \int_0^a \Big\{ &(\xi' - \xi_0)^T s^{(1)}(\theta) + (\beta' - \beta_0)^T s_2^{(1)-}(\zeta' \wedge \zeta_0, \alpha, \beta) \\ &+ (\gamma - \gamma_0)^T s_2^{(1)+}(\zeta' \vee \zeta_0, \alpha, \gamma) + (\beta - \gamma_0)^T s_2^{(1)}(]\zeta_0, \zeta'], \alpha, \gamma) \\ &+ (\gamma - \beta_0)^T s_2^{(1)}(]\zeta', \zeta_0], \alpha, \beta) - s^{(0)}(\theta) \log \frac{s^{(0)}(\theta')}{s^{(0)}(\theta_0)} \Big\} \, d\Lambda_\theta, \end{aligned}$$

it splits as the sum of a process $X_{1\theta}(\theta')$ defined by replacing ξ_0 with ξ in the expression of the regression function and $\log s^{(0)}(\theta') - \log s^{(0)}(\theta_0)$ with $\log s^{(0)}(\theta') - \log s^{(0)}(\theta)$ in $X_\theta(\theta')$, and a process $X_{2\theta}(\theta') = X_\theta(\theta') - X_{1\theta}(\theta')$ which depends on $\xi - \xi_0$ and $\log s^{(0)}(\theta) - \log s^{(0)}(\theta_0)$ but not on ξ'. The first term $X_{1\theta}(\theta')$ is maximum as $\theta' = \theta$, then the estimator $\widehat{\theta}_n$ converges a.s. to θ under P_θ. At θ such that the integral $X_\theta''(\theta)$ is not singular, the maximum likelihood estimator of η has the convergence rate $n^{-\frac{1}{2}}$ and the estimator of the change-point has the convergence rate n^{-1}, their asymptotic distributions are given by Theorem 6.5.

Local alternatives P_{θ_n} contiguous to the probability P_0 of the hypothesis H_0 with the rates a_n and b_n tending to infinity as n tends to infinity are defined by parameters $\zeta_n = \zeta_0 - n^{-1}u_n$, u_n converging to a non-zero limit u, $\xi_n = \xi_0 + n^{-\frac{1}{2}}v_n$, v_n converging to a non-zero limit v.

Proposition 6.2. *Under a fixed alternative P_θ, the statistic T_n tends to infinity. Under the local alternatives $P_{\theta_{n,u}}$ converging to P_0, the statistic T_n converges weakly to $T_0 + v_1^T I_0^{-1} v_1$.*

Proof. Under P_θ, $X_\theta'(\theta) = 0$ and the variable $X_n'(\theta)$ is centered. The process $W_{n,\theta}(\theta') = n^{\frac{1}{2}}\{X_n(\theta') - E_\theta X_n(\theta')\}$ is the sum of $W_{1n,\theta}(\theta') + W_{2n}(\theta)$ where

$$W_{1n,\theta}(\theta') = n^{\frac{1}{2}}\{X_n(\theta') - X_n(\theta) - E_\theta X_n(\theta') + E_\theta X_n(\theta)\},$$
$$W_{2n}(\theta) = n^{\frac{1}{2}}\{X_n(\theta) - E_\theta X_n(\theta)\},$$

the derivative $W_{1n,\theta,\xi}'(\theta)$ of the first term with respect to ξ, at θ, converges weakly under P_θ to a centered Gaussian process with a positive definite variance matrix I_θ and by an expansion in a neighborhood of ξ, it satisfies

$$W_{1n,\theta,\xi}'(\xi, \widehat{\zeta}_n) = -n^{\frac{1}{2}}(\widehat{\xi}_n - \xi)^T X_{n,\theta,\xi}''(\theta) + o_p(1)$$

and $n^{\frac{1}{2}}(\widehat{\xi}_n - \xi)$ converges weakly under P_θ to a centered Gaussian variable. A second order expansion of the process X_n entails

$$2n\{X_n(\widehat{\theta}_n) - X_n(\xi, \widehat{\zeta}_n)\} = W_{1n,\theta,\xi}'^T(\eta, \widehat{\zeta}_n) I_\theta^{-1} W_{1n,\theta,\xi}'(\xi, \widehat{\zeta}_n) + o_p(1),$$

it converges weakly to a χ^2 variable under P_θ. The variable

$$n\{X_n(\xi, \widehat{\zeta}_n) - X_n(\theta)\} = O_p(n\|\widehat{\zeta}_n - \zeta\|)$$

is a $O_p(1)$ and $n\{l_n(\theta) - \widehat{l}_{0n}\}$ tends to infinity, for $\widehat{\zeta}_n$ converging to ζ different from ζ_0, therefore T_n tends to infinity in probability.

Let P_{θ_n} be local alternatives with parameters θ_n such that $\tau_n = \tau_0 - n^{-1}u_n$ and $\xi_n = \xi_0 + n^{-\frac{1}{2}}v_n$ with sequences $(u_n)_n$ and respectively $(v_n)_n$ converging to non-zero limits u and respectively v. The estimators $\widehat{\xi}_{1n}$ and $\widehat{\xi}_{0n}$ converge a.s. to ξ_0 under P_{θ_n} and the variable $n^{\frac{1}{2}}(\widehat{\xi}_{1n} - \widehat{\xi}_{0n})$ converges in probability to a $p+q$-dimensional vector v_1. By a second order asymptotic expansion, the variable $2n\{X_{1n}(\widehat{\xi}_{1n}) - X_{1n}(\widehat{\xi}_{0n})\}$ converges in probability under P_n to $v_1^T I_0^{-1} v_1$, as n tends to infinity.

Under P_n, the variable $\widehat{\gamma}_n - \gamma_n$ has an expansion similar to the expansion of $\widehat{\gamma}_n - \gamma_0$ under P_0, $\widehat{\gamma}_n - \gamma_n = U_{2n}(\widehat{\xi}_{1n}, \gamma_n, \widehat{\zeta}_n) I_{2n}^{-1}(\widehat{\xi}_{1n}, \gamma_n, \widehat{\zeta}_n) + o_p(1)$ and it converges to a centered Gaussian variable. As $\gamma_n - \gamma_0$ converges to zero, it follows that $nX_{2n}(\widehat{\theta}_n)$ has the same limit as under H_0. □

6.5 Change-point at a time threshold

With a chronological change of the parameters at an unknown time, the proportional hazards regression model is defined by (6.1) under a probability P_θ with λ an unknown positive nonparametric function and the regression function (6.3)

$$r_\theta(Z(t)) = \alpha^T Z_1(t) + (\beta + \gamma 1_{\{t>\tau\}})^T Z_2(t),$$

where Z_1 and respectively Z_2 are left-continuous processes with right-hand limits with dimensions p and respectively q, and θ is the $(p+2q+1)$ dimensional parameter vector with components $\tau, \alpha, \beta, \gamma$. The parameters α, and respectively β and γ belong to open and bounded sets $A \subset \mathbb{R}^p$, and respectively $B \subset \mathbb{R}^q$ and $C \subset \mathbb{R}^q$, and $\xi = (\alpha^T, \beta^T, \gamma^T)^T$ belongs to their product Ξ, its value under P_0 is ξ_0. When a change-point occurs, C does not contain zero and the parameter τ belongs to an interval $[\tau_1, \tau_2]$ strictly included in the observation interval of the variable T, without specification we consider in this section that τ and τ_0 belong to $]0, a[$.

Let l_n be the logarithm of the partial likelihood (6.5) and let $\widehat{\theta}_n$ be the maximum likelihood estimator. At a fixed value of the parameter τ

$$\widehat{\xi}_n(\tau) = \arg\max_{\xi \in \Xi} l_n(\tau, \xi)$$

maximizes the function $l_n(\tau, \cdot)$. Let $\widehat{l}_n(\tau) = l_n(\tau, \widehat{\xi}_n(\tau))$, τ_0 is estimated by $\widehat{\tau}_n$ which maximizes $\widehat{l}_n$

$$\widehat{\tau}_n = \inf\Big\{\tau \in [\tau_1, \tau_2] : \max\{\widehat{l}_n(\tau^-), \widehat{l}_n(\tau)\} = \sup_{\tau \in [\tau_1, \tau_2]} \widehat{l}_n(\tau)\Big\}. \qquad (6.18)$$

The maximum likelihood estimator of ξ_0 is $\widehat{\xi}_n = \widehat{\xi}_n(\widehat{\tau}_n)$ and $\widehat{\theta}_n = (\widehat{\zeta}_n, \widehat{\xi}_n)$. The cumulative hazard function $\Lambda_0(t) = \int_0^T \lambda_0(s)\, ds$ is estimated by (6.6). The assumptions and notation for the asymptotic properties of the estimators are similar to those of Section (6.2). For t in $[0, a]$, θ in Θ and for $k = 0, 1, 2$, let

$$S_n^{(k)}(t; \alpha, \beta) = \sum_i Y_i(t) Z_i^{\otimes k}(t) \exp\{\alpha^T Z_{1i}(t) + \beta^T Z_{2i}(t)\},$$

$$s^{(k)}(t; \alpha, \beta) = E_0[Y_i(t) Z_i^{\otimes k}(t) \exp\{\alpha^T Z_{1i}(t) + \beta^T Z_{2i}(t)\}],$$

$$V_n(t; \alpha, \beta) = [S_n^{(2)} S_n^{(0)-1} - \{S_n^{(1)} S_n^{(0)-1}\}^{\otimes 2}](t; \alpha, \beta),$$

$$v(t; \alpha, \beta) = [s^{(2)} s^{(0)-1} - \{s^{(1)} s^{(0)-1}\}^{\otimes 2}](t; \alpha, \beta).$$

For every function or process $\varphi_n(t; \alpha, \beta)$, we denote

$$\varphi_n(t; \theta) = \varphi_n(t; \alpha, \beta) 1_{\{t \le \tau\}} + \varphi_n(t; \alpha, \beta + \gamma) 1_{\{t > \tau\}},$$

this notation defines the process $S_n^{(k)}(t;\theta)$, the functions $s^{(k)}(t;\theta)$ and

$$I(t;\alpha,\beta) = \int_0^t v(s;\theta)s^{(0)}(s;\theta)\lambda_0(s)\,ds,$$

$$X(\theta) = \int_0^a \{r_\theta(t) - r_{\theta_0}(t)\}^T s^{(1)}(t;\theta_0)\,d\Lambda_0(s) - \int_0^a \log\frac{s^{(0)}(t;\theta)}{s^{(0)}(t;\theta_0)} s^{(0)}(t;\theta_0)\,d\Lambda_0(t)$$

The process $X_n(\theta) = n^{-1}\{l_n(\theta) - l_n(\theta_0)\}$ is still defined by (6.9), M_i is the martingale of the compensated jumps of N_i, $\bar{N}_n = \sum_{i\le n} N_i$, and (6.8) defines the local martingales $\mathcal{M}_n^{(0)}$ and $\mathcal{M}_n^{(1)}$.

The asymptotic properties of the estimators will be established under the following conditions.

C1. The hazard function is continuous at τ_0, $\sup_{t\in[0,a]}\lambda_0(t)$ is finite, $\lambda_0(t)$ is strictly positive on $[0,a]$ and $P_0(T \ge a) > 0$.

C2. The variance $Var\,Z(t)$ is positive definite on $[0,a]$ and

$$E_0 \sup_{t\in[0,a]} \sup_{\alpha\in A} \sup_{\beta\in B\cup(B+C)} \{||Z(t)||^k e^{(\alpha^T Z_1(t)+\beta^T Z_2(t))}\}^2$$

is finite for $k = 0, 1, 2$.

Under the condition C2, the variables

$$\sup_{t\in[0,a]} \sup_{\alpha\in A} \sup_{\beta\in B\cup(B+C)} ||n^{-1}S_n^{(k)}(t;\alpha,\beta) - s^{(k)}(t;\alpha,\beta)||$$

converge a.s. to zero, for $k = 0, 1, 2$.

The function $s^{(0)}$ is strictly positive and we consider n large enough to ensure that the process $S_n^{(0)}$ is a.s. strictly positive.

The matrix $v(t;\alpha,\beta)$ is the limit of V_n and the conditions imply that for every θ, the matrix $I(t;\theta)$ is strictly positive definite for every $t > 0$ such that $\Lambda_0(t) > 0$. For $k = 0, 1$, the process

$$\mathcal{G}_n^{(k)} = n^{\frac{1}{2}}(n^{-1}S_n^{(k)} - s^{(k)}),$$

converges weakly to a centered Gaussian process with strictly positive definite variance functions.

6.6 Convergence of the estimators

We first prove the consistency of the estimators then we establish their convergence rates and their asymptotic distributions.

Theorem 6.7. *Under Conditions* C1–C2 *the estimator* $\widehat{\theta}_n$ *is a.s. consistent under* P_0.

Proof. The process X_n is expanded as a sum $X_n = X_{1n} + X_{2n} + X_{3n}$ according to the respective location of the variables T_i with respect to τ and τ_0

$$X_{1n}(\theta) = n^{-1} \sum_{i=1}^{n} \delta_i 1_{\{T_i \le \tau \wedge \tau_0\}} \Big\{ (\alpha - \alpha_0)^T Z_{1i}(T_i)$$

$$+(\beta - \beta_0)^T Z_{2i}(T_i) - \log \frac{S_n^{(0)}(T_i; \alpha, \beta)}{S_n^{(0)}(T_i; \alpha_0, \beta_0)} \Big\},$$

$$X_{2n}(\theta) = n^{-1} \sum_{i=1}^{n} \delta_i 1_{\{T_i > \tau \vee \tau_0\}} \Big\{ (\alpha - \alpha_0)^T Z_{1i}(T_i)$$

$$+(\beta - \beta_0 + \gamma - \gamma_0)^T Z_{2i}(T_i) - \log \frac{S_n^{(0)}(T_i; \alpha, \beta + \gamma)}{S_n^{(0)}(T_i; \alpha_0, \beta_0 + \gamma_0)} \Big\},$$

$$X_{3n}(\theta) = n^{-1} \sum_{i=1}^{n} \delta_i 1_{\{\tau < T_i \le \tau_0\}} \Big\{ (\alpha - \alpha_0)^T Z_{1i}(T_i)$$

$$+(\beta - \beta_0 + \gamma)^T Z_{2i}(T_i) - \log \frac{S_n^{(0)}(T_i; \alpha, \beta + \gamma)}{S_n^{(0)}(T_i; \alpha_0, \beta_0)} \Big\}$$

$$+n^{-1} \sum_{i=1}^{n} \delta_i 1_{\{\tau_0 < T_i \le \tau\}} \Big\{ (\alpha - \alpha_0)^T Z_{1i}(T_i)$$

$$+(\beta - \beta_0 - \gamma_0)^T Z_{2i}(T_i) - \log \frac{S_n^{(0)}(T_i; \alpha, \beta)}{S_n^{(0)}(T_i; \alpha_0, \beta_0 + \gamma_0)} \Big\}.$$

Under the conditions, they converge a.s. to functions of the parameters

$$X_1(\theta) = \int_0^{\tau \wedge \tau_0} \Big[\{(\alpha - \alpha_0)^T, (\beta - \beta_0)^T\} s^{(1)}(t; \alpha_0, \beta_0)$$

$$-s^{(0)}(t; \alpha_0, \beta_0) \log \frac{s^{(0)}(t; \alpha, \beta)}{s^{(0)}(t; \alpha_0, \beta_0)} \Big] \, d\Lambda_0(t),$$

$$
\begin{aligned}
X_2(\theta) = \int_{\tau\vee\tau_0}^{a} \Big[& \{(\alpha-\alpha_0)^T, (\beta-\beta_0+\gamma-\gamma_0)^T\} s^{(1)}(t;\alpha_0,\beta_0+\gamma_0) \\
& - s^{(0)}(t;\alpha_0,\beta_0+\gamma_0) \log \frac{s^{(0)}(t;\alpha,\beta+\gamma)}{s^{(0)}(t;\alpha_0,\beta_0+\gamma_0)} \Big] \, d\Lambda_0(t), \\
X_3(\theta) = 1_{\{\tau<\tau_0\}} \int_{\tau}^{\tau_0} \Big[& \{(\alpha-\alpha_0)^T, (\beta-\beta_0+\gamma)^T\} s^{(1)}(t;\alpha_0,\beta_0) \\
& - s^{(0)}(t;\alpha_0,\beta_0) \log \frac{s^{(0)}(t;\alpha,\beta+\gamma)}{s^{(0)}(t;\alpha_0,\beta_0)} \Big] \, d\Lambda_0(t) \\
& + 1_{\{\tau_0<\tau\}} \int_{\tau_0}^{\tau} \Big[\{(\alpha-\alpha_0)^T, (\beta-\beta_0-\gamma_0)^T\} s^{(1)}(t;\alpha_0,\beta_0+\gamma_0) \\
& - s^{(0)}(t;\alpha_0,\beta_0+\gamma_0) \log \frac{s^{(0)}(t;\alpha,\beta)}{s^{(0)}(t;\alpha_0,\beta_0+\gamma_0)} \Big] \, d\Lambda_0(t).
\end{aligned}
$$

For every θ, the matrices $I_1(\theta) = I(\tau\wedge\tau_0;\theta)$, $I_2(\theta) = I(\theta) - I(\tau\vee\tau_0;\theta)$ and $I_3(\theta) = I(\tau\wedge\tau_0;\theta) - I(\tau\vee\tau_0;\theta)$ are positive definite. The second order derivatives of X_1 and respectively $X_3 1_{\{\tau_0<\tau\}}$ with respect to $(\alpha^T, \beta^T)^T$ are $-I_1(\theta)$ and respectively $-I_3(\theta)1_{\{\tau_0<\tau\}}$, and the second order derivatives of X_2 and respectively $X_3 1_{\{\tau<\tau_0\}}$ with respect to $(\alpha^T, \beta^T+\gamma^T)^T$ are $-I_2(\theta)$ and respectively $-I_3(\theta)1_{\{\tau<\tau_0\}}$. It follows that for every θ, the functions X_k have the properties

1 $X_1(\theta) \le X_1(\theta_0) = 0$, with a strict inequality if $\alpha \ne \alpha_0$ or $\beta \ne \beta_0$;
2 $X_2(\theta) \le X_2(\theta_0) = 0$, with a strict inequality if $\xi \ne \xi_0$;
3 for every $\tau < \tau_0$, $X_3(\theta) \le X_3(\theta_0) = 0$, with a strict inequality if $\alpha \ne \alpha_0$ or $\beta+\gamma \ne \beta_0$;
4 for every $\tau > \tau_0$, $X_3(\theta) \le X_3(\theta_0) = 0$, with a strict inequality if $\alpha \ne \alpha_0$ or $\beta \ne \beta_0 + \gamma_0$.

The maximum value of the function $X = X_1 + X_2 + X_3$ is therefore zero and it cannot be reached if $\tau \ne \tau_0$ since the previous conditions cannot be simultaneously true with $\gamma_0 \ne 0$. The function X has therefore an unique maximum at θ_0. By maximization of the logarithm of the partial likelihood, the estimators satisfy the inequality

$$0 \le X_n(\widehat{\theta}_n) \le \sup_{\theta\in\Theta} |X_n(\theta) - X(\theta)| + X(\widehat{\theta}_n)$$

where $X(\widehat{\theta}_n) \le 0$ and $\sup_{\theta\in\Theta} |X_n(\theta) - X(\theta)|$ converges a.s. to zero, hence $\widehat{\theta}_n$ converges a.s. to θ_0. □

Lemma 6.5. *For $\varepsilon > 0$ sufficiently small, there exists a constant $\kappa_0 > 0$ such that $X(\theta) \le -\kappa_0\varepsilon$ for every θ in $V_\varepsilon(\theta_0)$.*

Proof. From the properties of the functions X_k, for $\varepsilon > 0$ sufficiently small, their Taylor expansions in $V_\varepsilon(\theta_0)$ provide the inequalities

$$\begin{aligned}
-X_1(\theta) &= \frac{1}{2}(\xi - \xi_0)^T I(\tau \wedge \tau_0; \theta_0)(\xi - \xi_0) + o(\|\xi - \xi_0\|^2) \\
&\geq \frac{1}{4}\|\xi - \xi_0\|^2 \inf_{\|x\|=1} x^T I_1(\theta_0)x + o(\|\xi - \xi_0\|^2) \\
-X_2(\theta) &\geq \frac{1}{4}\|\xi - \xi_0\|^2 \inf_{\|x\|=1} x^T I_2(\theta_0)x + o(\|\xi - \xi_0\|^2),
\end{aligned}$$

and the functions X_1 and X_2 satisfy the inequality of the proposition. By Jensen's inequality, the function X_3 is the integral of a concave function having a strict maximum at ξ_0. For $\varepsilon > 0$ sufficiently small and $\gamma_0 \neq 0$, we have

$$\begin{aligned}
\|\beta + \gamma - \beta_0\| &\geq \|\gamma_0\| - 2\varepsilon > \varepsilon, \\
\|\beta - \beta_0 - \gamma_0\| &\geq \|\gamma_0\| - \varepsilon,
\end{aligned}$$

and a Taylor expansion of the function $-X_3$ implies $X_3(\theta) \leq -c\varepsilon$ for a constant c depending on the matrix $I_3(\theta_0)$. □

The notations are similar to those of the previous sections, with the semi-norm $\rho(\theta, \theta_0) = (|\tau - \tau_0| + \|\xi - \xi_0\|^2)^{\frac{1}{2}}$ and a neighborhood $V_\varepsilon(\theta_0)$ defined with ρ. Let $W_n(\theta) = n^{\frac{1}{2}}\{X_n(\theta) - X(\theta)\}$.

Lemma 6.6. *For every $\varepsilon > 0$ there exists a constant $\kappa_1 > 0$ such that for n large enough, $E_0 \sup_{\theta \in V_\varepsilon(\theta_0)} |W_n(\theta)| \leq \kappa_1 \varepsilon$.*

Proof. The process W_n is the sum of centered processes $W_n = W_{1n} + W_{2n} + W_{3n}$ where

$$\begin{aligned}
W_{2n}(\theta) &= n^{-\frac{1}{2}} \sum_{i=1}^n \int_0^a \left[\log \frac{s^{(0)}(t;\theta)}{s^{(0)}(t;\theta_0)} - \log \frac{S_n^{(0)}(t;\theta)}{S_n^{(0)}(t;\theta_0)}\right] dN_i(t), \\
W_{3n}(\theta) &= -\int_0^a \log \frac{s^{(0)}(t;\theta)}{s^{(0)}(t;\theta_0)} n^{\frac{1}{2}} \{n^{-1} d\bar{N}_n(t) - s^{(0)}(t;\theta_0)\, d\Lambda_0(t)\}
\end{aligned}$$

and

$$\begin{aligned}
W_{1n}(\theta) = n^{-\frac{1}{2}} \sum_{i=1}^n \int_0^\tau & [\{r_\theta(t) - r_{\theta_0}(t)\}^T dN_i(t) \\
& - \xi^T \{s^{(1)}(t;\theta) - s^{(1)}(t;\theta_0)\}\, d\Lambda_0(t)] \\
+ n^{-\frac{1}{2}} \sum_{i=1}^n \int_\tau^a & [\{r_\theta(t) - r_{\theta_0}(t)\}^T dN_i(t) \\
& - (\alpha^T, \beta^T + \gamma^T)\{s^{(1)}(t;\theta) - s^{(1)}(t;\theta_0)\}\, d\Lambda_0(t)].
\end{aligned}$$

By Taylor expansions and under Condition C2, $s^{(0)}(t;\theta) - s^{(0)}(t;\theta_0) = (\theta-\theta_0)^T s^{(1)}(t;\theta_0) + o(\|\theta-\theta_0\|)$ for every t and

$$\log\{s^{(0)}(t;\theta)s^{(0)-1}(t;\theta_0)\} = (\theta-\theta_0)^T s^{(1)}(t;\theta_0)s^{(0)-1}(t;\theta_0) + o(\|\theta-\theta_0\|),$$

moreover the process $n^{\frac{1}{2}}\{n^{-1}\bar{N}_n(t) - \int_0^t s^{(0)}(s;\theta_0)\,d\Lambda_0(s)\}$ converges weakly to a centered Gaussian process with independent increments and the finite variance $\int_0^t s^{(0)}(s;\theta_0)\,d\Lambda_0(s)$ so the inequality is true for the process W_{3n}.

The process W_{2n} is also written as

$$\begin{aligned} W_{2n}(\theta) &= n^{-\frac{1}{2}}\sum_{i=1}^n \int_0^a \Big[\log\frac{n^{-1}S_n^{(0)}(t;\theta_0)}{s^{(0)}(t;\theta_0)} - \log\frac{n^{-1}S_n^{(0)}(t;\theta)}{s^{(0)}(t;\theta)}\Big]\,dN_i(t),\\ &= n^{-\frac{1}{2}}\sum_{i=1}^n \int_0^a \Big[\frac{n^{-1}S_n^{(0)}(t;\theta_0) - s^{(0)}(t;\theta_0)}{s^{(0)}(t;\theta_0)}\\ &\qquad - \frac{n^{-1}S_n^{(0)}(t;\theta) - s^{(0)}(t;\theta)}{s^{(0)}(t;\theta)}\Big]\{1+o_p(1)\}\,dN_i(t)\\ &= \int_0^a \Big[\frac{\mathcal{G}_n^{(0)}(t;\theta_0)}{s^{(0)}(t;\theta_0)} - \frac{\mathcal{G}_n^{(0)}(t;\theta)}{s^{(0)}(t;\theta)}\Big] s^{(0)}(t;\theta_0)\,d\Lambda_0(t) + o_p(1) \end{aligned}$$

where $S_n^{(0)}(t;\theta)$ and $s^{(0)}(t;\theta)$ are differentiable processes with respect to θ and such that for every θ in $V_\varepsilon(\theta_0)$

$$\begin{aligned} &|\mathcal{G}_n^{(0)}(t;\theta_0)s^{(0)-1}(t;\theta_0) - \mathcal{G}_n^{(0)}(t;\theta)s^{(0)-1}(t;\theta)|\\ &\quad \le \varepsilon\|\mathcal{G}_n^{(1)}(t;\theta_0)s^{(0)-1}(t;\theta_0) - \mathcal{G}_n^{(0)}(t;\theta_0)s^{(1)}(t;\theta_0)s^{(0)-2}(t;\theta_0)\|. \end{aligned}$$

Under P_0, the random variables $\int_0^a \mathcal{G}_n^{(0)}(\theta_0)s^{(1)}(\theta_0)s^{(0)-2}(\theta_0)d\Lambda_0$ and $\int_0^a \mathcal{G}_n^{(1)}(\theta_0)s^{(0)-1}(\theta_0)d\Lambda_0$ converge weakly to centered Gaussian variables with finite variances therefore there exists a strictly positive constant c_2 such that $\sup_{\theta\in V_\varepsilon(\theta_0)} E_0 W_{2n}^2 \le c_2\varepsilon^2$, this proves the lemma for W_{2n}.

The process W_{1n} is differentiable with respect to the components of (α,β) and γ, with partial derivatives

$$n^{-\frac{1}{2}}\sum_{i=1}^n \int_0^a \{Z_i(t)\,dN_i(t) - s^{(1)}(t;\theta_0)\,d\Lambda_0(t)\}$$

and, respectively

$$n^{-\frac{1}{2}}\sum_{i=1}^n \int_0^a \{Z_{2i}(t)\,dN_i(t) - s_2^{(1)}(t;\theta_0)\,d\Lambda_0(t)\},$$

they converge weakly to centered Gaussian variables with finite variances and the inequality is true for the process W_{1n}, by the same argument as for W_{2n}. □

Under the conditions, the convergence rates of the maximum likelihood estimators of the parameters and the likelihood are deduced from Lemmas (6.5) and (6.6), following the arguments of Theorem 5.1 in Ibragimov and Has'minskii (1981).

Theorem 6.8. *Under conditions* C1–C2, *for every* $\varepsilon > 0$

$$\overline{\lim}_{n\to\infty,A\to\infty} P_0(\sup_{u\in\mathcal{U}_{n,\varepsilon},\|u\|>A} X_n(\theta_{n,u}) \geq 0) = 0,1$$

$$\overline{\lim}_{n\to\infty,A\to\infty} P_0(n|\widehat{\tau}_n - \tau_0| > A) = 0,$$

$$\overline{\lim}_{n\to\infty,A\to\infty} P_0(n^{\frac{1}{2}}\|\widehat{\xi}_n - \xi_0\| > A) = 0.$$

Let $u_n = (n(\tau_n - \tau_0), n^{\frac{1}{2}}(\xi_n - \xi_0)^T)^T$ for sequences $(\tau_n)_n$ and $(\xi_n)_n$ converging to τ_0 and, respectively ξ_0, its first component is denoted u_{1n} and the vector $u_{2n} = n^{\frac{1}{2}}(\xi_n - \xi_0)$ has the components $u_{n,\alpha}, u_{n,\beta}$ and $u_{n,\gamma}$.

Let $\mathcal{U}_n^A = \{\|u_n\| \leq A\}$, where $\|u_n\|$ is related to the semi-norm ρ. The continuous part of the process $l_n(\theta) - l_n(\theta_0)$ is approximated on $\mathcal{U}_n^A$ by its second order expansion depending on the process

$$\widetilde{M}_n(t) = n^{-\frac{1}{2}} \sum_{i=1}^n \int_0^t \Big\{ Z_i(s) - \frac{S_n^{(1)}(s;\theta_0)}{S_n^{(0)}(s;\theta_0)} \Big\}\, dM_i(s)$$

where $M_i(t) = N_i(t) - \int_0^t s^{(0)}(s;\theta_0)\, d\Lambda_0(s)$, and the sum of its jumps is approximated by the process

$$Q_n(u_1) = \sum_{i=1}^n \delta_i 1_{\{\tau_{n,u} < T_i \leq \tau_0\}} \Big\{ \gamma_0^T Z_{2i}(T_i) - \log \frac{S_n^{(0)}(T_i;\alpha_0,\beta_0+\gamma_0)}{S_n^{(0)}(T_i;\alpha_0,\beta_0)} \Big\}$$
$$- \sum_{i=1}^n \delta_i 1_{\{\tau_0 < T_i \leq \tau_{n,u}\}} \Big\{ \gamma_0^T Z_{2i}(T_i) - \log \frac{S_n^{(0)}(T_i;\alpha_0,\beta_0)}{S_n^{(0)}(T_i;\alpha_0,\beta_0+\gamma_0)} \Big\},$$

where $\tau_{n,u} = \tau_0 + n^{-1}u_1$. Let $\widetilde{M}_{nq}$ be the last q components of the process $\widetilde{M}_n$, let

$$\widetilde{W}_n = \Big\{ \widetilde{M}_n^T(1), \widetilde{M}_{nq}^T(1) - \widetilde{M}_{nq}^T(\tau_0) \Big\}^T,$$

and let

$$\widetilde{I} = \begin{pmatrix} I(\theta_0) & I_q(\theta_0) - I_q(\tau_0;\theta_0) \\ I_q^T(\theta_0) - I_q^T(\tau_0;\theta_0) & I_{qq}(\theta_0) - I_{qq}(\tau_0;\theta_0) \end{pmatrix}$$

be the symmetric matrix of dimension $(p+2q)^{\otimes 2}$ defined by the last q rows I_q of the matrix I and its right lower $q \times q$ sub-matrix. The matrix $\widetilde{I}$ is the asymptotic variance of the variable $\widetilde{W}_n$ and it is positive definite.

Theorem 6.9. *Under Conditions* C1–C2

$$l_n(\theta) - l_n(\theta_0) = Q_n(u_1) + u_2^T \widetilde{W}_n - \frac{1}{2} u_2^T \widetilde{I} u_2 + o_p(1)$$

with an uniform $o_p(1)$ *in* $\mathcal{U}_n^A$, *for every* $A > 0$, *as* n *tends to infinity.*

Proof. Let $A > 0$, let $(u_n)_n$ a vector in $\mathcal{U}_n^A$ which converges to a vector u with strictly positive components and let $\tau_{n,u} = \tau_0 + n^{-1}u_{1n}$ and $\xi_{n,u} = \xi_0 + n^{-\frac{1}{2}}u_{2n}$ with components $\alpha_{n,u}$, $\beta_{n,u}$ and $\gamma_{n,u}$. The process

$$n^{-\frac{1}{2}}\sum_{i=1}^{n}\int_{\tau_0}^{\tau_{n,u}}\Big\{Z_i(s) - \frac{S_n^{(1)}(s;\theta_0)}{S_n^{(0)}(s;\theta_0)}\Big\}\,dN_i(s) = \int_{\tau_0}^{\tau_{n,u}} d\widetilde{M}_n$$

has the mean zero. Lenglart's inequality implies that for all $\eta > 0$ and $\varepsilon > 0$, the probability $P_0\Big\{\sup_{|u_1|\le A}\Big|\int_{\tau_0}^{\tau_{n,u}} d\widetilde{M}_{n,j}\Big| > \eta\Big\}$ is bounded by

$$\frac{\varepsilon}{\eta^2} + P_0\Big\{n^{-\frac{1}{2}}\sum_{i=1}^{n}\int_{\tau_0}^{\tau_{n,A}}\Big\{Z_{i,j}(s) - \frac{S_{n,j}^{(1)}(s;\theta_0)}{S_{n,j}^{(0)}(s;\theta_0)}\Big\}^2 Y_i(s)\,d\Lambda_0(s) > \varepsilon\Big\}$$

$$\le \frac{\varepsilon}{\eta^2} + \frac{1}{\varepsilon}E_0\int_{\tau_0}^{\tau_{n,A}} n^{-1}\Big|S_{n,j}^{(2)}(s;\theta_0) - \frac{S_{n,j}^{(1)2}(s;\theta_0)}{S_{n,j}^{(0)}(s;\theta_0)}\Big|\,d\Lambda_0(s)$$

as n tends to infinity, this inequality has the approximation

$$P_0\Big\{\sup_{|u_1|\le A}\Big|\int_{\tau_0}^{\tau_{n,u}} d\widetilde{M}_{n,j}\Big| > \eta\Big\}$$

$$\le \frac{\varepsilon}{\eta^2} + \frac{A}{n\varepsilon}\sup_{s\in V(\tau_0)}\Big|s_j^{(2)}(s;\theta_0) - \frac{s_j^{(1)2}(s;\theta_0)}{s_j^{(0)}(s;\theta_0)}\Big|\lambda_0(s) + o((n\varepsilon)^{-1}).$$

Let $\varepsilon = n^{-\frac{1}{2}}$, the bound converges to zero as n tends to infinity, for every $\eta > 0$. The discontinuous part of $l_n(\theta) - l_n(\theta_0)$ is therefore approximated by the process Q_n. By expansions of its logarithmic terms and because the derivative of the continuous part of $l_n(\theta) - l_n(\theta_0)$ with respect to α and β are centered, this continuous part is approximated by $u_2^T\widetilde{W}_n - \frac{1}{2}u_2^T\widetilde{I}u_2$. □

Theorem 6.9 splits the logarithm of the partial likelihood as the sum of the process Q_n depending only on the first component of u and a second order expansion of its continuous part depending only on its last $p+2q$ components. By the independence of the increments of the counting processes N_i, the estimators of $\widehat{\tau}_n$ and $\widehat{\xi}_n$ are independent.

The estimator of ξ has the expansion

$$n^{\frac{1}{2}}(\widehat{\xi}_n - \xi_0) = \widetilde{I}^{-1}\widetilde{W}_n + o_p(1) \tag{6.19}$$

and it converges weakly to a centered Gaussian variable with variance $\widetilde{I}^{-1}$. This is the limiting distribution of the variable $n^{\frac{1}{2}}(\widehat{\xi}_n - \xi_0)$ in a model with a known change-point at τ_0.

Theorem 6.10. *The process Q_n converges weakly to a Gaussian process with mean $E_0Q_n(u) = u_1\mu_0$ and with variance $Var_0Q_n(u) = u_1v_0$, and the* *estimator $\widehat{\tau}_n$ is such that $n(\widehat{\tau}_n - \tau_0)$ converges weakly to $(2\mu_0)^{-2}G_0^2$, where G_0 is a centered Gaussian variable.*

Proof. The process $Q_n = Q_n^- Q_n^+$ is such that

$$Q_n^-(u) = \sum_{i=1}^n \int_{\tau_{n,u}}^{\tau_0} \Big\{\gamma_0^T Z_{2i} - \log \frac{S_n^{(0)}(\alpha_0, \beta_0 + \gamma_0)}{S_n^{(0)}(\alpha_0, \beta_0)}\Big\}\, dM_i$$

$$+ \int_{\tau_{n,u}}^{\tau_0} \Big\{\gamma_0^T S_{n2}^{(1)}(\alpha_0, \beta_0) - S_n^{(0)}(\alpha_0, \beta_0) \log \frac{S_n^{(0)}(\alpha_0, \beta_0 + \gamma_0)}{S_n^{(0)}(\alpha_0, \beta_0)}\Big\}\, d\Lambda_0$$

where the second integral is approximated by

$$n^{-1} u_1 \lambda_0(\tau_0) \Big\{\gamma_0^T S_{n2}^{(1)}(\tau_0; \alpha_0, \beta_0) - S_n^{(0)}(\tau_0; \alpha_0, \beta_0) \log \frac{S_n^{(0)}(\tau_0; \alpha_0, \beta_0 + \gamma_0)}{S_n^{(0)}(\tau_0; \alpha_0, \beta_0)}\Big\}$$

and it converges a.s. to

$$u_1 \lambda_0(\tau_0) \Big\{\gamma_0^T s_2^{(1)}(\tau_0; \alpha_0, \beta_0) - s^{(0)}(\tau_0; \alpha_0, \beta_0) \log \frac{s^{(0)}(\tau_0; \alpha_0, \beta_0 + \gamma_0)}{s^{(0)}(\tau_0; \alpha_0, \beta_0)}\Big\}.$$

The first integral is centered and its variance is

$$v_n(u_1) = \sum_{i=1}^n \int_{\tau_{n,u}}^{\tau_0} \Big\{\gamma_0^T Z_{2i} - \log \frac{S_n^{(0)}(\alpha_0, \beta_0 + \gamma_0)}{S_n^{(0)}(\alpha_0, \beta_0)}\Big\}^2 Y_i e^{r_{\theta_0}(Z_i)}\, d\Lambda_0$$

$$= \int_{\tau_{n,u}}^{\tau_0} \Big\{\gamma_0^T S_{n2}^{(2)}(\alpha_0, \beta_0)\gamma_0 - 2\gamma_0^T S_{n2}^{(1)}(\alpha_0, \beta_0) \log \frac{S_n^{(0)}(\alpha_0, \beta_0 + \gamma_0)}{S_n^{(0)}(\alpha_0, \beta_0)}$$

$$+ S_n^{(0)}(\alpha_0, \beta_0) \log^2 \frac{S_n^{(0)}(\alpha_0, \beta_0 + \gamma_0)}{S_n^{(0)}(\alpha_0, \beta_0)}\Big\}\, d\Lambda_0$$

which is asymptotically equivalent to

$$v(u_1) = \Big\{\gamma_0^T s_2^{(2)}(\tau_0; \alpha_0, \beta_0)\gamma_0 - 2\gamma_0^T s_2^{(1)}(\tau_0; \alpha_0, \beta_0) \log \frac{s^{(0)}(\tau_0; \alpha_0, \beta_0 + \gamma_0)}{s^{(0)}(\tau_0; \alpha_0, \beta_0)}$$

$$+ s^{(0)}(\tau_0; \alpha_0, \beta_0) \log^2 \frac{s^{(0)}(\tau_0; \alpha_0, \beta_0 + \gamma_0)}{s^{(0)}(\tau_0; \alpha_0, \beta_0)}\Big\}\, u_1 \lambda_0(\tau_0) = \mu_0^- u_1.$$

By the same argument for Q_n^+ and by Rebolledo's convergence theorem for local martingales, the process Q_n converges weakly to the sum Q_0 of a mean $u_1\mu_0$ and a centered Gaussian variable G_0 with variance $u_1 v_0$

$$Q_0(u_1) = u_1\mu_0 + u_1^{\frac{1}{2}} G_0.$$

Its maximum satisfies $\mu_0 u_1^{\frac{1}{2}} + \frac{1}{2}G_0 = 0$ i.e. $u_1 = (2\mu_0)^{-2}G_0^2$. □

The weak convergence of the estimator of the cumulative baseline intensity is deduced from the expansion of the second term of the sum

$$n^{\frac{1}{2}}\{\widehat{\Lambda}_n(t) - \Lambda_0(t)\} = \int_0^t \frac{dM_{0n}(s)}{n^{-1}S_n^{(0)}(s;\widehat{\theta}_n)} - \int_0^t \frac{n^{\frac{1}{2}}\{S_n^{(0)}(s;\widehat{\theta}_n) - S_n^{(0)}(s;\theta_0)\}}{S_n^{(0)}(s;\widehat{\theta}_n)}\, d\Lambda_0(s). \tag{6.20}$$

Theorem 6.11. *The process*

$$n^{\frac{1}{2}}\{\widehat{\Lambda}_n(t) - \Lambda_0(t)\} + n^{\frac{1}{2}}(\widehat{\theta}_n - \theta_0)^T \int_0^t \frac{s^{(1)}(s;\theta_0)}{s^{(0)}(s;\theta_0)}\, d\Lambda_0(s)$$

is asymptotically independent of $\widehat{\theta}_n$ and it converges weakly under P_0 to a centered Gaussian process with independent increments and with variance $\int_0^t s^{(0)-1}(\theta_0)\, d\Lambda_0$.

Proof. The first term in the expansion (6.20) is a local martingale and it converges weakly to a centered Gaussian process with independent increments and with variance $\int_0^t s^{(0)-1}(\theta_0)\, d\Lambda_0$. The limiting distribution of the second term is obtained from an expansion of $n^{\frac{1}{2}}\{S_n^{(0)}(s;\widehat{\theta}_n) - S_n^{(0)}(s;\theta_0)\}$ as

$$\begin{aligned}
&1_{[0,\widehat{\tau}_n\wedge\tau_0]}(s)n^{\frac{1}{2}}\{(\widehat{\alpha}_n-\alpha_0)^T,(\widehat{\beta}_n-\beta_0)^T\}S_n^{(1)}(s;\alpha_0,\beta_0)\\
&+1_{[\widehat{\tau}_n,\tau_0]}(s)n^{\frac{1}{2}}\{(\widehat{\alpha}_n-\alpha_0)^T,(\widehat{\beta}_n-\beta_0)^T,\widehat{\gamma}_n^T\}S_n^{(1)}(s;\alpha_0,\beta_0+\gamma_0)\\
&+1_{[\tau_0,\widehat{\tau}_n]}(s)n^{\frac{1}{2}}\{(\widehat{\alpha}_n-\alpha_0)^T,(\widehat{\beta}_n-\beta_0)^T,-\gamma_0^T\}S_n^{(1)}(s;\alpha_0,\beta_0)\\
&+1_{[\tau_0\vee\widehat{\tau}_n,a]}(s)n^{\frac{1}{2}}\{(\widehat{\alpha}_n-\alpha_0)^T,(\widehat{\beta}_n-\beta_0)^T,(\widehat{\gamma}_n-\gamma_0)^T\}S_n^{(1)}(s;\alpha_0,\beta_0+\gamma_0)\\
&+o_p(\|\widehat{\xi}_n-\xi_0\|)
\end{aligned}$$

and the restriction of the second term of (6.20) to the intervals of length $|\widehat{\tau}_n - \tau_0| = O(n^{-1})$ tends to zero. The sum of the restrictions to the other intervals is asymptotically equivalent to

$$n^{\frac{1}{2}}(\widehat{\theta}_n - \theta_0)^T \int_0^a s^{(1)}(\theta_0)s^{(0)-1}(\theta_0)\, d\Lambda_0.$$

By centering the process $n^{\frac{1}{2}}\{\widehat{\Lambda}_n - \Lambda_0\}$ with this limit, it is asymptotically equivalent to the first term of the expansion (6.20). □

The properties of the estimators are the same in Model (6.1) with a parametric regression function $r_\theta(z)$ of $C^2(\Xi)$, for every z in $\mathcal{Z}$. The notations of the partial derivatives of the processes $S_n^{(0)}(t;\theta)$ and l_n are modified according to those of the regression function, the process X_n is now the sum of the processes

$$X_{1n} = n^{-1}\sum_{i=1}^n \delta_i 1_{\{T_i\le\tau\wedge\tau_0\}}\Big\{r_{\alpha,\beta}(Z_i) - r_{\alpha_0,\beta_0}(Z_i) - \log\frac{S_n^{(0)}(T_i;\alpha,\beta)}{S_n^{(0)}(T_i;\alpha_0,\beta_0)}\Big\},$$

$$X_{2n} = n^{-1}\sum_{i=1}^n \delta_i 1_{\{T_i>\tau\vee\tau_0\}}\Big\{r_{\alpha,\beta+\gamma}(Z_i) - r_{\alpha_0,\beta_0+\gamma_0}(Z_i) - \log\frac{S_n^{(0)}(T_i;\alpha,\beta+\gamma)}{S_n^{(0)}(T_i;\alpha_0,\beta_0+\gamma_0)}\Big\},$$

$$X_{3n} = n^{-1}\sum_{i=1}^n \delta_i 1_{\{\tau<T_i\le\tau_0\}}\Big\{r_{\alpha,\beta+\gamma}(Z_i) - r_{\alpha_0,\beta_0}(Z_i) - \log\frac{S_n^{(0)}(T_i;\alpha,\beta+\gamma)}{S_n^{(0)}(T_i;\alpha_0,\beta_0)}\Big\} + n^{-1}\sum_{i=1}^n \delta_i 1_{\{\tau_0<T_i\le\tau\}}\Big\{r_{\alpha,\beta}(Z_i) - r_{\alpha_0,\beta_0+\gamma_0}(Z_i) - \log\frac{S_n^{(0)}(T_i;\alpha,\beta)}{S_n^{(0)}(T_i;\alpha_0,\beta_0+\gamma_0)}\Big\}$$

and the continuous part of the process $l_n(\theta) - l_n(\theta_0)$ is approximated on $\mathcal{U}_n^A$ using the processes

$$\widetilde{W}_n(t) = n^{-\frac12}\sum_{i=1}^n \int_0^t \Big\{\dot r_{\theta_0,\xi}(Z_i(s)) - \frac{S_n^{(1)}(s;\theta_0)}{S_n^{(0)}(s;\theta_0)}\Big\}\,dM_i(s),$$

$$\widetilde{l}_n(t) = n^{-1}\sum_{i=1}^n \int_0^t \{\ddot r_{\theta_0,\xi}(Z_i(s)) - V_n(s;\theta_0)\}\,dM_i(s).$$

6.7 Likelihood ratio test of models without change

The log-likelihood ratio test of the hypothesis H_0 of regression without change of parameter against the alternative of a change at an unknown index vector k_0, with distinct parameters β and γ, is performed with the statistic

$$T_n = 2\{l_n(\widehat{\theta}_n) - \widehat{l}_{0n}\},$$

where $\widehat{l}_{0n} = l_n(\widehat{\alpha}_{0n}, \widehat{\beta}_{0n})$ is the maximum of the log-partial likelihood under the hypothesis H_0. Let $\xi_1 = (\alpha^T, \beta^T)^T$ be the parameter vector under P_0, and let

$$U_{0n}(\xi_1) = l'_{0n}(\xi_1) = \sum_{i=1}^n \int_0^a \{Z_i - S_n^{(1)}(\xi_1) S_n^{(1)-1}(\xi_1)\}\, dM_i,$$

the variable $n^{-\frac{1}{2}} U_{0n}(\xi_{01})$ converges weakly under P_0 to a centered Gaussian variable with variance I_0^{-1} where $I_0 = -\lim_n n^{-1} l''_n(\xi_1)$ and $\widehat{l}_{0n}$ converges weakly to a χ^2_{p+q} variable, by a second order expansion.

Under P_0, $\tau_0 = a$ and the process $X_n(\theta)$ reduces to the sum of the processes

$$X_{1n}(\theta) = n^{-1} \sum_{i=1}^n \delta_i 1_{\{T_i \le \tau\}} \Big\{ (\alpha - \alpha_0)^T Z_{1i}(T_i) + (\beta - \beta_0)^T Z_{2i}(T_i) - \log \frac{S_n^{(0)}(T_i; \alpha, \beta)}{S_n^{(0)}(T_i; \alpha_0, \beta_0)} \Big\},$$

$$X_{3n}(\theta) = n^{-1} \sum_{i=1}^n \delta_i 1_{\{\tau < T_i\}} \Big\{ (\alpha - \alpha_0)^T Z_{1i}(T_i) + (\beta - \beta_0 + \gamma)^T Z_{2i}(T_i) - \log \frac{S_n^{(0)}(T_i; \alpha, \beta + \gamma)}{S_n^{(0)}(T_i; \alpha_0, \beta_0)} \Big\}.$$

Under the conditions, they converge a.s. uniformly under P_0 to their expectation functions

$$X_1(\theta) = \int_0^\tau \Big[\{(\alpha - \alpha_0)^T, (\beta - \beta_0)^T\}^T s^{(1)}(t; \alpha_0, \beta_0) - s^{(0)}(t; \alpha_0, \beta_0) \log \frac{s^{(0)}(t; \alpha, \beta)}{s^{(0)}(t; \alpha_0, \beta_0)} \Big] \, d\Lambda_0(t),$$

$$X_3(\theta) = \int_\tau^a \Big[\{(\alpha - \alpha_0)^T, (\beta - \beta_0 + \gamma)^T\}^T s^{(1)}(t; \alpha_0, \beta_0) - s^{(0)}(t; \alpha_0, \beta_0) \log \frac{s^{(0)}(t; \alpha, \beta + \gamma)}{s^{(0)}(t; \alpha_0, \beta_0)} \Big] \, d\Lambda_0(t),$$

the function $X = X_1 + X_3$ is maximum at θ_0 and $\widehat{\theta}_n$ converges a.s. under P_0 to θ_0, with $\gamma_0 = 0$.

Proposition 6.3. *The statistic T_n converges weakly under H_0 to a χ^2_q variable.*

Proof. The process $W_{1n}(\theta, \xi_0) = nX_n(\theta) - nX_n(\xi, \tau_0)$ develops as

$$W_{1n}(\theta) = nX_{3n}(\theta) - \sum_{i=1}^{n} \delta_i 1_{\{T_i > \tau\}} \Big\{ (\alpha - \alpha_0)^T Z_{1i}(T_i) + (\beta - \beta_0)^T Z_{2i}(T_i) - \log \frac{S_n^{(0)}(T_i; \alpha, \beta)}{S_n^{(0)}(T_i; \alpha_0, \beta_0)} \Big\}$$

$$= \sum_{i=1}^{n} \delta_i 1_{\{T_i > \tau\}} \Big\{ \gamma^T Z_{2i}(T_i) - \log \frac{S_n^{(0)}(T_i; \alpha, \beta + \gamma)}{S_n^{(0)}(T_i; \alpha, \beta)} \Big\}$$

and the process $W_{2n}(\theta) = nX_n(\xi, \tau_0) - nX_{0n}(\xi_0)$ is written as

$$W_{2n}(\xi_1, \xi_0) = \sum_{i=1}^{n} \delta_i \Big\{ (\alpha - \alpha_0)^T Z_{1i}(T_i) + (\beta - \beta_0)^T Z_{2i}(T_i) - \log \frac{S_n^{(0)}(T_i; \alpha, \beta)}{S_n^{(0)}(T_i; \alpha_0, \beta_0)} \Big\}.$$

Let $U_{0n}(\xi_1)$ and $-I_{0n}(\xi_1)$ be the first two derivatives of the process W_{2n} with respect to ξ_1, I_n is the asymptotic variance of U_{0n} under P_0 and it converges to a strictly positive definite matrix I_0. By a first order asymptotic expansion of $U_{0n}(\widehat{\xi}_{1n})$, we have

$$n^{\frac{1}{2}}(\widehat{\xi}_{1n} - \widehat{\xi}_{0n}) = I_0^{-1} U_{1n}(\widehat{\xi}_{0n}) + o_p(1) = o_p(1)$$

and, by a second order asymptotic expansion, the variable $W_{2n}(\widehat{\xi}_{1n}, \widehat{\xi}_{0n})$ converges in probability to zero.

Let $U_{1n}(\theta)$ and $-I_{1n}(\theta)$ be the first two derivatives of the process W_{1n} with respect to θ

$$U_{n,\alpha}(\theta) = \sum_{i=1}^{n} \delta_i 1_{\{\tau < T_i\}} \Big\{ \frac{S_{1n}^{(1)}(T_i; \alpha, \beta)}{S_n^{(0)}(T_i; \alpha, \beta)} - \frac{S_{1n}^{(1)}(T_i; \alpha, \beta + \gamma)}{S_n^{(0)}(T_i; \alpha, \beta + \gamma)} \Big\},$$

$$U_{n,\beta}(\theta) = \sum_{i=1}^{n} \delta_i 1_{\{T_i \le \tau\}} \Big\{ \frac{S_{2n}^{(1)}(T_i; \alpha, \beta)}{S_n^{(0)}(T_i; \alpha, \beta)} - \frac{S_{2n}^{(1)}(T_i; \alpha, \beta + \gamma)}{S_n^{(0)}(T_i; \alpha, \beta + \gamma)} \Big\},$$

$$U_{n,\gamma}(\theta) = \sum_{i=1}^{n} \delta_i 1_{\{T_i \le \tau\}} \Big\{ Z_{2i}(T_i) - \frac{S_{2n}^{(1)}(T_i; \alpha, \beta + \gamma)}{S_n^{(0)}(T_i; \alpha, \beta + \gamma)} \Big\},$$

I_n is the asymptotic variance of U_n, it converges to a strictly positive definite matrix I_0. Under P_0, the function W_2, limit of the process W_{2n}, is a $O(|\tau - \tau_0|)$, Lemma 6.5 is therefore satisfied, the proof of Lemma 6.6 is still valid and the convergence rate of $\widehat{\tau}_n$ is n, by Theorem 6.8. The variable $U_n(\widehat{\theta}_n)$ depends only on the time variables larger than $\widehat{\tau}_n$ and by the convergence of $\widehat{\gamma}_n$ to zero, the variables $U_{n,\alpha}(\widehat{\theta}_n)$ and $U_{n,\beta}(\widehat{\theta}_n)$ converge to zero in probability. The expectation of the variable $U_n(\theta_n)$ converges to zero and its variance converges to a strictly positive definite matrix I_{γ_0}.

By a first order expansion of the variable $U_n(\widehat{\theta}_n)$ for $\widehat{\gamma}_n$ in a neighborhood of $\gamma_0 = 0$, we have

$$\widehat{\gamma}_n = I_{n,\gamma}^{-1}(\widehat{\theta}_n)U_{n,\gamma}(\widehat{\xi}_{1n}, 0, \widehat{\tau}_n) + o_p(1)$$

where $U_{n,\gamma}(\widehat{\xi}_{1n}, 0, \widehat{\tau}_n)$ converges to a centered Gaussian variable with variance I_{γ_0}. The limit of the test statistic follows from a second order asymptotic expansion of $W_{1n}(\widehat{\theta}_n)$. □

Local alternatives P_{θ_n} are defined by parameters $\tau_n = \tau_0 - n^{-1}u_n$ in $]0, a[$ and $\xi_n = (\alpha_n^T, \beta_n^T, \gamma_n^T)^T$ such that $\xi_n = \xi_0 + n^{-\frac{1}{2}}v_n$, where u_n and v_n converge to finite limits u, and respectively v, as n tends to infinity.

Proposition 6.4. *Under fixed alternatives, the statistic T_n tends to infinity as n tends to infinity and under local alternatives P_{θ_n} H_0, the statistic T_n converges weakly to $T_0 + v_1^T I_0^{-1} v_1$.*

Proof. Under a fixed alternative P_θ with τ in $]0, a[$ and a non-zero parameter vector γ, the log-partial likelihood is the sum $l_n(\widehat{\theta}_n) - l_n(\theta) - \{l_n(\widehat{\theta}_{0n}) - l_n(\theta_0)\} + l_n(\theta) - l_n(\theta_0)$ where the first two differences are is asymptotically equivalent to χ^2 variables and the third one tends to infinity.

Under local alternatives $P_n = P_\theta$, the variable $n^{\frac{1}{2}}(\widehat{\xi}_{1n} - \widehat{\xi}_{0n})$ converges in probability to v_1 and $W_{2n}(\widehat{\xi}_{1n}, \widehat{\xi}_{0n})$ converges in probability to $v_1^T I_0^{-1} v_1$. From the convergence rate of the estimators, the variable $W_{1n}(\widehat{\theta}_n)$ has the same limiting distribution under P_n as under H_0. □

Chapter 7

Change-points for auto-regressive series

Abstract. This chapter studies auto-regressive models with a random initial value and with or without change-points at unknown sampling indices. The convergence rates of the series and the empirical estimators are established according to the the domain of the parameters and their weak convergences are proved.

7.1 Series of AR(p) models

Let $(X_t)_{t=0}^{\infty}$ be a time series and let $X_{(t)} = (X_1, \ldots, X_t)^T$ be the t-dimensional vector of the series starting from X_1. An auto-regressive series $(X_t)_t$ of order p is defined by a vector $X_0^{(p)} = (X_0, \ldots, X_{-(p-1)})^T$ of p random initial values with finite means and variances, and by the model

$$X_t = \mu + \sum_{k=1}^{p} \alpha_k X_{t-k} + \varepsilon_t,\ t \in \mathrm{N}, \tag{7.1}$$

with parameters μ and $\alpha = (\alpha_1, \ldots, \alpha_k)^T$, and $(\varepsilon_t)_t$ is a sequence of independent, centred and identically distributed errors with variance σ^2 and unknown distribution. Let m be the expectation of X_0, when $\sum_{k=0}^{p-1} \alpha_k \neq 1$, all parameters of model (7.1) are identifiable and the model is equivalently written with $\mu = m(1 - \sum_{k=1}^{p} \alpha_k)$ as

$$X_t = \left(1 - \sum_{k=1}^{p} \alpha_k\right) m + \sum_{k=1}^{p} \alpha_k X_{t-k} + \varepsilon_t,$$

the series $Y_t = X_t - m$ satisfies

$$Y_t = \sum_{k=1}^{p} \alpha_k Y_{t-k} + \varepsilon_t,\ t \in \mathrm{N}. \tag{7.2}$$

In model AR(1) defined as

$$X_t = \mu + \alpha X_{t-1} + \varepsilon_t, \tag{7.3}$$

$$Y_t = \alpha Y_{t-1} + \varepsilon_t, \tag{7.4}$$

the series X_t and Y_t are expressed as sums since their initial values

$$X_t = \mu \sum_{k=0}^{t-1} \alpha^k + \alpha^t X_0 + \sum_{k=1}^{t} \alpha^{t-k} \varepsilon_k,$$

$$Y_t = \alpha^t Y_0 + \sum_{k=1}^{t} \alpha^{t-k} \varepsilon_k.$$

If $|\alpha| < 1$, X_t and the mean $\bar{X}_t = t^{-1} \sum_{k=1}^t X_k$ converge to m, the series $\sigma^{-1}(1-\alpha^2)^{\frac{1}{2}}(X_t - m)$ and $t\sigma^{-1}(1-\alpha^2)^{\frac{1}{2}}(\bar{X}_t - m)$ converge weakly to free distributions, as t tends to by infinity. If $|\alpha| \geq 1$, the processes are a.s. explosive and other normalization coefficients must be introduced for their weak convergence.

If $\alpha = 1$, $X_t = \mu + X_{t-1} + \varepsilon_t = t\mu + X_0 + \sum_{k=1}^t \varepsilon_k$, the series $t^{-1}(X_t - X_0)$ converges a.s. to μ and $\sigma^{-1} t^{\frac{1}{2}} \{t^{-1}(X_t - X_0) - \mu\}$ converges weakly to a Gaussian process.

If $\alpha = -1$, the sub-series $X_{2t} = X_0 + \sum_{k=1}^{2t} (-1)^k \varepsilon_k$ and $X_{2t+1} = \mu - X_0 + \sum_{k=1}^{2t+1} (-1)^{k+1} \varepsilon_k$ converge if the distribution of the error is symmetric, X_t is the sum of means and random walks with parameter $\alpha = 1$.

If $|\alpha| > 1$, $(X_t - m) = \alpha^t (X_0 - m) + \sum_{k=1}^t \alpha^{t-k} \varepsilon_k$, with two diverging subsequences when $\alpha < -1$, according to the behaviour of α^{2t} which tends to $+\infty$ and α^{2t+1} which tends to $-\infty$. The process

$$\alpha^{-t}(X_t - m) - (X_0 - m) \sim \alpha^{-t} \sum_{k=0}^{t-1} \alpha^k \varepsilon_{t-k}$$

is asymptotically Gaussian.

Estimators are defined for every finite sample by minimization of quadratic norm of the error vector $E_t = (\varepsilon_k)_{k=1}^T$ of the series and the behaviour of the estimator $\widehat{\alpha}_t$ of α is an important question since the domain of α determines the behaviour of the series. When the initial value X_0 is constant, the process $t^{\frac{1}{2}}(\widehat{\alpha}_t - \alpha)$ is asymptotically Gaussian if $|\alpha| < 1$. In this case and with a Gaussian error, White (1958) established an approximation of $\alpha^t(\widehat{\alpha}_t - \alpha)/(\alpha^2 - 1)$ for $|\alpha| > 1$, and an approximation of $t(\widehat{\alpha}_t - 1)$ by the ratio of two integrated Wiener processes for $|\alpha| = 1$.

For $|\alpha| \geq 1$, the results depend on the distribution of ε_t. Other models with an unknown error distribution have also been studied at the critical value $|\alpha| = 1$, with independent or mixing errors: Phillips (1978, 1987), Haldrup (1994) with mixing errors, Dickey and Fuller (1979) with independent and identically distributed errors. In random walk models with a drift of the mean, the observations have other limits according to the model for the mean, see for example Haldrup and Hylleberg (1995). Martingale techniques are used for the convergence and the limits are transformed Brownian motions.

In Section 7.2, the asymptotic distributions of the sample-paths and the estimators are established for every distribution of the error. The estimators of the parameters of model (7.3) are asymptotically free of X_0 when $|\alpha| < 1$ or $\alpha = 1$ and the asymptotic distribution of the estimators of α is free of σ.

The convergence rate of $\widehat{\alpha}_t$ is $t^{-\frac{1}{2}}$ when $|\alpha| < 1$, t^{-1} when $\alpha = 1$ and $\mu \neq 0$, or $|\alpha| > 1$; it becomes $t^{-3/2}$ when $\alpha = 1$ and $\mu = 0$. For the estimator $\widehat{\mu}_t$ of the mean, the convergence rate is 1 when $|\alpha| < 1$, $t^{-\frac{1}{2}}$ when $\alpha = 1$ and $t^{-1}\alpha^t$ when $|\alpha| > 1$.

7.2 Convergence in the AR(1) model

The least squares estimators of the parameters α, μ and σ^2 minimize the Euclidean norm of the error $\|E_t\|_t^2 = \sum_{k=1}^t \varepsilon_k^2$. By the equivalence between (7.3) and (7.4), the least squares estimators are

$$\begin{aligned}
\widehat{\mu}_t &= (1 - \widehat{\alpha}_t)\bar{X}_t + t^{-1}(\widehat{\alpha}_t X_t - X_0), \\
\widehat{m}_t &= (1 - \widehat{\alpha}_t)^{-1}\widehat{\mu}_t, \\
\widehat{\alpha}_t &= \frac{\sum_{k=1}^t (X_{k-1} - \widehat{m}_t)(X_k - \widehat{m}_t)}{\sum_{k=1}^t (X_{k-1} - \widehat{m}_t)^2}.
\end{aligned} \tag{7.5}$$

Then with the residuals $\widehat{E}_t = X_t - \widehat{\mu}_t - \widehat{\alpha}_t X_{t-1}$, the variance of the error is estimated by

$$\widehat{\sigma}_t^2 = \frac{1}{t}\|\widehat{E}_t\|_t^2 = \frac{1}{t}\sum_{k=1}^t \{X_k - (1 - \widehat{\alpha}_t)\bar{X}_t - \widehat{\alpha}_t X_{k-1}\}^2.$$

For $|\alpha| \neq 1$, $\widehat{\mu}_t = (1 - \widehat{\alpha}_t)\bar{X}_t + O_p(t^{-1})$ and $\widehat{m}_t = \bar{X}_t$. For $\alpha = 1$, another estimator of μ is defined. Let W denote the standard Brownian motion and let $\bar{W} = \int_0^{\cdot} W(t)\, dt$.

Partial sums $S_{k,\alpha}$ are defined for weighted errors by $S_{0,\alpha} = 0$ and

$$S_{k,\alpha} = \sum_{j=1}^{k} \alpha^{k-j}\varepsilon_j = \sum_{j=0}^{k-1} \alpha^j \varepsilon_{k-j}, \; k \geq 1,$$

$$\widetilde{S}_{k,\alpha} = \sum_{j=1}^{k} \varepsilon_j \sum_{l=0}^{j-1} \alpha^l \varepsilon_{j-l},$$

$$\bar{S}_{k,\alpha} = \sum_{j=1}^{k} S_{j,\alpha}.$$

Their variances are

$$VarS_k = \sigma^2 \sum_{l=0}^{k-1} \alpha^{2l} = \sigma^2(1-\alpha^{2k})(1-\alpha^2)^{-1}, \;\text{ if } \alpha \neq 1,$$

$$VarS̄_{t,\alpha} = \sigma^2 2 \sum_{l=1}^{t} (t-l) \sum_{k=0}^{l-1} \alpha^{2k} + o(1)$$

$$= \frac{\sigma^2}{1-\alpha^2} t^2 + o(t^2), \;\text{ if } |\alpha| < 1,$$

$$Var\bar{S}_{t,\alpha} = 2\sigma^2 \sum_{l=1}^{t} (t-l)(l-1) = \frac{1}{3}\sigma^2 t^3 \{1+o(1)\}, \;\text{ if } \alpha = 1,$$

$$Var\bar{S}_{t,\alpha} = 2\sigma^2 \left\{ \frac{\alpha^2}{(\alpha^2-1)^2} t\alpha^{2t} - \frac{\alpha^2}{\alpha^2-1} \sum_{l=0}^{t-1} l\alpha^{2l} \right\} \{1+o(1)\},$$

$$= 2\sigma^2 \frac{\alpha^{2(t+1)}}{(\alpha^2-1)^3} \{1+o(1)\}, \;\text{ if } |\alpha| > 1.$$

The convergence of the partial sums is proved using the martingale property of $S_{k,\alpha}$ and the expression of their variances. Let

$$\bar{W}_\alpha(s) = \sqrt{2}\,\sigma \left(\int_0^s x^2 \alpha^x \, dx \right)^{-\frac{1}{2}} \int \alpha^s \, d\bar{W}_s (\alpha^2-1)^{-\frac{1}{2}} (\alpha^4-1)^{-1}, \quad (7.6)$$

the variance of $\bar{W}_s$ is $\frac{s^3}{3}$.

Lemma 7.1.

When $|\alpha| < 1$, the process

$$(S_{[ns],\alpha}, n^{-1} s^{\frac{1}{2}} \bar{S}_{[ns],\alpha}, n^{-\frac{1}{2}} s^{\frac{1}{2}} \widetilde{S}_{[ns],\alpha})_{s\in[0,1]}$$

converges in distribution to

$$\frac{\sigma}{(1-\alpha^2)^{\frac{1}{2}}} \left(W, \sqrt{3}\bar{W}, \sigma\sqrt{2} \int_0^s W \, dW \right)_{s\in[0,1]},$$

and $(n^{-1}s\sum_{k=1}^{[ns]} S_{k,\alpha}^2)_{s\in[0,1]}$ *converges weakly to* $2\sigma^2(1-\alpha^2)^{-1}\int_0^s W_x^2\,dx$.

When $\alpha = 1$, *the process* $(n^{-\frac{1}{2}}S_{[ns],1}, n^{-3/2}\bar{S}_{[ns],1}, n^{-1}\widetilde{S}_{[ns],1})_{s\in[0,1]}$ *converges weakly to*

$$\sigma\Big(W, \bar{W}, \sigma\int_0^{\cdot} W\,dW\Big),$$

$n^{-2}\sum_{k=1}^{[ns]} S_{k,\alpha}^2$ *converges weakly to to* $\sigma^2\int_0^s W_x^2\,dx$.

When $|\alpha| > 1$, *the process* $\alpha^{-[ns]}(S_{[ns],\alpha}, s^{3/2}\bar{S}_{[ns],\alpha}, s\widetilde{S}_{[ns],\alpha})_{s\in[0,1]}$ *converges weakly to*

$$\sigma\Big(\frac{W}{(\alpha^2-1)^{\frac{1}{2}}}, \frac{\sqrt{6}\alpha\bar{W}}{(\alpha^2-1)^{3/2}}, \frac{\sqrt{2}\alpha\sigma}{(\alpha^2-1)}\int_0^{\cdot} W\,dW\Big),$$

the process $(\alpha^{-2[ns]}\sum_{k=0}^{[ns]}\alpha^k S_{k,\alpha})_{s\in[0,1]}$ *converges weakly to* $\bar{W}_\alpha$, $(\alpha^{-2[ns]}s^2\sum_{k=0}^{[ns]} S_{k,\alpha}^2)_{s\in[0,1]}$ *converges weakly to* $2\sigma^2(\alpha^2-1)^{-2}\int_0^{\cdot} W_x^2\,dx$,

and the process $(n^{-1}s^2\alpha^{-2[ns]}\sum_{k=0}^{[ns]} kS_{k,\alpha}^2)_s$ *converges weakly to* $3\sigma^2(\alpha^2-1)^{-2}\int_0^s xW_x^2\,dx$.

Proof. As the variance of $\bar{W}_s$ is $\frac{s^3}{3}$, $n^{-2}Var\bar{S}_{[ns],\alpha}$ converges to $\frac{3}{s}\sigma^2(1-\alpha^2)^{-1}Var\bar{W}_s$ if $|\alpha| < 1$ and the limit of $s^2\alpha^{-2[ns]}Var\bar{S}_{[ns],\alpha}$ is $\frac{6}{s^3}Var\bar{W}_s\sigma^2\alpha^2(\alpha^2-1)^{-3}$ if $|\alpha| > 1$. The convergence of the weighted sums of the $S_{k,\alpha}$ is obtained by similar arguments, the variances are asymptotically equivalent to $t\sigma^2\alpha^2(1-\alpha^2)^{-1}$ if $|\alpha| < 1$ and $2\sigma^2\alpha^{4t}(\alpha^2-1)^{-1}(\alpha^4-1)^{-2}$ if $|\alpha| > 1$, with $Var\int_0^x \alpha^s\,d\bar{W}_s = \int_0^x s^2\alpha^{2s}\,ds$.

The convergence of $\widetilde{S}_{[ns],\alpha}$ is deduced from the behaviour of $S_{[ns],\alpha}$ and the weak convergence of $n^{-1/2}\sum_{k=}^{[ns]}\varepsilon_k$ to σW. The variance of the process $\int_0^s W\,dW = \frac{1}{2}(W_s^2 - s)$ is the limiting variance of $\sigma^{-2}\widetilde{S}_{[ns],1}$ which equals $\frac{1}{2}t^2$. The variance of $\widetilde{S}_{[ns],\alpha}$ is $\sigma^4 t(1-\alpha^2)^{-1}$ if $|\alpha| < 1$, $\frac{1}{2}\sigma^4 t^2$ when $\alpha = 1$, and $\sigma^4\alpha^{2(t+1)}(\alpha^2-1)^{-2}$ when $|\alpha| > 1$.

If $|\alpha| > 1$, the expectation of the process $t^{-1}\alpha^{-2t}\sum_{k=1}^t kS_{k,\alpha}^2$ is

$$\sigma^2(\alpha^2-1)^{-1}t^{-1}\alpha^{-2t}\sum_{k=1}^t k\alpha^{2k}$$

and it converges to $\sigma^2\alpha^2(\alpha^2-1)^{-2}$. □

By definition of the estimator $\widehat{\alpha}_t$, we have

$$\begin{aligned}
\widehat{\alpha}_t - \alpha &= \frac{\sum_{k=1}^t (X_{k-1} - \bar{X}_t)(X_k - \bar{X}_t - \alpha(X_{k-1} - \bar{X}_t))}{\sum_{k=1}^t (X_{k-1} - \bar{X}_t)^2} \\
&= \frac{\sum_{k=1}^t (X_{k-1} - \bar{X}_t)((1-\alpha)(m - \bar{X}_t) + \varepsilon_k)}{\sum_{k=1}^t (X_{k-1} - \bar{X}_t)^2} \\
&= \frac{-(X_t - X_0)(1-\alpha)t^{-1}\bar{S}_{t,\alpha} + \sum_{k=1}^t \varepsilon_k X_{k-1} - t\bar{X}_t\bar{\varepsilon}_t}{\sum_{k=1}^t (X_{k-1} - \bar{X}_t)^2},
\end{aligned}$$

with $\bar{\varepsilon}_t = t^{-1}\sum_{k=1}^t \varepsilon_k$. The asymptotic behaviour of the sample path of the series and the estimators according to the domain of α are deduced from Lemma 7.1 (Pons, 2008).

Proposition 7.1. *When $|\alpha| < 1$, the process $(X_{[ns]} - m, \bar{X}_{[ns]} - m)_{s\in[0,1]}$ converges weakly to a Brownian motion $\sigma(1-\alpha^2)^{-\frac{1}{2}}(W_s, s^{-\frac{1}{2}}\sqrt{3}\bar{W}_s)_s$. The process $(\widehat{m}_{[ns]} - m, n^{\frac{1}{2}}(\widehat{\alpha}_{[ns]} - \alpha))_{s\in[0,1]}$ converges weakly to*

$$\left\{\frac{\sigma}{(1-\alpha^2)^{\frac{1}{2}}} s^{-3/2}\sqrt{3}\bar{W}_s, (1-\alpha^2)^{\frac{1}{2}} \frac{\int_0^s (\sqrt{2}W_x - \sqrt{3}s^{-1}\bar{W}_s)\, dW_x}{\int_0^s (2W_x^2 - 3s^{-2}\bar{W}_s^2)\, dx}\right\}_s.$$

When $\alpha = 1$, the equivalence between (7.3) and (7.4) is no more satisfied if μ is different from zero. By equation (7.3), we have

$$\begin{aligned}
X_t &= \mu + X_{t-1} + \varepsilon_t = X_0 + (t-1)\mu + S_t, \\
\bar{X}_t &= X_0 + \frac{1}{2}(t+1)\mu + t^{-1}\sum_{k=1}^t S_t.
\end{aligned}$$

The behaviour of the observation processes X_t and $\bar{X}_t$ is now characterized by the following approximations

$$\begin{aligned}
n^{\frac{1}{2}}\{n^{-1}(X_{[ns]} - X_0) - \mu\}_s &= n^{-\frac{1}{2}}(S_{[ns]})_s + o_{L_2}(1), \\
n^{\frac{1}{2}}\{n^{-1}(\bar{X}_{[ns]} - X_0) - \mu/2\}_s &= n^{-\frac{3}{2}}(\bar{S}_{[ns]})_s + o_{L_2}(1)
\end{aligned}$$

the first process is asymptotically equivalent to σW and the second one converges weakly to $\sigma\bar{W}$.

When it is not zero, the parameter μ is estimated by

$$\widehat{\mu}_t = t^{-1}\sum_{k=1}^t (X_k - X_{k-1}) = \mu + t^{-1}\bar{\varepsilon}_t$$

and $n^{\frac{1}{2}}(\widehat{\mu}_{[ns]} - \mu)_s$ converges weakly to σW, its asymptotic variance is $\sigma^2 t$. Let $\widehat{\alpha}_t$ be the estimator defined as in Proposition 7.1.

Proposition 7.2. *If $\alpha = 1$ and $\mu = 0$, the process $n(\widehat{\alpha}_{[ns]} - \alpha)_{s\in[0,1]}$ converges weakly to*

$$\Big\{\int_0^{\cdot} (W_x - \bar{W})^2\,dx\Big\}^{-1}\Big\{\int_0^{\cdot} W\,dW - W\bar{W}\Big\}.$$

As $\alpha = 1$ and $\mu \neq 0$, the process $n^{\frac{3}{2}} s^3(\widehat{\alpha}_{[ns]} - \alpha)_s$ converges weakly to

$$\mu^{-1}\sigma 2\sqrt{3}\Big\{\int_0^{\cdot} x\,dW - \frac{1}{2}W\Big\}.$$

The limit

$$\Big\{\int_0^1 (W - \bar{W}(1))^2\,dx\Big\}^{-1}\Big\{\int_0^1 W\,dW - W(1)\bar{W}(1)\Big\}$$

of the variable $t(\widehat{\alpha}_t - \alpha)$ for $\mu = 0$ has been given by Dickey and Fuller (1979) and Haldrup and Hylleberg (1995) in other models and Proposition 7.2 extends their results. For μ different from zero, the above developments still hold when $|\alpha| < 1$ with $\widehat{\mu}_t$ instead of $\bar{X}_t$ and the asymptotic behaviour of the estimator (7.5) of α is modified.

Proposition 7.3. *As $\alpha = 1$ and $\widehat{\mu}_t = t^{-1}\sum_{k=1}^t (X_k - X_{k-1})$, $n(\widehat{\alpha}_{[ns]} - \alpha)_s$ is defined by (7.5) and it converges weakly to*

$$\Big[\Big\{\int_0^s \{(x-s)W_x + xW(s)\}\,dx\Big\}^{-1}\Big\{\frac{s^3}{\sqrt{3}}\int_0^s x\,dW_x - sW(s)\Big\}\Big]_s.$$

When $|\alpha| > 1$, the variable $X_k - m$ develops as a sum of independent centred variables $X_k - m = \alpha^k(X_0 - m) + \sum_{j=0}^{k-1}\alpha^j\varepsilon_{k-j}$. The process $\alpha^{-t}(X_t - m) - (X_0 - m) = \alpha^{-t}S_{t-1,\alpha}$ converges weakly to $\sigma(\alpha^2-1)^{-\frac{1}{2}}W(1)$ and

$$t\alpha^{-t}(\bar{X}_t - m) = (X_0 - m)\frac{\alpha}{\alpha - 1} + \alpha^{-t}\bar{S}_{t,\alpha} + o_p(1). \tag{7.7}$$

The estimator of m is asymptotically equivalent to $\widehat{m}_t = \bar{X}_t$, it has the approximation $\bar{X}_t = m + o_p(1)$ from (7.7), and the weak convergence of $t\alpha^{-t}(\bar{X}_t - m)$ is a consequence of Lemma 7.1.

Proposition 7.4. *As $|\alpha| > 1$ and $m \neq 0$, $\alpha^{-[ns]}[ns](\widehat{m}_{[ns]} - m)_s$ converges weakly to*

$$\frac{\alpha}{\alpha - 1}(X_0 - m) + \frac{\alpha\sigma\sqrt{6}}{\{s(\alpha^2-1)\}^{\frac{3}{2}}}\bar{W}$$

and $n(\widehat{\alpha}_{[ns]} - \alpha)$ converges weakly to

$$G_{\alpha,X_0,m} = -\Big\{\frac{(X_0 - m)^2}{\alpha^2 - 1} + 2(X_0 - m)\bar{W}_\alpha + 2\frac{\sigma^2\int_0^s W_x^2\,dx}{(\alpha^2-1)^2}\Big\}^{-1} \times\Big\{X_0 - m + \frac{W\sigma}{(\alpha^2-1)^{\frac{1}{2}}}\Big\}\Big\{X_0 - m + \frac{\sqrt{6}\alpha\bar{W}\sigma s^{-\frac{3}{2}}}{(\alpha^2-1)^{\frac{3}{2}}}\Big\}.$$

Remark 1. Tho behaviour of the observation processes X_t and $\bar{X}_t$ vary in the following way: when $|\alpha| < 1$, $X_t - m$ and $\bar{X}_t - m$ are of order $O_p(1)$; when $\alpha = 1$, $X_t - X_0$ and $\bar{X}_t - X_0$ are $O_p(t^{\frac{1}{2}})$ if $\mu = 0$, $X_t - X_0$ and $\bar{X}_t - X_0$ are $O_p(t)$ if $\mu \neq 0$; $X_t - m = O_p(\alpha^t)$ and $\bar{X}_t - m = O_p(t^{-1}\alpha^t)$ when $|\alpha| > 1$.

Remark 2. When μ is known, $Y_t = \alpha^t Y_0 + S_{t,\alpha}$ and the estimator of α becomes

$$\widehat{\alpha}_t = \frac{\sum_{k=1}^t Y_k Y_{k-1}}{\sum_{k=1}^t Y_{k-1}^2},$$

it satisfies

$$\widehat{\alpha}_t - \alpha = \frac{\sum_{k=1}^t Y_{k-1}\varepsilon_k}{\sum_{k=1}^t Y_{k-1}^2}$$

and, when $\alpha = 1$, the limiting distribution of Proposition 7.2 is modified in the form $\int_0^{\cdot} W\,dW\{\int_0^{\cdot} W^2\,dx\}^{-1}$. For $\alpha \neq 1$,

$$\widehat{\alpha}_t - \alpha = \frac{\frac{\alpha^t - 1}{\alpha - 1} Y_0 + \sum_{k=1}^t \varepsilon_k S_{k,\alpha}}{\sum_{k=1}^t Y_{k-1}^2}.$$

When $|\alpha| < 1$,

$$t^{\frac{1}{2}}(\widehat{\alpha}_t - \alpha) = \frac{t^{-\frac{1}{2}} \sum_{k=1}^t \varepsilon_k S_{k,\alpha}}{t^{-1} \sum_{k=1}^t Y_{k-1}^2} + o_p(1)$$

and its asymptotic distribution is $(1-\alpha^2)^{\frac{1}{2}} \int_0^{\cdot} W\,dW\{\sqrt{2}\int_0^{\cdot} W^2\,dx\}^{-1}$. When X_0 is deterministic, it is replaced by the mean m and as $|\alpha| > 1$, $t(\widehat{\alpha}_t - \alpha)$ converges weakly to G_α defined by

$$G_\alpha(s) = -\sqrt{3}\alpha\, W_s \bar{W}_s \left\{ \sqrt{2} s^{\frac{3}{2}} \int_0^s W_x^2\,dx \right\}^{-1}.$$

In each case, a normalization replacing the terms α by their estimators provides a statistic which convergence to a free distribution.

Test statistics for the hypothesis $H_0 : \alpha = 1$ or $H_0 : \alpha = \alpha_0$, against the alternative that α belongs to domains which do not include 1, were studied by Pons (2009), their asymptotic distributions are deduced from Propositions 7.1 and 7.4.

A test of the hypothesis H_0 that α_0 belongs to domain $\mathcal{D} = \{\alpha; |\alpha| < 1\}$, against the alternative $H_1 : \alpha_0 = 1$ or $H_2 : |\alpha_0| > 1$ relies on the estimator of

$\widehat{\alpha}_t$, its convergence rate under H_0 is β_t^{-1}, and the limit under the hypothesis and the alternatives of the statistic

$$T_\alpha = \beta_t^{-1}\widehat{\sigma}_{t;\widehat{\mu}_t,\widehat{\alpha}_t}^{-1}\{\widehat{\alpha}_t - 1\}$$

is deduced from the propositions.

The statistic

$$T_{\mathcal{D}} = \sup_{\alpha\in\mathcal{D}} \beta_t^{-1}\widehat{\sigma}_{t;\widehat{\mu}_t,\widehat{\alpha}_t}^{-1}(\widehat{\alpha}_t - 1)$$

converges weakly to $\sigma^{-1}\sup_{\alpha\in\mathcal{D}}\lim_{t\to\infty}\beta_t^{-1}(\widehat{\alpha}_t - 1)$ defined according to the domain of α, then a test based on T_α is consistent, its asymptotic power is deduced from the propositions.

7.3 Convergence in the AR(p) model

In the AR(p) model (7.1), the mean squares estimators of the mean and variance parameters are

$$\begin{aligned}
\widehat{\mu}_t &= \bar{X}_t - (\bar{X}_{t-1},\ldots,\bar{X}_{t-p})\widehat{\alpha}_t,\\
\widehat{m}_t &= \bar{X}_t,\\
\widehat{\sigma}_t^2 &= \|\widehat{E}_t\|_t^2 = \frac{1}{t}\sum_{k=1}^{t}\{X_k - \widehat{\mu}_t - \mathrm{X}_{k-1}^{(p)'}\widehat{\alpha}_t\}^2.
\end{aligned}$$

The estimator $\widehat{\mu}_t$ has the same asymptotic behaviour as (7.5). The estimators of the regression parameters are dependent and they are usually estimated by an orthogonal projection into the space generated by the regressors. In model (7.1), they depend on all past observations and $E(X_t - m)(X_s - m) = ES_{t\wedge s}^2$ is defined by Lemma 7.1. The marginal estimators

$$\widehat{\alpha}_{j,t} = \frac{\sum_{k=1}^t (X_{k-j} - \bar{X}_t)(X_{k-j+1} - \bar{X}_t)}{\sum_{k=1}^t (X_{k-j} - \bar{X}_t)^2}, j = 1,\ldots,p-1,$$

are consistent and their joint behaviour depends on the domain of the parameters. When the norm of a coefficient is larger than 1, the sample-paths of X_t and $\bar{X}_t$ diverge, the behaviour of the estimators is the same as in Proposition 7.4. Let $W^{(p)}$ denote the p-dimensional Brownian motion with covariance function

$$EW_j^{(p)}(s)W_{j'}^{(p)}(s') = (s\wedge s')\alpha^{|j-j'|}.$$

Proposition 7.5. *When $|a_j| < 1$ for every $j = 1, \ldots, p$, $(X^{(p)}_{[ns]} - m\mathbb{1}_p)_s$ converges weakly to $r\sigma(1-\alpha^2)^{-\frac{1}{2}}W^{(p)}$, $n^{\frac{1}{2}}(\bar{X}^{(p)}_{[ns]} - m\mathbb{1}_p)_s$ converges weakly to $\sigma(1-\alpha^2)^{-\frac{1}{2}}\bar{W}^{(p)}$, and $n^{\frac{1}{2}}(\widehat{\alpha}_{[ns]} - \alpha)_{s\in[0,1]}$ converges weakly to*

$$(1-\alpha^2)^{\frac{1}{2}}\left\{\frac{\int_0^\cdot W_j^{(p)}\,dW_j^{(p)} - W_j^{(p)}\bar{W}_j^{(p)}}{\int_0^\cdot (W_j^{(p)}(x) - \bar{W}_j(x))^2\,dx}\right\}_{j=1,\ldots,p-1}.$$

If $\alpha_j \geq 1$ for every j, the marginal convergence of the estimators described in Section 7.3 and the results of Proposition 7.2 and 7.4 extend to a joint convergence, as in Proposition 7.5.

A mixture of two series of order p and q yields a series X_t such that

$$\begin{cases} X_{t+1} = \mu + \alpha^T X_t^{(p)} + \beta^T Y_t^{(q)} + \varepsilon_{t+1} \\ Y_{t+1} = m + a^T Y_t^{(s)} + e_{t+1}. \end{cases} \tag{7.8}$$

When both series are observed and such that (X_t, Y_t) is independent of a bivariate error (ε_t, e_t) with null expectation, the regression parameters α, β and $\mu = (1-\alpha)\mu_X + (1-\beta)\mu$ are identifiable. The convergence of the series depends on the domain of the parameters α, β and a. Asymptotically Gaussian estimators are explicitly defined by

$$\widehat{\alpha}_{j,t} = \frac{\sum_{k=1}^t (X_{k-j} - \bar{X}_t)(X_{k-j+1} - \bar{X}_t)}{\sum_{k=1}^t (X_{k-j} - \bar{X}_t)^2},\ j = 1, \ldots, p, \tag{7.9}$$

$$\widehat{\beta}_{j,t} = \frac{\sum_{k=1}^t (Y_{k-j} - \bar{Y}_t)(X_{k-j+1} - \bar{X}_t)}{\sum_{k=1}^t (X_{k-j} - \bar{X}_t)^2},\ j = 1, \ldots, q, \tag{7.10}$$

$$\widehat{a}_{j,t} = \frac{\sum_{k=1}^t (Y_{k-j} - \bar{Y}_t)(Y_{k-j+1} - \bar{Y}_t)}{\sum_{k=1}^t (Y_{k-j} - \bar{Y}_t)^2},\ j = 1, \ldots, s. \tag{7.11}$$

The mean μ is estimated by $\widehat{\mu}_t = (1-\widehat{\alpha}_t)\bar{X}_t + (1-\widehat{\beta}_t)\bar{Y}_t$ if $|\alpha| \neq 1$ and $|\beta| \neq 1$, and the means are modified by a relevant estimator $\widehat{\mu}_t$ otherwise. In the absence of mixture, the parameter β is zero and the Student statistic for $\widehat{\beta}_t$ allows for testing the hypothesis $\beta = 0$ against an alternative $|\beta| < 1$ when the parameter β belongs to the domain $]-1,1[^q$.

In a random coefficients model

$$X_{t+1}^{(p)} = \mu + A^T X_t^{(p)} + \varepsilon_{t+1},$$

the vector A is a random variable independent of $X_t^{(p)}$, such that $EA = \alpha$ and $Var A = V$. Then

$$E(X_{t+1}^{(p)} - m | X_t^{(p)}) = \alpha^T (X_t^{(p)} - m),$$

$$E\{(X_{t+1}^{(p)} - m)^2 | X_t^{(p)}\} = V(X_t^{(p)} - m)^2,$$

$$E(X_t^{(p)} - m)^2 = V^{\otimes t} Var X_0^{(p)},$$

the model of the mean is still 7.1 and it is similar for the variance but with a quadratic series. In a random coefficients model, the convergence rates for the mean and the variance of the process are the same as for the mean of the process of a simple auto-regressive series. The estimator of α is still 7.1 and for the variance V it is written with the squares of the observations.

7.4 Change-points in AR(p) models

A change-point in model (7.1) occurs at an unknown time τ or at an unknown threshold η of the series. In both models, X_t is divided in

$$X_{1,t} = X_t I_t \quad \text{and} \quad X_{2,t} = X_t(1 - I_t)$$

with a random indicator $I_t = I_{t,\eta} = 1\{X_t \le \eta\}$ in a model with a change-point at a threshold of the series and $I_t = I_{t,\tau} = 1\{t \le \tau\}$ in a model with a time threshold. The p-dimensional vector α is replaced by two vectors α and β. The model with a change-point at a threshold η of the series is related to a random stopping time $\tau_\eta = \sup\{t; X_t \le \eta\}$ and the maximum value of series in the model with a time-dependent threshold t is $\eta_t = \sup\{X_s : s \in [0,t]\}$. With a change-point, the model (7.1) is modified as

$$X_t = \mu_1 I_t + \mu_2(1 - I_t) + \alpha^T X_{1,t}^{(p)} + \beta^T X_{2,t}^{(p)} + \varepsilon_t \tag{7.12}$$

or $X_t = \mu + \alpha^T X_{1,t}^{(p)} + \beta^T X_{2,t}^{(p)} + \varepsilon_t$ for a model without change-point in the mean. The initial variable X_0 has the mean $EX_0 = m_\alpha$ and the variance $Var X_0 = \sigma_0^2$. The parameters are μ, or μ_1 and μ_2, α, β and σ^2. For the limits of the series $(X_k)_{k \le n}$ and its estimators, as n tends to infinity, $n^{-1}\tau$ is also denoted $\gamma = \gamma_\tau$ and $\tau = [\gamma n]$. Before a fixed change-point τ, a series $(X_k)_{k \le n}$ is asymptotically neglectable as n tends to infinity. The parameter vectors are denoted ξ with components $\alpha, m_\alpha, \beta, m_\beta$ and $\theta = (\xi^T, \tau)^T$.

For the auto-regressive model of order 1 with a time change-point, this equation is still denoted

$$X_{t,\alpha} = m_\alpha + \alpha^T(X_0 - m_\alpha) + \sum_{k=1}^{t} \alpha^{t-k}\varepsilon_k,\ t \le \tau,$$

$$X_{t,\beta} = m_\beta + \beta^{t-\tau}(X_{\tau,\alpha} - m_\alpha) + \sum_{k=1}^{t-\tau} \beta^{t-\tau-k}\varepsilon_{k+\tau},\ t > \tau,$$

or $m_\beta = \mu(1-\beta)^{-1}$. With $\alpha = 1$, $X_{t,\alpha} = X_0 + (t-1)\mu + \sum_{k=1}^{T} \varepsilon_k$ and with $\beta = 1$ and $t > \tau$, $X_{t,\beta} = X_{\tau,\alpha} + (t-k-1)\mu + \sum_{k=1}^{t-\tau} \varepsilon_{k+\tau}$.

Consider an AR(1) model with a change-point at time τ, under a probability distribution P_θ, where θ is the vector of all parameters $\alpha, \beta, m_\alpha, m_\beta, \tau$, with a sequence of independent and identically distributed error variables $(\varepsilon_t)_t$. The true parameter vector of the model is θ_0. The time τ corresponds either to a change-point of the series or a stopping time for a change-point at a threshold of the process X. The mean square estimators of α, μ and σ^2 are defined by minimization of the Euclidean norm of the error vector $\|E_t\|_t^2 = \sum_{k=1}^t \varepsilon_k^2$

$$\begin{aligned}
\widehat{\alpha}_{t,\tau} &= \frac{\sum_{k=1}^t (I_{k-1,\tau}X_{k-1} - \widehat{m}_{\alpha,\tau})(I_{k,\tau}X_k - \widehat{m}_{\alpha,\tau})}{\sum_{k=1}^t (I_{k-1}X_{k-1} - \widehat{m}_{\alpha,\tau})^2}, \ t \le \tau, \\
\widehat{\beta}_{t,\tau} &= \frac{\sum_{k=1}^t ((1-I_{k-1,\tau})X_{k-1} - \widehat{m}_{\beta,t})((1-I_{k,\tau})X_k - \widehat{m}_{\beta,t})}{\sum_{k=1}^t \{(1-I_{k-1,\tau})X_{k-1} - \widehat{m}_{\beta,t}\}^2}, \ t > \tau, \\
\widehat{E}_t &= \widehat{\mu}_{1,t}I_t + \widehat{\mu}_{2,t}(1-I_t) + \widehat{\alpha}_{t,\tau}^T X_{1,t} + \widehat{\beta}_{t,\tau}^T X_{2,t}, \\
\widehat{\sigma}_t^2 &= t^{-1}\|\widehat{E}_t\|_t^2
\end{aligned}$$

where the estimators of $m_\alpha = (1-\alpha)^{-1}\mu$ and $m_\beta = (1-\beta)^{-1}\mu$ are asymptotically equivalent to

$$\widehat{m}_{\alpha,\tau} = \bar{X}_\tau,$$

$$\widehat{m}_{\beta,\tau+k} = \bar{X}_{\tau,k} := k^{-1}\sum_{j=1}^{k} X_{\tau+j}, \quad \text{for } t = \tau + k \ge \tau,$$

and $\widehat{\mu}_t = \bar{X}_\tau - \widehat{\alpha}_\tau \bar{X}_\tau - \widehat{\beta}_t \bar{X}_{\tau,t-\tau}$ as $|\alpha|$ and $|\beta| \neq 1$.

When $\alpha = 1$ or $\beta = 1$, the parameters m_α and m_β have no meaning and in the expression $\widehat{\alpha}_t$ and $\widehat{\beta}_t$, the variables are now centred by the empirical means of the differences between the observed values of the process

$$\widehat{\mu}_{\alpha,t} = \frac{1}{t}\sum_{k=1}^{t}(X_k - X_{k-1}), \ t \le \tau, \qquad \text{if } \alpha = 1$$

$$\widehat{\mu}_{\beta,t} = \frac{1}{t-\tau}\sum_{k=\tau+1}^{t}(X_k - X_{k-1}), \ t > \tau, \quad \text{if } \beta = 1$$

in the model with a time change-point and by

$$\widehat{\mu}_{t,\eta} = \frac{\sum_{k=1}^t (X_k - X_{k-1})1_{\{X_k \le \eta, X_{k-1} \le \eta\}}}{\sum_{k=1}^t 1_{\{X_k \le \eta, X_{k-1} \le \eta\}}}, \ \text{if } \alpha = 1$$

and

$$\widehat{\mu}_{t,\eta} = \frac{\sum_{k=1}^{t}(X_k - X_{k-1})1_{\{X_k>\eta, X_{k-1}>\eta\}}}{\sum_{k=1}^{t} 1_{\{X_k>\eta, X_{k-1}>\eta\}}}, \text{ if } \beta = 1$$

in the model with a change-point at a threshold. Let $\tau = [\gamma t]$ for a value γ belonging to $]0,1[$. All estimators change according to the threshold that is finally estimated by minimization of $\widehat{\sigma}_t^2$ with respect to γ in the model with a time change-point or η in the model with a threshold in the series.

For $t \leq \tau$ the sample-paths are

$$\begin{aligned}
X_t &= m_\alpha + S_{t,\alpha} + o_p(1), && \text{if } |\alpha| < 1,\\
X_t &= X_0 + (t-1)\mu + S_{t,1}, && \text{if } \alpha = 1,\\
X_t &= m_\alpha + \alpha^T(X_0 - m_\alpha) + S_{t,\alpha}, && \text{if } |\alpha| > 1
\end{aligned}$$

and

$$\begin{aligned}
\bar{X}_t &= m_\alpha + t^{-1}\bar{S}_{t,\alpha} + o_p(1), && \text{if } |\alpha| < 1,\\
\bar{X}_t &= X_0 + \frac{1}{2}(t+1)\mu + t^{-1}\bar{S}_{t,1}, && \text{if } \alpha = 1,\\
\bar{X}_t &= m_\alpha + t^{-1}\frac{\alpha^{t+1}-1}{\alpha-1}(X_0 - m_\alpha) + t^{-1}\bar{S}_{t,\alpha}, && \text{if } |\alpha| > 1.
\end{aligned}$$

The expectation of X_t and $\bar{X}_t$ are therefore always m_α if $\mu = 0$, as $\alpha = 1$. The processes X_t and $\bar{X}_t$ have the same convergence to m_α depending on to $|\alpha|$, as t tends to infinity. For $t < \tau$, the limits of the processes are described in Section 7.2.

The sample-paths of $(X_t)_{t \leq \tau}$ and $\bar{X}_{\tau,k}$ have similar expressions for $t \leq \tau$ and $t > \tau$, where X_τ replace X_0 as initial value at τ. Let $k = t - \tau$ and the partial sums

$$S_{\tau,k,\beta} = \sum_{j=1}^{k} \beta^{k-j}\varepsilon_{\tau+j} = \sum_{j=0}^{k-1} \beta^j \varepsilon_{\tau+k-j}, \tag{7.13}$$

$$\widetilde{S}_{\tau,k,\beta} = \sum_{j=1}^{k} \varepsilon_j S_{\tau,j,\beta},$$

$$\bar{S}_{\tau,k,\beta} = \sum_{j=1}^{k} S_{\tau,j,\beta}.$$

The equation $X_{t,\beta} = m_\beta + \beta^{t-\tau}(X_{\tau,\alpha} - m_\alpha) + \sum_{k=1}^{t-\tau}\beta^{t-\tau-k}\varepsilon_{k+\tau}$, for $t > \tau$, provides expressions for the sample-paths of X_t and $\bar{X}_t$ according to the sums (7.13) as

$$\begin{aligned}
X_t &= m_\beta + S_{\tau,t-\tau,\beta} + o_p(1), && \text{if } |\beta| < 1,\\
X_t &= X_\tau + (t-\tau-1)\mu + S_{\tau,t-\tau,1}, && \text{if } \beta = 1,
\end{aligned}$$

and

$$\begin{aligned}
X_t &= m_\beta + \beta^{t-\tau}(X_\tau - m_\alpha) + S_{\tau,t-\tau,\beta} \\
&= m_\beta + \beta^{t-\tau}\{\alpha^\tau(X_0 - m_\alpha) + S_{\tau,\alpha}\} + S_{\tau,t-\tau,\beta} + o_p(1), \text{ if } |\beta| > 1.
\end{aligned}$$

Furthermore

$$\begin{aligned}
\bar{X}_{\tau,t-\tau} &= m_\beta + (t-\tau)^{-1}\bar{S}_{\tau,t-\tau,\beta} + o_p(1), & \text{if } |\beta| < 1, \\
\bar{X}_{\tau,t-\tau} &= X_\tau + \frac{1}{2}(t-\tau+1)\mu + (t-\tau)^{-1}\bar{S}_{\tau,t-\tau,1}, & \text{if } \beta = 1,
\end{aligned}$$

$$\bar{X}_{\tau,t-\tau} = m_\beta + (t-\tau)^{-1}\frac{\beta^{t-\tau+1}-1}{\beta-1}(X_\tau - m_\alpha) + (t-\tau)^{-1}\bar{S}_{\tau,t-\tau,\beta}, \text{ if } |\beta| > 1.$$

The behaviour of $X_\tau - m_\alpha$ still differs according to the three domains of α if $\beta = 1$ and $|\beta| > 1$. As $|\alpha| > 1$ and $|\beta| > 1$

$$\bar{X}_{\tau,t-\tau} = m_\beta + (t-\tau)^{-1}\frac{\beta^{t-\tau+1}}{\beta-1}\{\alpha^\tau(X_0 - m_\alpha) + S_{\tau,\alpha}\} + o_p(1).$$

The limits for $|\beta| < 1$ and $|\alpha| < 1$ are similar on the interval $]\gamma, 1]$, for $\beta = 1$ and $\alpha = 1$ they are still similar starting from the limit of X_τ which depend on the domain of α. When $t > \tau$ and $|\beta| > 1$, they are different and depend on both X_τ and the sums (7.13), with results similar to Proposition 7.4 on the interval $]\gamma, 1]$ if $|\alpha| < 1$.

For every β, the limits of the partial sums and the sample-paths of the processes are given by the lemma and proposition below. We define limiting processes as

$$\begin{aligned}
S_{\tau,\beta} &= \sigma(\beta^2 - 1)^{-\frac{1}{2}} W, \\
\widetilde{S}_{\tau,\beta} &= \sigma^2\beta(\beta^2 - 1)^{-\frac{1}{2}} \int_\gamma^\cdot W\, dW, \\
\bar{S}_{\tau,\beta} &= \sigma\{\beta(\beta - 1)\}^{\frac{1}{2}}(\beta^2 - 1)^{-1}(\bar{W} - \bar{W}_\gamma).
\end{aligned}$$

Proposition 7.6. *Let $\gamma = \gamma_\tau$, in the model with a change-point at τ. As n tends to infinity*

if $|\beta| < 1$, the process

$$(S_{\tau,[ns],\beta}, n^{-1}(s-\gamma)^{\frac{1}{2}}\bar{S}_{\tau,[ns],\beta}, n^{-\frac{1}{2}}(s-\gamma)^{\frac{1}{2}}\widetilde{S}_{\tau,[ns],\beta})_{s\in]\gamma,1]}$$

converges weakly to

$$\sigma(1-\beta^2)^{-\frac{1}{2}}\left(W, \sqrt{3}(\bar{W} - \bar{W}_\gamma), \sqrt{2}\sigma\int_\gamma^\cdot W\, dW\right)_{]\gamma,1]}.$$

if $\beta = 1$, *the process* $(n^{-\frac{1}{2}}S_{\tau,[ns],1}, n^{-\frac{3}{2}}\bar{S}_{\tau,[ns],1}, n^{-1}\widetilde{S}_{\tau,[ns],1})_{s\in]\gamma,1]}$ *converges weakly* $\sigma(W, \bar{W} - \bar{W}_\gamma, \sigma\int_\gamma^\cdot W\,dW)$ *and* $n^{-2}\sum_{k=[n\gamma]+1}^{[n\cdot]} S_{k,\alpha}^2$ *converges weakly to* $\sigma^2\int_\gamma^\cdot W_x^2\,dx$ *on* $]\gamma,1]$.

if $|\beta| > 1$, *the process*

$$\beta^{-[ns]}((s-\gamma)^{\frac{1}{2}}S_{\tau,[ns],\beta}, \widetilde{S}_{\tau,[ns],\beta}, n^{\frac{1}{2}}\bar{S}_{\tau,[ns],\beta})_{s\in]\gamma,1]}$$

converges weakly to $(S_{\tau,\beta}, \widetilde{S}_{\tau,\beta}, \bar{S}_{\tau,\beta})_{s\in]\gamma,1]}$.

When $|\beta| > 1$, the main term of the expression of X_k and $\bar{X}_k$, for $\tau < k \leq t$, depends on the observations at τ if $|\alpha| > 1$. In the other cases, the value during the first phase are neglectable.

Proposition 7.7. *In the model of change-point of order 1, if* $\gamma = \gamma_\tau$ *and* n *tends to infinity*

- *for* $|\beta| < 1$, $(1-\beta^2)^{\frac{1}{2}}(X_{[ns]} - m_\beta, \bar{X}_{\tau,[ns]-\tau} - m_\beta)_{s\in]\gamma,1]}$ *converges weakly under* P_θ *to* $\sigma(W - W_\gamma, (s-\gamma)^{-\frac{1}{2}}\sqrt{3}(\bar{W} - \bar{W}_\gamma))_{]\gamma,1]}$.
- *for* $\beta = 1$, $n^{\frac{1}{2}}\{n^{-1}(X_{\tau,[ns]-\tau} - X_\tau) - (s-\gamma)\mu, n^{-1}(\bar{X}_{\tau,[ns]-\tau} - \bar{X}_\tau) - (s-\gamma)\mu/2\}_{s\in]\gamma,1]}$ *converges weakly* P_θ *to* $\sigma(W - W_\gamma, (s-\gamma)^{-1}(\bar{W}_s - \bar{W}_\gamma))_{]\gamma,1]}$.
- *for* $|\beta| > 1$ *and* $|\alpha| < 1$, $\beta^{-([ns]-\tau)}(X_{\tau,[ns]-\tau} - m_\beta, ([ns]-\tau)(\bar{X}_{[ns]} - m_\beta))_{s\in]\gamma,1]}$ *converges weakly* P_θ *to*

$$\left(\frac{\sigma}{(\beta^2-1)^{\frac{1}{2}}}(W - W_\gamma), \frac{\beta\sigma\sqrt{6}}{\{(s-\gamma)(\beta^2-1)\}^{\frac{3}{2}}}(\bar{W} - \bar{W}_\gamma)\right)_{]\gamma,1]}.$$

- *for* $|\beta| > 1$ *and* $|\alpha| > 1$, $\alpha^{-\tau}\beta^{-([ns]-\tau)}(X_{\tau,[ns]-\tau} - m_\beta, ([ns]-\tau)(\bar{X}_{[ns]} - m_\beta))_{s\in]\gamma,1]}$ *converges weakly* P_θ *to* $(X_0 - m_\alpha + \sigma W_\tau(\alpha^2-1)^{-\frac{1}{2}})(1,1)^T$.

The change-point is estimated by minimization of the mean square error defined for $t > \tau$ by

$$\widehat{\sigma}_{t,\tau}^2 = t^{-1}\left(\sum_{k=1}^t \widehat{E}_{k;\widehat{m}_{\alpha,\tau},\widehat{\alpha}_{k,\tau}}^2 + \sum_{k=\tau+1}^t \widehat{E}_{k;\widehat{m}_{\beta,\tau},\widehat{\beta}_{k,\tau}}^2\right)$$

where the k-th error term is defined, for $k \leq \tau$, by

$$\widehat{E}_{k;\widehat{m}_{\alpha,\tau},\widehat{\alpha}_{k,\tau}} = X_k - (1-\widehat{\alpha}_{k,\tau})\bar{X}_\tau - \widehat{\alpha}_{k,\tau}X_{k-1},\ k \leq \tau,$$

$$\widehat{E}_{k;\widehat{m}_{\beta,\tau},\widehat{\beta}_{k,\tau}} = X_k - (1-\widehat{\beta}_{k,\tau})\bar{X}_{\tau,t-\tau} - \widehat{\beta}_{k,\tau}X_{k-1},\ \tau < k \leq t.$$

When the variables ε_t are independent and identically distributed, the estimator $\widehat{\sigma}_{t,\tau}^2$ converges in probability to the variance σ^2 as t tends to infinity.

When the variables ε_t are dependent, this limit and the previous limits also depend of their mixing coefficient. In the following they are supposed to be independent, otherwise the estimator of σ^2 has to be modified.

7.5 Convergence in models with change-points

The estimators of α and β are sums on disjoint sets of indices, $\widehat{\alpha}_{t,\tau}$ and $\widehat{\beta}_{t,\tau}$ are therefore asymptotically independent and their limits are those of the auto-regressive model without change-point, with time variables modified as in Lemma 7.6 and Proposition 7.7 for the residual sums and the sample-paths. The thresholds η and τ are first supposed to be known and the distribution governing the model is P_θ.

The asymptotic behaviour of the estimators of the first phase are deduced from Section 7.2, then we study the asymptotic behaviour of the empirical variance of the series. The process $\bar{W}_\alpha$ is defined by (7.6), let $G_{\alpha,X_0,m_\alpha}(s)$ be the process

$$-\left\{\frac{(X_0-m_\alpha)^2}{\alpha^2-1}+2(X_0-m_\alpha)\bar{W}_\alpha+2\frac{\sigma^2\int_0^\cdot W_x^2\,dx}{(\alpha^2-1)^2}\right\}^{-1} \cdot\left\{X_0-m_\alpha+\frac{W\sigma}{(\alpha^2-1)^{\frac{1}{2}}}\right\}\left\{X_0-m_\alpha+\frac{\sqrt{6}\alpha\bar{W}\sigma s^{-\frac{3}{2}}}{(\alpha^2-1)^{\frac{3}{2}}}\right\}.$$

Theorem 7.1. *In the model of order 1 and under P_θ*

- *when $|\alpha|<1$, the process $(\widehat{m}_{\alpha,[ns]}-m_\alpha, n^{\frac{1}{2}}(\widehat{\alpha}_{[ns]}-\alpha))_{s\in[0,\gamma]}$ converges weakly to*

$$\left\{\frac{\sigma}{(1-\alpha^2)^{\frac{1}{2}}}s^{-\frac{3}{2}}\sqrt{3}\bar{W}_s,(1-\alpha^2)^{\frac{1}{2}}\frac{\int_0^s(\sqrt{2}W_x-\sqrt{3}s^{-1}\bar{W}_s)\,dW_x}{\int_0^s(2W_x^2-3s^{-2}\bar{W}_s^2)\,dx}\right\}_{s\in[0,\gamma]}.$$

- *When $\alpha=1$ and $\mu=0$, the process $n(\widehat{\alpha}_{[ns]}-\alpha)_{s\in[0,\gamma]}$ converges weakly to*

$$\left\{\int_0^\cdot(W_x-\bar{W})^2\,dx\right\}^{-1}\left\{\int_0^\cdot W\,dW-W\bar{W}\right\}.$$

- *When $\alpha=1$ and $\mu\neq 0$, the process $n^{\frac{3}{2}}s^3(\widehat{\alpha}_{[ns]}-\alpha)_{s\in[0,\gamma]}$ converges weakly to*

$$\mu^{-1}\sigma 2\sqrt{3}\left\{\int_0^\cdot x\,dW-\frac{1}{2}W\right\}.$$

- *When* $\alpha > 1$, *the process* $\alpha^{-[ns]}(\widehat{m}_{\alpha,[ns]} - m_\alpha)_{s\in[0,\gamma]}$ *converges weakly to*
$$\frac{\alpha}{\alpha-1}(X_0 - m_\alpha) + \Big[\frac{\alpha\sigma\sqrt{6}}{\{s(\alpha^2-1)\}^{\frac{3}{2}}}\bar{W}_s\Big]_{s\in[0,\gamma]}$$
and $n(\widehat{\alpha}_{[ns]} - \alpha)_{s\in[0,\gamma]}$ *converges weakly to the process* G_{α,X_0,m_α}.

For a test of the hypothesis H_0 of a parameter belonging to a domain $\mathcal{D}$ in a model of order 1, against the alternative H_1 of a domain $\bar{\mathcal{D}}$, the Student statistic is
$$T_{\tau,\alpha} = \sup_{\alpha\in\mathcal{D}} v_{\tau,\alpha}(\widehat{\alpha}_\tau - \alpha)\widehat{\sigma}_{\tau,\mathcal{D}}^{-1}$$
where $v_{\tau,\mathcal{D}}$ is the convergence rate of $\widehat{\alpha}_\tau$ in the domain $\mathcal{D}$ and $\widehat{\sigma}_{\tau,\mathcal{D}}$ is the estimator of the variance in $\mathcal{D}$. It converges weakly, as τ tends to infinity to the limit $\sup_{\alpha\in\mathcal{D}}\lim_\tau T_{\tau,\alpha}$ defined according to $\mathcal{D}$. The convergences are modified for the second phase of the model, after τ.

Theorem 7.2. *In the model of order 1, and under* P_θ

when $|\beta| < 1$, *the process* $(\widehat{m}_{\beta,[ns]} - m_\beta, n^{\frac{1}{2}}(\widehat{\beta}_{[ns]} - \beta))_{s\in]\gamma,1]}$ *converges weakly to*
$$\Big\{\frac{\sigma}{(1-\beta^2)^{\frac{1}{2}}}s^{-\frac{3}{2}}\sqrt{3}(\bar{W}_s - \bar{W}_\gamma), (1-\beta^2)^{\frac{1}{2}}\frac{\int_\gamma^s(\sqrt{2}W_x - \sqrt{3}s^{-1}\bar{W}_s)\,dW_x}{\int_\gamma^s(2W_x^2 - 3s^{-2}\bar{W}_s^2)\,dx}\Big\}.$$
When $\beta = 1$ *and* $\mu = 0$, *the process* $n(\widehat{\beta}_{[ns]} - \beta)_{s\in]\gamma,1]}$ *converges weakly to*
$$\Big\{\int_\gamma^\cdot (W_x - \bar{W})^2\,dx\Big\}^{-1}\Big\{\int_\gamma^\cdot W\,dW - W\bar{W}\Big\}.$$
When $\beta = 1$ *and* $\mu \neq 0$, *the process* $n^{\frac{3}{2}}\{s^3(\widehat{\beta}_{[ns]} - \beta)\}_{s\in]\gamma,1]}$ *converges weakly to the process*
$$\mu^{-1}\sigma 2\sqrt{3}\Big\{\int_\gamma^\cdot x\,dW - \frac{1}{2}W\Big\}.$$
When $|\beta| > 1$, *the process* $\{\beta^{-[ns]}(\widehat{m}_{\beta,[ns]} - m_\beta)\}_{s\in]\gamma,1]}$ *converges weakly to the process*
$$\frac{\beta}{\beta-1}(X_\tau - m_\beta) + \frac{\beta\sigma\sqrt{6}}{\{s(\beta^2-1)\}^{\frac{3}{2}}}\bar{W}_s$$
and $n(\widehat{\beta}_{[ns]} - \beta)_{s\in]\gamma,1]}$ *converges weakly to the process* $G_{\beta,X_\tau,m_\beta}(s)$
$$-\Big\{\frac{(X_\tau - m_\beta)^2}{\beta^2-1} + 2(X_\tau - m_\beta)\bar{W}_\beta + 2\frac{\sigma^2\int_0^s W_x^2\,dx}{(\beta^2-1)^2}\Big\}^{-1}$$
$$\times\Big\{X_\tau - m_\beta + \frac{W\sigma}{(\beta^2-1)^{\frac{1}{2}}}\Big\}\Big\{X_\tau - m_\beta + \frac{\sqrt{6}\beta\bar{W}\sigma s^{-\frac{3}{2}}}{(\beta^2-1)^{\frac{3}{2}}}\Big\}.$$

The criterion for the estimation of the change-point parameter is the minimization of the mean square error of estimation $\widehat{\sigma}_t^2(\tau)$, the estimator $\widehat{\tau}_n$ is therefore consistent. After the estimation of the change-point parameter τ, the regression parameter are estimated replacing τ by its estimator in their expression at fixed τ and they are consistent. Let

$$\widehat{C}_t(\tau) = \widehat{\sigma}_{0t}^2 - \widehat{\sigma}_t^2(\tau) = \widehat{\sigma}_{0t}^2 - \widehat{\sigma}_{1t}^2(\tau) - \widehat{\sigma}_{2t}^2(\tau) \tag{7.14}$$

where $\widehat{\sigma}_{1t}^2$ and $\widehat{\sigma}_{2t}^2$ are the respective estimators of the variance on the two phases of the model at unknown τ and $\widehat{\sigma}_{0t}^2$ is the estimator of the variance when the value τ_0 of the change-point parameter is known. The process $\widehat{C}_t$ is the sum of $\widehat{E}_{k;\widehat{\mu}_{\alpha,\tau},\widehat{\alpha}_\tau}$, if $|\alpha| \neq 1$, and $\widehat{E}_{k;\widehat{\mu}_{\beta,t},\widehat{\beta}_t}$, if $|\beta| \neq 1$.

When $|\alpha| \neq 1$ and $|\beta| \neq 1$, the error in the prediction of X_k are

$$\widehat{E}_{k;\widehat{\mu}_{\alpha,\tau},\widehat{\alpha}_\tau} = \varepsilon_k - (\widehat{\alpha}_\tau - \alpha)(X_{k-1} - m_\alpha) - (1 - \widehat{\alpha}_\tau)\bar{X}, \; k \leq \tau$$
$$\widehat{E}_{k;\widehat{\mu}_{\beta,\tau},\widehat{\beta}_t} = \varepsilon_k - (\widehat{\beta}_t - \beta)(X_{k-1} - m_\beta) - (1 - \widehat{\beta}_t)(\bar{X}_t - \bar{X}_\tau),$$

for $\tau < k \leq t$. The variance estimators are defined by

$$\begin{aligned}
\widehat{\sigma}^2_{\tau;\widehat{\mu}_{\alpha,\tau},\widehat{\alpha}_\tau} &= \tau^{-1}\sum_{k=1}^{\tau}\{X_k - (1-\widehat{\alpha}_\tau)\bar{X}_\tau - \widehat{\alpha}_\tau X_{k-1}\}^2 \\
&= \widehat{\sigma}^2_{\tau;\widehat{\alpha}_\tau} + (1-\widehat{\alpha}_\tau)^2(\bar{X}_\tau - m_\alpha)^2 \\
&\quad -2(1-\widehat{\alpha}_\tau)(\bar{X}_\tau - m_\alpha)\{\bar{\varepsilon}_\tau - (\widehat{\alpha}_\tau - \alpha)(\bar{X}_{\tau-1} - m_\alpha)\}, \\
\widehat{\sigma}^2_{\tau;\widehat{\alpha}_\tau} &= \tau^{-1}\sum_{k=1}^{\tau}\varepsilon_k^2 + (\widehat{\alpha}_\tau - \alpha)^2\tau^{-1}\sum_{k=1}^{\tau}(X_{k-1} - m_\alpha)^2 \\
&\quad -2(\widehat{\alpha}_\tau - \alpha)\tau^{-1}\sum_{k=1}^{\tau}\varepsilon_k(X_{k-1} - m_\alpha)
\end{aligned}$$

where the variables $X_k - m_\alpha$ and $\bar{X}_k - m_\alpha$ develop into sums according to the domain of α. When $|\alpha| < 1$,

$$X_k - m_\alpha = S_{k,\alpha} + o_p(1), \quad \bar{X}_\tau - m_\alpha = \tau^{-1}\bar{S}_{\tau,\alpha} + o_p(1),$$

and $\tau^{-1}\sum_{k=1}^{\tau}\varepsilon_k(X_{k-1} - m_\alpha) = \tau^{-1}\widetilde{S}_{\tau,\alpha} + o_p(1)$.

If $m_\alpha = 0$, the variance estimator is

$$\widehat{\sigma}^2_{\tau;\widehat{\alpha}_\tau} = \tau^{-1}\sum_{k=1}^{\tau}\{X_k - \widehat{\alpha}_\tau X_{k-1}\}^2 = \tau^{-1}\sum_{k=1}^{\tau}\{\varepsilon_k - (\widehat{\alpha}_\tau - \alpha)X_{k-1}\}^2.$$

For the true parameter values, $\widehat{\sigma}^2_{\tau;0} = \tau^{-1}\sum_{k=1}^{\tau}\varepsilon_k^2$. In each case, the estimator $\widehat{\sigma}^2_{\widehat{\tau}_t}$ converges to σ^2 as τ tends to infinity.

In the second phase of the model, the estimators have similar expressions

$$\begin{aligned}\widehat{\sigma}^2_{t;\widehat{\mu}_{\beta,t},\widehat{\beta}_t} &= (t-\tau)^{-1}\sum_{k=\tau+1}^{t}\{X_k-(1-\widehat{\beta}_t)\bar{X}_{\tau,t-\tau}-\widehat{\beta}_tX_{k-1}\}^2\\ &= \widehat{\sigma}^2_{t;\widehat{\beta}_t}+(1-\widehat{\beta}_t)^2(\bar{X}_{\tau,t-\tau}-m_\beta)^2\\ &\quad -2(1-\widehat{\beta}_t)(\bar{X}_t-m_\beta)\{\bar{\varepsilon}_{\tau,t-\tau}-(\widehat{\beta}_t-\beta)(\bar{X}_{\tau,t-1-\tau}-m_\beta)\},\\ \widehat{\sigma}^2_{t;\widehat{\beta}_t} &= (t-\tau)^{-1}\sum_{k=\tau+1}^{t}\varepsilon^2_{\tau+k}+\frac{(\widehat{\beta}_t-\beta)^2}{t-\tau}\sum_{k=\tau+1}^{t}(X_{k-1}-m_\beta)^2\\ &\quad -2\frac{\widehat{\beta}_t-\beta}{t-\tau}\sum_{k=\tau+1}^{T}\varepsilon_{\tau+k}(X_{k-1}-m_\beta).\end{aligned}$$

The variables $X_{\tau+k}-m_\beta$ and $\bar{X}_{\tau+k}-m_\beta$ develop as sums according to $S_{\tau,\tau+k,\beta}$ and $\bar{S}_{\tau,\tau+k,\beta}$, depending on the domain of β.

When $|\beta|<1$,

$X_{\tau+k}-m_\beta = S_{\tau,\tau+k,\beta}+o_p(1)$, $\bar{X}_{\tau,t-\tau}-m_\beta=(t-\tau)^{-1}\bar{S}_{\tau,t-\tau,\beta}+o_p(1)$, and $(t-\tau)^{-1}\sum_{k=1}^t\varepsilon_{\tau+k}(X_{\tau+k-1}-m_\beta)=(t-\tau)^{-1}\widetilde{S}_{\tau,t-\tau,\beta}+o_p(1)$.

If $m_\beta=0$, an estimator of the variance on $]\tau,t]$ is

$$\begin{aligned}\widehat{\sigma}^2_{\tau,t-\tau;\widehat{\beta}_\tau} &= (t-\tau)^{-1}\sum_{k=\tau+1}^{t}\{X_k-\widehat{\beta}_tX_{k-1}\}^2\\ &= (t-\tau)^{-1}\sum_{k=\tau+1}^{t}\{\varepsilon_k-(\widehat{\beta}_t-\beta)X_{k-1}\}^2\end{aligned}$$

and for the true parameter values $\widehat{\sigma}^2_{\tau,t;0}=(t-\tau)^{-1}\sum_{k=\tau+1}^t\varepsilon^2_k$.

Proposition 7.8. *Let $v_\tau=1$ if $|\alpha|<1$ and $v_\tau=\alpha^{-2\tau}$ if $|\alpha|>1$. In the model of order 1 with a change-point at $\tau=[n\gamma]$ and under P_θ, the process $v_{[ns]}(\widehat{\sigma}^2_{[ns]}-\sigma^2)_{s\in[0,\gamma]}$ converges weakly on $[0,\tau]$ to*

$3\sigma^2(1-\alpha)\{(1+\alpha)s\}^{-1}\bar{W}^2_s$, *if* $|\alpha|<1$,

$2\sigma^2\{(\alpha^2-1)s\}^{-2}\int_0^s W_x^2\,dx+2(X_0-m_\alpha)\bar{W}_\alpha-\frac{(X_0-m_\alpha)^2}{1-\alpha^2}$, *if* $|\alpha|>1$.

The results are similar for the observations after the change-point. Let $v_t=1$ if $|\beta|<1$ and let $v_t=\beta^{-2t}$ if $|\beta|>1$.

Proposition 7.9. *In the change-point model of order 1, and under P_θ, the process $v_{[ns]}(\widehat{\sigma}^2_{\tau,[ns]-\tau,\widehat{\beta}_t}-\sigma^2)_{s\in]\gamma,1]}$ converges weakly on $]\tau,t]$ to*

$$3\sigma^2(1-\beta)(1+\beta)^{-1}(s-\gamma)^{-1}(\bar{W}^2_s-\bar{W}^2_\gamma)\ \textit{if}\ |\beta|<1,$$

if $|\beta| > 1$ and $|\alpha| < 1$, it converges weakly to

$$\lim_n \beta^{-2[ns]} \sum_{k=\tau+1}^{[ns]-\tau} S_{k-1,\beta}^2 = 2\sigma^2(\alpha^2-1)(s-\gamma)\}^{-2} \int_\gamma^s W_x^2\,dx,\ s \in]\gamma, 1],$$

if $|\alpha| > 1$ and $|\beta| > 1$, with $|\alpha| < |\beta|^{\frac{1}{2}}$, $\alpha^{-\tau} v_{[ns]}(\widehat{\sigma}^2_{\tau,[ns]-\tau,\widehat{\beta}_t} - \sigma^2)_{s\in]\gamma,1]}$ converges weakly to the process defined on $]\gamma, 1]$ by

$$2\sigma^2\Big\{(X_0 - m_\alpha) + \frac{\sigma^2}{(\alpha^2-1)}(W - W_\gamma)\Big\}(\bar{W}_\beta - \bar{W}_\beta(\gamma)).$$

These convergences extend to the models of order p with the notation (7.12).

7.6 Tests in models with change-points

A test statistic for an AR(1) model without change-point against the alternative of a model with a change relies on the same criterion as the construction of the estimators. Let

$$C_t(\theta) = \widehat{\sigma}_t^2(\theta) - \widehat{\sigma}_t^2(\theta_0)$$

be defined by the empirical variance $\widehat{\sigma}_t^2(\theta)$ of the model with parameter $\theta = (\xi, \tau)$. When the change-point is unknown, it is estimated by maximization of $\widehat{C}_t(\tau) = C_t(\widehat{\xi}_t, \tau)$ and the parameter θ is estimated by

$$\widehat{\theta}_t = \arg\max_\theta C_t(\theta).$$

The empirical variances are

$$\widehat{\sigma}_t^2(\theta) = \tau^{-1}\sum_{k=1}^{\tau}\{X_k - (1-\alpha)m_\alpha - \alpha X_{k-1}\}^2$$
$$+(t-\tau)^{-1}\sum_{k=\tau+1}^{t}\{X_k - (1-\beta)m_\beta - \beta X_{k-1}\}^2$$

and $\widehat{\sigma}_t^2(\theta)$ converges in probability under P_0 to

$$\sigma_{0t}^2(\theta) = E_0\{X_k - r_{\mu,\alpha}(X_{k-1})\}^2 + E_0\{X_k - r_{\mu,\beta}(X_{k-1})\}^2.$$

The convergence rates of the estimators of the parameters define the convergence rate of C_t according to the domain of α and β. Let φ_t be the convergence rate of the empirical variance

$$\begin{aligned} \varphi_t &= 1 \quad \text{if } |\alpha_0| < 1, |\beta_0| < 1, \\ &= \alpha_0^{2t} \quad \text{if } |\alpha_0| > 1 \wedge |\beta_0|, \\ &= \beta_0^{2t} \quad \text{if } |\beta_0| > 1 \wedge |\alpha_0|. \end{aligned} \tag{7.15}$$

Theorem 7.3. *With $\gamma_0 < 1$, the estimator of the change-point parameter is such that*

$$\widehat{\tau}_t = \arg \inf_{\tau \in [0,t]} t^{\frac{1}{2}}(\tau - \tau_0)\left[\frac{1}{(t-\tau)(t-\tau_0)} \sum_{k=\tau_0+1}^{t} \varepsilon_k^2 - \frac{1}{\tau\tau_0} \sum_{k=1}^{\tau_0} \varepsilon_k^2\}\right] - \gamma_0 + o_p(1),$$

$\widehat{\gamma}_t - \gamma_0$ is independent of the estimators $\widehat{\xi}_t$ of the regression parameters and it converges weakly to

$$\arg \inf_{u \in [-\gamma_0, 1-\gamma_0]} u\left\{\frac{G_{\varepsilon^2}(1) - G_{\varepsilon^2}(\gamma_0)}{(1-\gamma_0-u)(1-\gamma_0)^{\frac{1}{2}}} - \frac{G_{\varepsilon^2}(\gamma_0)}{(\gamma_0+u)\gamma_0^{\frac{1}{2}}}\right\},$$

the variable $\varphi_t C_t$ is asymptotically independent of $\widehat{\gamma}_t$ and it converges weakly under P_0 to a squared Gaussian process with mean zero.

Furthermore

$$\begin{aligned} C_t(\theta) &= O_p(\|\xi - \xi_0\| + t^{-\frac{1}{2}}|\gamma - \gamma_0|), \text{ if } |\alpha_0| < 1, |\beta_0| < 1, \\ &= O_p(\|\alpha - \alpha_0\|\alpha_0^{2t} + t^{-\frac{1}{2}}|\gamma - \gamma_0|), \text{ if } |\alpha_0| > 1 \wedge |\beta_0|, \\ &= O_p(\|\beta - \beta_0\|\beta_0^{2t} + t^{-\frac{1}{2}}|\gamma - \gamma_0|), \text{ if } |\beta_0| > 1 \wedge |\alpha_0|, \end{aligned} \quad (7.16)$$

The asymptotic distribution of $\varphi_t C_t$ defined by the limit of

$$\begin{aligned} \varphi_t \Big[& \tau^{-1} \sum_{k=1}^{\tau} (r_{\mu,\alpha}^2 - r_{\mu_0,\alpha_0}^2)(X_{k-1}) + (t-\tau)^{-1} \sum_{k=\tau+1}^{t} (r_{\mu,\beta}^2 - r_{\mu_0,\beta_0}^2)(X_{k-1}) \\ & -2\{\bar{X}_\tau(\mu_\alpha - \mu_{0,\alpha}) + \bar{X}_{\tau;t-\tau}(\mu_\beta - \mu_{0,\beta}) \\ & +(\alpha - \alpha_0)\tau^{-1} \sum_{k=1}^{\tau} X_k X_{k-1} + (\beta - \beta_0)(t-\tau)^{-1} \sum_{k=\tau+1}^{t} X_k X_{k-1}\} \Big] \end{aligned}$$

and the arguments for its convergence ar similar to those of the linear regression function, using the asymptotic distributions of the estimator $\widehat{\theta}_t$. By continuity, as t tends to infinity, $\inf_{\theta \in \Theta} \varphi_t C_t(\theta) = \varphi_t C_t(\widehat{\theta}_t)$ converges weakly to the minimum of the limit of the process $\varphi_t C_t$.

A test for the hypothesis H_0 of the absence of change-point in the model of order 1, as $\gamma_0 = 1$, relies on the statistic

$$T_t = \varphi_t |C_t(\widehat{\theta}_t) - C_{0t}(\widehat{\xi}_{0t})|$$

where

$$C_{0t}(\theta) = \widehat{\sigma}_{0t}^2(\theta) - \widehat{\sigma}_{0t}^2(\xi_0)$$

and $\widehat{\sigma}^2_{0t}(\xi_0)$ is the estimator of the variance in the model without change-point where the unknown parameter ξ_0 under H_0 reduces to the first components ξ_1 of the parameter ξ, with components μ and α

$$\widehat{\sigma}^2_{0t}(\xi_0) = t^{-1} \sum_{k=1}^{t} \{X_k - \mu - \alpha X_{k-1}\}^2,$$

it converge in probability to $\sigma_0^2(\xi_0) = E_0\{X_k - r_{\mu,\alpha}(X_{k-1})\}^2$.

Propositions 7.8 and 7.9 provide the limiting distribution of $C_{0t}(\widehat{\theta}_{0t})$ under $P_0 = P_{\theta_0}$, with the same rates as $C_t(\widehat{\theta}_t)$. From Theorem 7.3, T_t is asymptotically independent of $\widehat{\tau}_t$ under P_0 but the asymptotic distribution of $\widehat{\gamma}_t$ of Theorem 7.3 is not valid at γ_0.

With the notations (7.16), the tests statistic is the estimator of the process

$$\begin{aligned}
\varphi_t|C_t(\theta) - C_{0t}(\widehat{\theta}_{0t})| &= \varphi_t|C_{2t}(\mu, \beta, \tau) - C_{2t}(\widehat{\mu}_{0t}, \widehat{\alpha}_{0t}, \tau)| \\
&= \varphi_t \tau^{-1} \Big| \sum_{k=1}^{\tau} [\{X_k - r_{\mu,\beta}(X_{k-1})\}^2 - \{X_k - r_{\mu_0,\beta_0}(X_{k-1})\}^2 \\
&\quad - \{X_k - r_{\widehat{\mu}_{0t},\widehat{\alpha}_{0t}}(X_{k-1})\}^2 + \{X_k - r_{\mu_0,\alpha_0}(X_{k-1})\}^2] \Big| + o_p(1).
\end{aligned}$$

Under H_0

$$\begin{aligned}
T_t = \inf_{\tau} \Big[\varphi_t \tau^{-1} \Big| \sum_{k=1}^{\tau} [\{X_k - r_{\widehat{\mu}_{\tau,t-\tau},\widehat{\alpha}_{\tau,t-\tau}}(X_{k-1})\}^2 \\
- \{X_k - r_{\widehat{\mu}_{0t},\widehat{\alpha}_{0t}}(X_{k-1})\}^2] \Big| \Big] + o_p(1)
\end{aligned}$$

and T_t converges in probability to a squared Gaussian variable T_0 with mean zero, depending on the asymptotic distribution of $(\widehat{\xi}_{1t} - \widehat{\xi}_{0t})$.

Under a fixed alternative of parameter θ_0, T_t diverges. Under local alternatives P_{θ_n} converging to the probability P_{θ_0} of the hypothesis, the asymptotic behaviour of T_t depends on the limit of $\theta_n - \theta_0$ and it converges weakly to a limit T different from T_0. The statistic T_t provides a consistent test and its asymptotic distribution is characterized by the limits of the estimators, the level of the test and its power under local alternatives follow.

These convergences extend to the models of order p with the notation (7.12). A change-point may occur on any component of $X^{(p)}$ and the parameter γ is a p-dimensional vector. The approach for a test of the hypothesis H_0 of no change-point in the model of order p, as γ_0 is a vector with components 1, is similar to that of the linear regression on p regression variables.

7.7 Change-points at a threshold of series

Consider the model (7.12) of order 1 with a change-point at a threshold η of the series, with the equivalence between the chronological change-point model and the model with random stopping time

$$\tau_1 = \min\{k : I_k = 0\}.$$

It extends recursively to a sequence of stopping times

$$\tau_j = \min\{k > \tau_{j-1} : I_k = 0\}, j > 1.$$

The series have similar asymptotic behaviour starting from the first value of the series which goes across the threshold η at time

$$s_j = \min\{k > \tau_{j-1} : I_k = 1\}, j > 1.$$

The estimators of the parameters in the first phase of the model are restricted to the set of random intervals $[s_j, \tau_j]$ where X_t stands below η, for the second phase the observations are restricted to the set of random intervals $]\tau_{j-1}, s_j[$ where X remains above η.

The variables τ_j are stopping times of the series defined recursively for $t > s_{j-1}$ as

$$\begin{aligned} X_t &= m_\alpha + S_{s_{j-1},t-s_{j-1},\alpha} + o_p(1), \text{ if } |\alpha| < 1, \\ &= X_{s_{j-1}} + (t - s_{j-1} - 1)\mu + S_{s_{j-1},t-s_{j-1},1}, \text{ if } \alpha = 1, \\ &= m_\alpha + \alpha^{t-s_{j-1}}(X_{s_{j-1}-1} - m_\beta) + S_{s_{j-1},t-s_{j-1},\alpha}, \text{ if } |\alpha| > 1, \end{aligned}$$

and the s_j are stopping times defined for $t > \tau_{j-1}$ by

$$\begin{aligned} X_t &= m_\beta + S_{\tau_{j-1},t-\tau_{j-1},\beta} + o_p(1), \text{ if } |\beta| < 1, \\ &= X_{\tau_{j-1}} + (t - \tau_{j-1} - 1)\mu + S_{\tau_{j-1},t-\tau_{j-1},1}, \text{ if } \beta = 1, \\ &= m_\beta + \beta^{t-\tau_{j-1}}(X_{\tau_{j-1}} - m_\alpha) + S_{\tau_{j-1},t-\tau_{j-1},\beta} \text{ if } |\beta| > 1. \end{aligned}$$

The sequences $t^{-1}\tau_j$ and $t^{-1}s_j$ converge to the corresponding stopping times of the limit of X_t as t tends to infinity. The partial sums are therefore defined as sums over indices belonging to countable union of intervals $[s_j, \tau_j]$

and $]r_j, s_{j+1}[$, respectively, for the two phases of the model. As previously, theirs limits are deduced from integrals on the corresponding sub-intervals The estimators of the parameters are still expressions of the partial sums. The results generalize to processes of order p with a change-point in each p component.

Chapter 8

Change-points in nonparametric models

Abstract. This chapter studies nonparametric models for densities, regressions and autoregressive series with unknown change-points according to thresholds of regression variables or at sampling indices. The convergence rates of the estimators and their weak convergences are proved. Mean squares and likelihood ratio tests of models without change-points are considered, their asymptotic behaviour under the null hypothesis and alternatives are determined.

8.1 Nonparametric models

A nonparametric model with change-points defines a partition of the probability space into classes according to thresholds of the regression variable or the observation index. On a probability space $\left(\Omega, \mathcal{A}, P\right)$, let $(A_k)_{0\leq k\leq K}$ be a partition of Ω into disjoint subsets. A regression X for a response variable Y splits according to the partition into a set of variables $X_1, \ldots, X_{K_1}$ with respective density functions $f_1, \ldots, f_{K_1}$ on a metric space $(\mathcal{X}, d)$. Models with random classes independent of the variable Y are also conditional mixture models, the density of the variable

$$X = \sum_{k=1}^{K_1} X_k 1_{A_k} \tag{8.1}$$

is a finite mixture of the densities $f_1, \ldots, f_{K_1}$ with the respective mixture probabilities $p_1, \ldots, p_{K_1}$ such that $p_k = P(A_k)$ and f_k is the density of X conditionally on A_k, then $f_X = \sum_{k=1}^{K_1} p_k f_k$.

Consider a regression model for a random variable Y on the variable X defined by (8.1), $Y = m(X) + \varepsilon$, with an error ε such that $E(\varepsilon|X) = 0$

and a variance σ^2. In a regression model with a change-point, the mean m develops as

$$E(Y|X) = \sum_{k=1}^{K_1} \frac{E(Y1_{A_k}|X_k)}{P(A_k|X_k)} = \sum_{k=1}^{K_1} m_k(X_k).$$

If for every $k = 1, \ldots, K_1$, the class A_k of a mixture is independent of the variable X_k, the regression functions are

$$m_k(x) = P(A_k)^{-1} E(Y1_{A_k}|X_k = x).$$

The mixture of the variable X is transposed into a mixture for the regression model of the variable Y.

Consider the regression model of (X, Y) belonging to $\mathbb{R}^2$ with a change-point according to a threshold γ of the variable X, it is defined by the equation

$$Y = (1 - \delta_\gamma)m_1(X) + \delta_\gamma m_2(X) + \varepsilon, \tag{8.2}$$

where the variables X and ε are independent, $E\varepsilon = 0$ and $Var\varepsilon = \sigma^2$ and the indicator is $\delta_\gamma = 1_{\{X>\gamma\}}$, γ belonging to Γ a compact subset of $\mathbb{R}$. A change-point of mean such as (8.2) is a consequence of a change-point of the density de (X, Y) according to the sets $\{X \leq \gamma\}$ and $\{X > \gamma\}$. We assume that the functions m_1 and m_2 are continuous and the change-point parameter has a true value γ_0 under the probability measure P_0 of the observations, such that $p_0 = P(X \leq \gamma_0)$ is different from 0 and 1 and $m_1(\gamma_0) \neq m_2(\gamma_0)$. The approach of Chapter 4 adapts to the nonparametric regression models, with nonparametric estimators of the regression functions.

For a continuous model with different regularity properties with different left and right- derivatives of some order, the nonparametric estimators and the estimator of the change-point are still consistent but their limiting distributions are modified. Discontinuities of a derivative of the regression function may therefore be estimated and the results of this chapter apply to their estimators.

8.2 Nonparametric density with a change

On a probability space (Ω, A, P_γ), let Y be a real variable with two-phase density function on a bounded interval I_Y

$$f_\gamma = (1 - \delta_\gamma)f_{1,\gamma} + \delta_\gamma f_{2,\gamma}, \tag{8.3}$$

with sub-densities $f_1 = f_{1,\gamma}$ and $f_2 = f_{2,\gamma}$ of $C^2(I_Y)$ such that γ is the first point of discontinuity of the density f_γ of the variable Y

$$f_{1,\gamma}(\gamma) \neq f_{2,\gamma}(\gamma^+)$$

and with $\delta_\gamma(y) = 1_{\{y>\gamma\}}$. Under the probability measure P_0 of a n-sample of observations $(Y_i)_{i=1,\dots,n}$, the density with a change-point at γ_0 is denoted $f_0 = (1-\delta_0)f_{01} + \delta_0 f_{02}$. For a sequence $(\gamma_n)_n$ converging to γ_0, the density $f_n = (1-\delta_{\gamma_n})f_{1,\gamma_n} + \delta_{\gamma_n} f_{2,\gamma_n}$ converges to f_0. The parameter of the model is $\theta = (f_1, f_2, \gamma)$ under P_γ, with value θ_0 under P_0.

The sub-densities f_1 and f_2 are estimated using a symmetric kernel K in $L^2 \cap C^2$ on a compact support, with bandwidth $h = h_n$ converging to zero as n tends to infinity and satisfying the Conditions 2.1 of Chapter 2, with the optimal bandwidth $h_n = O(n^{-\frac{1}{5}})$ for the quadratic norm. At every fixed γ of I, the estimators are defined from the sub-sample of the observations below γ for $f_{1,\gamma}$, and respectively from the observations above γ for $f_{2,\gamma}$, for every x of I

$$\widehat{f}_{1nh,\gamma}(y) = n^{-1}\sum_{i=1}^n K_h(y-Y_i)(1-\delta_{\gamma,i}),\ y \le \gamma,$$

$$\widehat{f}_{2nh,\gamma}(y) = n^{-1}\sum_{i=1}^n K_h(y-Y_i)\delta_{\gamma,i},\ y > \gamma.$$

The estimators $\widehat{f}_{1nh,\gamma}$ and $\widehat{f}_{2nh,\gamma}$ are independent, for a kernel with a compact support, their expectation under P_0 are

$$\begin{aligned}
f_{1nh,\gamma}(y) &= \int K(u)1_{\{y+hu\le\gamma, y\le\gamma\}} f_0(y+hu)\,du,\ y\le\gamma\\
&= f_{01}(y)1_{\{y\le\gamma_0\wedge\gamma\}} + f_{02}(y)1_{\{\gamma_0<y\le\gamma\}} + O(h^2),\ y\le\gamma,\\
f_{1nh,\gamma}(y) &= \int K(u)1_{\{y>\gamma, y+hu>\gamma\}} f_0(y+hu)\,du = f_{02}(y) + O(h^2),\ y\ge\gamma,\\
f_{2nh,\gamma}(y) &= \int K(u)1_{\{y>\gamma, y+hu>\gamma\}} f_0(y+hu)\,du\\
&= f_{02}(y)1_{\{y>\gamma_0\vee\gamma\}} + f_{01}(y)1_{\{\gamma<y\le\gamma_0\}} + O(h^2),\ y>\gamma,\\
f_{2nh,\gamma}(y) &= \int K(u)1_{\{y\le\gamma, y+hu>\gamma\}} f_0(y+hu)\,du\\
&= f_{01}(y) + O(h^2),\ y\le\gamma,
\end{aligned}$$

so $\widehat{f}_{1nh,\gamma}(y)$ converges a.s. to $f_{1\gamma}(y)$, for $y \le \gamma$, and $\widehat{f}_{2nh,\gamma}(y)$ converges a.s. to $f_{2,\gamma}(y)$, for $y > \gamma$. For a sequence $(\gamma_n)_n$ converging to γ_0, f_{nh,γ_n}

converges to f_0 and $\widehat{f}_{nh,\widehat{\gamma}_n}$ converges a.s. to f_0. Their variances have the approximations

$$\begin{aligned}
v_{1nh,\gamma}(y) &= (nh)^{-1} \int 1_{\{u \geq h^{-1}(\gamma - y)\}} 1_{\{y \leq \gamma\}} \{f_0(y+hu)K^2(u)\}\, du \\
&= (nh)^{-1} k_2 f_{01}(y) 1_{\{y \leq \gamma\}} + o((nh)^{-1}), \\
v_{2nh,\gamma}(y) &= (nh)^{-1} \int 1_{\{u > h^{-1}(\gamma - y)\}} 1_{\{y > \gamma\}} \{f_0(y+hu)K^2(u)\}\, du \\
&= (nh)^{-1} k_2 f_{02}(y) 1_{\{y > \gamma\}} + o((nh)^{-1})
\end{aligned}$$

the convergence rate of the estimators $\widehat{f}_{1nh,\gamma}$ and $\widehat{f}_{2nh,\gamma}$ is therefore $(nh)^{-\frac{1}{2}}$.

Like in Section 3.3 for a parametric density, the parameter γ is estimated by $\widehat{\gamma}_{nh}$, the first value which maximizes the estimated likelihood of the sample

$$\widehat{l}_{nh}(\gamma) = \sum_{k=1}^{n} \Big\{ \log \widehat{f}_{1nh,\gamma}(Y_k) 1_{\{Y_k \leq \gamma\}} + \log \widehat{f}_{2nh,\gamma}(Y_k) 1_{\{Y_k > \gamma\}} \Big\}$$

and the estimators of the sub-densities are $\widehat{f}_{jnh} = \widehat{f}_{jnh,\widehat{\gamma}_n}$, for $j = 1, 2$.

The log-likelihood ratio process $X_n(\theta) = n^{-1}\{l_n(\theta) - l_n(\theta_0)\}$ of the sample under P_θ and P_0 is

$$\begin{aligned}
X_n(\theta) = n^{-1} \sum_{k=1}^{n} \Big[& 1_{\{\gamma_0 < \gamma\}} \Big\{ \log \frac{f_{1,\gamma}(Y_k)}{f_{01}(Y_k)} 1_{\{Y_k \leq \gamma_0\}} \\
& + \log \frac{f_{1,\gamma}(Y_k)}{f_{02}(Y_k)} 1_{]\gamma_0,\gamma]}(Y_k) + \log \frac{f_{2,\gamma}(Y_k)}{f_{02}(Y_k)} 1_{\{Y_k > \gamma\}} \Big\} \\
& + 1_{\{\gamma < \gamma_0\}} \Big\{ \log \frac{f_{1,\gamma}(Y_k)}{f_{01}(Y_k)} 1_{\{Y_k \leq \gamma\}} \\
& + \log \frac{f_{2,\gamma}(Y_k)}{f_{01}(Y_k)} 1_{]\gamma,\gamma_0]}(Y_k) + \log \frac{f_{2,\gamma}(Y_k)}{f_{02}(Y_k)} 1_{\{Y_k > \gamma_0\}} \Big\} \Big],
\end{aligned}$$

under P_0, it converges a.s. uniformly on I to its expectation

$$\begin{aligned}
X(\theta) = 1_{\{\gamma_0 < \gamma\}} \Big\{ & \log \int_{-\infty}^{\gamma_0} \frac{f_{1,\gamma}}{f_{01}}\, dF_0 + \int_{\gamma_0}^{\gamma} \log \frac{f_{1,\gamma}}{f_{02}}\, dF_0 \\
& + \int_{\gamma}^{\infty} \log \frac{f_{2,\gamma}}{f_{02}}\, dF_0 \Big\} + 1_{\{\gamma < \gamma_0\}} \Big\{ \int_{-\infty}^{\gamma} \log \frac{f_{1,\gamma}}{f_{01}}\, dF_0 \\
& + \int_{\gamma}^{\gamma_0} \log \frac{f_{2,\gamma}}{f_{01}}\, dF_0 + \int_{\gamma_0}^{\infty} \log \frac{f_{2,\gamma}}{f_{02}}\, dF_0 \Big\}.
\end{aligned}$$

The maximum of the function X is zero at θ_0 and the a.s. consistency of the estimator $\widehat{\gamma}_{nh}$ follows like in Theorem 3.1, then the estimators $\widehat{f}_{1nh}$ and

$\widehat{f}_{2nh}$ are a.s. consistent. Under Conditions C of Section 2.5 and from the expression of the variances v_{1nh,γ_0} and v_{2nh,γ_0}, it follows that the kernel estimator $\widehat{f}_{nh}$ is $(nh)^{\frac{1}{2}}$ consistent and asymptotically Gaussian.

Let $\rho(\theta, \theta_0) = (\|f_1 - f_{01}\|_I^2 + \|f_2 - f_{02}\|_I^2 + |\gamma - \gamma_0|)^{\frac{1}{2}}$ and let $V_\varepsilon(\theta_0)$ be an ε-neighborhood of θ_0, for $\varepsilon > 0$. The process

$$W_n(\theta) = n^{\frac{1}{2}}\{X_n(\theta) - X(\theta)\}$$

has a finite variance in $V_\varepsilon(\theta_0)$ under the next condition.

Condition C2. The expectation $E_0\{\sup_{\theta \in V_\varepsilon(\theta_0)} \log^2 f_\gamma(Y)\}$ is finite.

Lemma 8.1. *For $\varepsilon > 0$ sufficiently small, there exists a constant $\kappa_0 > 0$ such that $E_0 X(\theta) \leq -\kappa_0 \rho(\theta, \theta_0)$.*

This is a consequence of the expansion of X according to the logarithmic expansion $\log f_{k,\gamma}(y) - \log f_{0k}(y) = f_{0k}^{-1}(y)\{f_{k,\gamma}(y) - f_{0k}(y)\}\{1 + o_p(1)\}$ and it is a $O(|\gamma - \gamma_0|)$.

Lemma 8.2. *Under Condition C2, for every $\varepsilon > 0$ there exists a constant $\kappa_1 > 0$ such that for n large enough, $E_0 \sup_{\rho(\theta,\theta_0) \leq \varepsilon} |W_n(\theta)| \leq \kappa_1 \varepsilon$.*

Proof. The inequality for the process W_n follows from the Cauchy–Schwarz inequality. For θ in an ε neighborhood $V_\varepsilon(\theta_0)$ of θ_0, the process $W_n(\theta)$ has an expansion according to the logarithmic expansion of $\log f_{k,\gamma}(y) - \log f_{0k}(y)$ which implies that $Var \sup_{\theta \in V_\varepsilon(\theta_0)} W_n(\theta)$ is a $O(\rho^2(\theta, \theta_0))$. □

Theorem 8.1. *Under the conditions* C *and* C2

$$\overline{\lim}_{n,A \to \infty} P_0((nh)^{\frac{1}{2}} \rho(\widehat{\theta}_{nh}, \theta_0) > A) = 0.$$

This convergence rate is deduced from Lemmas 8.1 and 8.2 and from the convergence rate of the kernel estimators of the sub-densities, so the estimator $\widehat{\gamma}_{nh}$ has the convergence rate nh.

The log-likelihood ratio process X_n is estimated by

$$\begin{aligned}
\widehat{X}_{nh}(\gamma) = n^{-1}\sum_{k=1}^{n}\Big[1_{\{\gamma_0<\gamma\}}\Big\{\log\frac{\widehat{f}_{1nh,\gamma}(Y_k)}{\widehat{f}_{1nh,\gamma_0}(Y_k)}1_{\{Y_k\le\gamma_0\}} \\
+\log\frac{\widehat{f}_{1nh,\gamma}(Y_k)}{\widehat{f}_{2nh,\gamma_0}(Y_k)}1_{]\gamma_0,\gamma]}(Y_k)+\log\frac{\widehat{f}_{2nh,\gamma}(Y_k)}{\widehat{f}_{2nh,\gamma_0}(Y_k)}1_{\{Y_k>\gamma\}}\Big\} \\
+1_{\{\gamma<\gamma_0\}}\Big\{\log\frac{\widehat{f}_{1nh,\gamma}(Y_k)}{\widehat{f}_{1nh,\gamma_0}(Y_k)}1_{\{Y_k\le\gamma\}} \qquad (8.4)\\
+\log\frac{\widehat{f}_{2nh,\gamma}(Y_k)}{\widehat{f}_{1nh,\gamma_0}(Y_k)}1_{]\gamma,\gamma_0]}(Y_k)+\log\frac{\widehat{f}_{2nh,\gamma}(Y_k)}{\widehat{f}_{2nh,\gamma_0}(Y_k)}1_{\{Y_k>\gamma_0\}}\Big\}\Big],
\end{aligned}$$

under P_0, it converges a.s. to the function $X(\gamma)$, as n tends to infinity and h tends to zero, according to the limits of the expectation functions of the kernel estimators.

The process X_n is the sum of the processes

$$\begin{aligned}
X_{1n}(\theta) &= n^{-1}\sum_{k=1}^{n}\Big\{\log\frac{f_{1,\gamma}(Y_k)}{f_{01}(Y_k)}1_{\{Y_k\le\gamma_0\}}+\log\frac{f_{2,\gamma}(Y_k)}{f_{02}(Y_k)}1_{\{Y_k>\gamma_0\}}\Big\} \\
X_{2n}(\theta) &= n^{-1}\sum_{k=1}^{n}\Big\{1_{\{\gamma_0<\gamma\}}\log\frac{f_{1,\gamma}(Y_k)}{f_{2,\gamma}(Y_k)}1_{]\gamma_0,\gamma]}(Y_k) \\
&\qquad +1_{\{\gamma<\gamma_0\}}\log\frac{f_{2,\gamma}(Y_k)}{f_{1,\gamma}(Y_k)}1_{]\gamma,\gamma_0]}(Y_k)\Big\}.
\end{aligned}$$

Let $\mathcal{U}_{nh}=\{u=nh(\gamma-\gamma_0),\gamma\in I\}$ and let $\mathcal{U}_{nh}^A=\{u\in\mathcal{U}_{nh}:|u|\le A\}$, with $A>0$. For u in $\mathcal{U}_{nh}^A$, let $\gamma_{nh,u}=\gamma_0+(nh)^{-1}u$ and let

$$\begin{aligned}
\widetilde{X}_{2n}(u) &= X_{2n}(f_{01},f_{02},\gamma_{nh,u}) \\
&= n^{-1}\sum_{k=1}^{n}\Big\{1_{\{\gamma_0<\gamma\}}\log\frac{f_{01}(Y_k)}{f_{02}(Y_k)}1_{]\gamma_0,\gamma]}(Y_k) \\
&\qquad +1_{\{\gamma<\gamma_0\}}\log\frac{f_{02}(Y_k)}{f_{01}(Y_k)}1_{]\gamma,\gamma_0]}(Y_k)\Big\}.
\end{aligned}$$

Theorem 8.2. *For every $A>0$, on $\mathcal{U}_{nh}^A$, the process nhX_n has the uniform asymptotic expansion*

$$nhX_n(f_1,f_2,\gamma_{nh,u})=nhX_{1n}(f_1,f_2)+nh\widetilde{X}_{2n}(u)+o_p(1).$$

Proof. On $\mathcal{U}^A_{nh}$, the remainder term $nh\{X_{2n}(\theta_{nh,u}) - \widetilde{X}_{2n}(u)\}$ of the approximation of the process nhX_n by $nh\{X_{1n} + \widetilde{X}_{2n}\}$ is

$$h\sum_{k=1}^{n}\Big[1_{\{\gamma_0<\gamma_{nh,u}\}}\Big\{\log\frac{f_{1,\gamma_{nh,u}}(Y_k)}{f_{01}(Y_k)}1_{]\gamma_0,\gamma_{nh,u}]}(Y_k)$$
$$-\log\frac{f_{2,\gamma_{nh,u}}(Y_k)}{f_{02}(Y_k)}1_{\{\gamma_0<Y_k\leq\gamma_{nh,u}\}}\Big\}$$
$$+1_{\{\gamma_{nh,u}<\gamma_0\}}\Big\{\log\frac{f_{2,\gamma_{nh,u}}(Y_k)}{f_{02}(Y_k)}1_{]\gamma_{nh,u},\gamma_0]}(Y_k)$$
$$-\log\frac{f_{1,\gamma_{nh,u}}(Y_k)}{f_{01}(Y_k)}1_{\{\gamma_{nh,u}<Y_k\leq\gamma_0\}}\Big\}\Big],$$

and it converges uniformly to zero in probability due to the convergence rate of the parameter. □

Theorem 8.3. *Under P_0 and Conditions C and C2, the variable $n(\widehat{\gamma}_{nh} - \gamma_0)$ is bounded in probability and it converges weakly to the location u_0 of the maximum of an uncentered Gaussian process.*

Proof. For A and n sufficiently large, the variable $\widehat{u}_{nh} = nh(\widehat{\gamma}_{nh} - \gamma_0)$ belongs to $\mathcal{U}^A_{nh}$ with a probability converging to one. It maximizes the process $nh\widetilde{X}_{2n}$ which is a sum of the variations of processes on intervals of length $(nh)^{-1}$, therefore it is tight by Billingsley's criterion (15.21) for processes of $D(I)$. The mean of $nh\widetilde{X}_{2n}(u)$ is

$$\mu = u\{f_{02}(\gamma_0)1_{\{u>0\}} + f_{01}(\gamma_0)1_{\{u<0\}}\}\log\frac{f_{01}(\gamma_0)}{f_{02}(\gamma_0)}$$

and its variance is

$$v = |u|\{f_{02}(\gamma_0)1_{\{u>0\}} + f_{01}(\gamma_0)1_{\{u<0\}}\}\log^2\frac{f_{01}(\gamma_0)}{f_{02}(\gamma_0)},$$

therefore it converges weakly to a Gaussian process with mean μ and with variance v. □

8.3 Likelihood ratio test of no change

The likelihood ratio test for the hypothesis H_0 of a density without change-point against the alternative of model (8.3), is performed with the statistic

$$T_n = \sup_{\gamma\in I} 2nh\{\widehat{X}_{nh}(\gamma) - \widehat{X}_{0nh}\}$$

with the process $\widehat{X}_{nh}$ defined by (8.4) and $\widehat{X}_{0n}$ is the estimator of the process X_n under H_0, with γ_0 the end-point of I_Y, for a density f_0 of $C^2(I_Y)$. Under an alternative with a change-point at $\gamma < \gamma_0$, the density f_γ is right continuous and has a left-hand limit at γ where it is discontinuous, $f_{2\gamma}(\gamma^+)$ is distinct from $f_{1\gamma}(\gamma)$.

Proposition 8.1. *Under the hypothesis H_0, Conditions* C *and* C2, *the statistic T_n converges in probability to*

$$T_0 = -u_0 \int Var_0 \widehat{f}_{0nh}(y) f'^2(y) f^{-1}(y)\, dy.$$

Proof. Under H_0, the density f_0 est estimated by a kernel estimator $\widehat{f}_{0nh}$ and the process $\widehat{X}_{nh} - \widehat{X}_{0nh}$ reduces to

$$\begin{aligned}\widehat{X}_{nh}(\gamma) - \widehat{X}_{0nh} &= \int 1_{\{y \le \gamma\}} \log \frac{\widehat{f}_{1nh,\gamma}(y)}{\widehat{f}_{0nh}(y)}\, d\widehat{F}_n(y)\\ &+ \int 1_{\{y > \gamma\}} \log \frac{\widehat{f}_{2nh,\gamma}(y)}{\widehat{f}_{0nh}(y)}\, d\widehat{F}_n(y).\end{aligned}$$

The process $\widehat{X}_{nh} - \widehat{X}_{0nh}$ converges a.s. uniformly in I_Y, under P_0, to the function

$$X(\gamma) = \int 1_{\{y \le \gamma\}} \log \frac{f_{1,\gamma}(y)}{f_0(y)}\, dF_0(y) + \int 1_{\{y > \gamma\}} \log \frac{f_{2,\gamma}(y)}{f_0(y)}\, dF_0(y),$$

the first integral is negative and maximum with the value zero at γ_0 where $f_{01} = f_{1,\gamma_0} = f_0$, then the second integral is zero and $f_{02} = 0$ beyond γ_0. The estimators are therefore a.s. consistent of under H_0 by the same arguments as in Theorem 3.1, from the uniform a.s. convergence of X_n to X. The estimators $\widehat{f}_{0nh}$ and $\widehat{f}_{1nh}$ converge a.s. to f_0 and $\widehat{f}_{2nh}$ converges a.s. to zero under H_0. As the variances of the estimators $\widehat{f}_{knh}$ and $\widehat{f}_{1nh,\gamma}$ have the order $(nh)^{-1}$, the processes $(nh)^{\frac{1}{2}}(\widehat{f}_{knh} - f_{knh})$ converge to Gaussian processes, for $k = 0, 1$.

The first integral $\mathcal{I}_{1nh}(\gamma)$ of $\widehat{X}_{nh}(\gamma) - \widehat{X}_{0nh}$ has a second order expansion

$$\begin{aligned}\mathcal{I}_{1nh}(\gamma) &= \int 1_{\{y \le \gamma\}} \frac{(\widehat{f}_{1nh,\gamma} - \widehat{f}_{0nh})(y)}{\widehat{f}_{0nh}(y)}\, d\widehat{F}_n(y)\\ &- \int 1_{\{y \le \gamma\}} \frac{(\widehat{f}_{1nh,\gamma} - \widehat{f}_{0nh})^2(y)}{2\widehat{f}^2_{0nh}(y)}\, dF_0(y) + o_p((nh)^{-1}),\end{aligned}$$

its first term is asymptotically equivalent under P_0 to the integral of $\widehat{f}_{0nh} - \widehat{f}_{1nh,\gamma} = \widehat{f}_{2nh,\gamma}$ which is zero below γ, then $\mathcal{I}_{1nh}(\gamma) = o_p((nh)^{-1})$.

On a bounded interval, the inequalities of Lemmas 8.1 and 8.2 are still satisfied and, by Theorem 8.1, the estimator $\widehat{\gamma}_{nh}$ is $(nh)^{-1}$ consistent. Let $\gamma_{nh,u} = \gamma_0 - (nh)^{-1}u$, the second integral $\mathcal{I}_{2nh}(\gamma_{nh,u})$ of the expression of $(\widehat{X}_{nh} - \widehat{X}_{0nh})(\gamma_{nh,u})$, the estimators $1_{\{y>\gamma_{nh,u}\}}\widehat{f}_{0nh}(y)$ and $\widehat{f}_{2nh,\gamma_{nh,u}}(y)$ converge a.s. under P_0 to the same limit. The variables $(nh)^{\frac{1}{2}}\{\widehat{f}_{0nh}(Y_k) - \widehat{f}_{2nh}(Y_k)\}1_{\{Y_k>\gamma_{nh,u}\}}$ converges weakly to a centered Gaussian variable with a strictly positive variance $uVar_0\widehat{f}_{0nh}(Y_k)$. A second order expansion of the process $2nh(\widehat{X}_{nh} - \widehat{X}_{0nh})(\gamma_{nh,u})$ as $n^{-1}\sum_{i=1}^n \xi_{inh} + o_p(1)$ with

$$\xi_{inh} = -nh\{\widehat{f}_{0nh}(Y_k) - \widehat{f}_{2nh}(Y_k)\}^2 \frac{f'^2}{f}(Y_k)1_{\{Y_k>\widehat{\gamma}_{nh}\}}$$

implies that T_n converge a.s. to T_0. □

Under fixed alternatives with γ strictly lower than the value γ_0 of the hypothesis H_0 and $f_{1\gamma}$ distinct from $f_{2\gamma}$ the processes $(nh)^{\frac{1}{2}}(\widehat{f}_{1nh} - f_{1,\gamma})$ and $(nh)^{\frac{1}{2}}(\widehat{f}_{2nh} - f_{2,\gamma})$ converge weakly to Gaussian processes and T_n tends to infinity.

Let K_n be local alternatives defined by a change point at $\gamma_{nh,u} = \gamma_0 - (nh)^{-1}u_n$, where u_n converges to u in $\mathcal{U}^A_{nh}$, and distinct sub-densities

$$f_{1nh,\gamma_{nh,u}} = f_0 + (nh)^{-\frac{1}{2}} g_{1n},\ y \leq \gamma_n,$$
$$f_{2nh,\gamma_{nh,u}} = f_0 + (nh)^{-\frac{1}{2}} g_{2n},\ y > \gamma_n,$$

where the functions g_{kn} converges uniformly to g_k, for $k = 1, 2$.

Proposition 8.2. *Under the local alternatives K_n, Conditions* C *and* C2, *the statistic T_n converges weakly to $T_0 - T$ where T is a strictly positive constant.*

Proof. Under K_n, the process $(nh)^{\frac{1}{2}}(\widehat{f}_{1nh} - \widehat{f}_{0nh})$ is the sum $(nh)^{\frac{1}{2}}(\widehat{f}_{1nh} - f_{1nh}) + g_{1n} - (nh)^{\frac{1}{2}}(\widehat{f}_{0nh} - f_0)$ where $(nh)^{\frac{1}{2}}(\widehat{f}_{1nh} - f_{1nh})$ and $(nh)^{\frac{1}{2}}(\widehat{f}_{0nh} - f_0)$ converge weakly to the same distribution and the integral $nh\mathcal{I}_{1nh}(\widehat{\gamma}_{nh})$ converges in probability to $\frac{1}{2}\int g_1^2(y)\,dy$. The variable $nh(\gamma_0 - \widehat{\gamma}_{nh})$ converges weakly to $u + u_0$ under the alternatives and the asymptotic distribution of the integral $nh\mathcal{I}_{2nh}(\widehat{\gamma}_{nh})$ is similar to its limit under H_0 except u_0 which is replaced by $u + u_0$. □

Proposition 8.2 entails that the likelihood ratio test is consistent under local alternatives.

8.4 Mixture of densities

Let (X, Y) be a variable on a probability space (Ω, A, P), such that X induces a change-point for the distribution function of Y at a threshold γ. The variable X has a mixture distribution function with the probabilities $p_\gamma = P(X \le \gamma)$ and $1 - p_\gamma$

$$F_{X,\gamma} = p_\gamma F_{1X,\gamma} + (1 - p_\gamma) F_{2X,\gamma} \tag{8.5}$$

and $F_{1X,\gamma} = P(X \le x \wedge \gamma)$ and $F_{2X,\gamma} = P(\gamma < X \le x)$.

The variable Y has the conditional distribution functions

$$F_{1Y,\gamma}(y) = P(Y \le y | X \le \gamma), \quad F_{2Y,\gamma}(y) = P(Y \le y | X > \gamma).$$

Under the probability measure of the observed sample $(X_i, Y_i)_{i=1,\dots,n}$, the change-point occurs at γ_0, $p_0 = p_{\gamma_0}$ and the (sub-) distribution functions are F_{01X}, F_{02X} and F_{0X} for the variable X, F_{01Y}, F_{02Y} and F_{0Y} for Y.

Estimators of the mixture probabilities in this model are deduced from the partition in disjoint sub-populations according to the change-point estimator for the observed auxiliary variable X, when the change-point of the density of X is unknown, in model (8.3). The probability p_0 is estimated by

$$\widehat{p}_n = n^{-1} \sum_{i=1}^n 1_{\{X_i \le \widehat{\gamma}_n\}} = \widehat{F}_{X,n}(\widehat{\gamma}_n), \tag{8.6}$$

with the estimator $\widehat{\gamma}_n$ of Section 8.2. The distribution functions F_{01Y} and F_{02Y} are estimated by

$$\widehat{F}_{1n,X}(x) = \widehat{p}_n^{-1} n^{-1} \sum_{i=1}^n 1_{\{X_i \le x \wedge \widehat{\gamma}_n\}},$$

$$\widehat{F}_{2nX}(x) = (1 - \widehat{p}_n)^{-1} n^{-1} \sum_{i=1}^n 1_{\{\widehat{\gamma}_n < X_i \le x\}}.$$

The conditional distribution functions F_{01Y} and F_{02Y} are estimated by

$$\widehat{F}_{1nY}(y) = \widehat{p}_n^{-1} n^{-1} \sum_{i=1}^n 1_{\{X_i \le \widehat{\gamma}_n\}} 1_{\{Y_i \le y\}},$$

$$\widehat{F}_{2nY}(y) = (1 - \widehat{p}_n)^{-1} n^{-1} \sum_{i=1}^n 1_{\{X_i > \widehat{\gamma}_n\}} 1_{\{Y_i \le y\}},$$

and they are independent. Their asymptotic behaviour is a direct consequence of the convergence of empirical processes. Let W be the standard Brownian bridge and let $W_X = W \circ F_X$. Let W_2 be the standard Brownian bridge on $[0,1]^2$ and let $W_{1,Y} = W_2 \circ F_{X,Y}(\gamma_0, \cdot)$ and $W_{2,Y} = W_2 \circ F_{X,Y}(]\gamma_0, \infty[, \cdot)$.

Proposition 8.3. *Under P_0, the estimator $\widehat{p}_{0n}$ converge a.s. to p_0 and the variable $n^{\frac{1}{2}}(\widehat{p}_n - p_0)$ converges weakly to $W_X(\gamma_0)$. The estimators $\widehat{F}_{1nY}$ and $\widehat{F}_{2nY}$ converge a.s. uniformly to F_{1Y}, and respectively F_{2Y}. The processes $n^{\frac{1}{2}}(\widehat{F}_{1nY} - F_{01Y}, \widehat{F}_{2nY} - F_{02Y})$ converges weakly to the process*

$$(p_0^{-1}\{W_{1,Y} - F_{01Y}W_X(\gamma_0)\}, (1-p_0)^{-1}\{W_{2,Y} + F_{02Y}W_X(\gamma_0)\}).$$

8.5 Nonparametric regression with a change

Let (X,Y) be a variable defined on a probability space (Ω, A, P), with values in $(\mathbb{R}^2, \mathcal{B})$ and with distribution function $F_{X,Y}$. Let $(X_i, Y_i)_{i \le n}$ be a n-sample of (X,Y) on a bounded support I_{XY}. The nonparametric continuous functions m_1 and m_2 of the expectation of Y conditionally on $\{X = x\}$ are distinct and they define the conditional mean of Y under a probability distribution $P = P_{\gamma, m_1, m_2}$ as $E_\gamma(Y|X) = m_\gamma(X)$ where

$$m_\gamma(x) = \{1 - \delta_\gamma(x)\}m_1(x) + \delta_\gamma(x)m_2(x) \tag{8.7}$$

and $\delta_\gamma(x) = 1_{\{x>\gamma\}}$. They are estimated from the sub-samples $(X_i\delta_{\gamma,i}, Y_i)_{i\le n}$ with the regression variable below γ and $(X_i(1-\delta_{\gamma,i}), Y_i)_{i\le n}$ with the regression variable above γ, for every γ such that $p_\gamma = P(X \le \gamma)$ does not belong to $\{0,1\}$. Let $F_{Y|X}$ be the distribution function of Y conditionally on X and let F_X be the marginal distribution function of X, $I_{Y|X}$ and I_X are their respective supports. They have densities $f_{X,Y}$ for (X,Y) and f_X is the marginal density of X.

Let K be a symmetric kernel in $L^2 \cap C^2$ on a compact support, let $h = h_n$ be a bandwidth converging to zero as n tends to infinity, and let K_h be the kernel with bandwidth h. The estimators of the functions $m_{1,\gamma}$, and respectively $m_{2,\gamma}$, are defined at every fixed γ of Γ from the sub-sample of the observations below γ for $m_{1,\gamma} = (1-\delta_\gamma)m_1$, and respectively from the observations above γ for $m_{2,\gamma} = \delta_\gamma m_2$, for every x of I_X

$$\widehat{m}_{1nh}(x,\gamma) = \frac{\sum_{i=1}^n K_h(x - X_i)(1-\delta_{\gamma,i})Y_i}{\sum_{i=1}^n K_h(x-X_i)(1-\delta_{\gamma,i})}, \ x \le \gamma,$$

$$\widehat{m}_{2nh}(x,\gamma) = \frac{\sum_{i=1}^n K_h(x - X_i)\delta_{\gamma,i}Y_i}{\sum_{i=1}^n K_h(x-X_i)\delta_{\gamma,i}}, \ x > \gamma.$$

The estimators are independent since the observations that define them are independent, they are consistent. Let $f_{knh}(x,\gamma)$ be the expectations of the denominator of $\widehat{m}_{knh}(x,\gamma)$, for $k = 1,2$. The expectations of $\widehat{m}_{1nh}(x,\gamma)$ and $\widehat{m}_{2nh}(x,\gamma)$ are asymptotically equivalent to

$$\begin{aligned} m_{1nh}(x,\gamma) &= \frac{\int K_h(x-t)1_{\{t\le\gamma\}}m_0(t)\,dF_{0X}(t)}{f_{1nh}(x,\gamma)},\ x\le\gamma,\\ &= m_{01}(x)1_{\{x\le\gamma\wedge\gamma_0\}} + m_{02}(x)1_{\{\gamma_0<x\le\gamma\}} + O(h^2),\\ m_{2nh}(x,\gamma) &= \frac{\int K_h(x-t)1_{\{t>\gamma\}}m_0(t)\,dF_{0X}(t)}{f_{2nh}(x,\gamma)},\ x>\gamma\\ &= m_{02}(x)1_{\{x>\gamma\vee\gamma_0\}} + m_{01}(x)1_{\{\gamma<x\le\gamma_0\}} + O(h^2), \end{aligned}$$

using a kernel with a compact support. The variances of the estimators $\widehat{m}_{knh}(x,\gamma)$ have the bias

$$b_{knh}(x,\gamma) = \frac{m_{2K}}{2nh} f_k^{-1}(x,\gamma)\{(m_k f_X)^{(2)}(x,\gamma) - m_k(x) f_X^{(2)}(x,\gamma)\} + o((nh)^{-1})$$

and the estimators are asymptotically Gaussian with the convergence rate $(nh)^{-\frac{1}{2}}$. The size of the jump of the expectation of the estimator $\widehat{m}_{nh}$ at γ_0 is $E_0\{\widehat{m}_{nh}(\gamma_0^+) - \widehat{m}_{nh}(\gamma_0^-)\}$ and it is asymptotically equivalent to the jump $m_{02}(\gamma_0^+) - m_{01}(\gamma_0)$ of m_0 at γ_0.

The variance σ_γ^2 of Y in the model (8.7) with a change-point at γ is estimated by the process

$$\widehat{\sigma}^2_{\gamma,nh} = n^{-1}\sum_{i=1}^n \{Y_i - (1-\delta_{\gamma,i})\widehat{m}_{1nh}(X_i,\gamma) + \delta_{\gamma,i}\{\widehat{m}_{2nh}(X_i,\gamma)\}^2.$$

The change-point parameter γ is estimated by minimization of the variance estimator

$$\widehat{\gamma}_{nh} = \arg\inf_{\gamma\in\Gamma} \widehat{\sigma}^2_{\gamma,nh}$$

and the estimators of the functions m_1 and m_2 are

$$\widehat{m}_{knh}(x) = \widehat{m}_{knh}(x,\widehat{\gamma}_{nh}),\ k=1,2.$$

The true values of the parameters and functions of the model under P_0 are denoted γ_0, $m_0 = (m_{01}, m_{02})^T$ and $\sigma_0^2 = \sigma_{\gamma_0}^2$ which is estimated by $\widehat{\sigma}^2_{nh} = \widehat{\sigma}^2_{\widehat{\gamma}_{nh},nh}$.

Conditions C

C1 The function K is a symmetric density of $C^2([-1;1])$, such that the integrals $m_{2K} = \int v^2K(v)dv$, $k_j = \int K^j(v)dv$ and $\int |K'(v)|^j dv$, for $j = 1,2$ are finite;

C2 The regression functions $m_{1,\gamma}$ and $m_{2,\gamma}$, m_{01} and m_{02} belong to the class of functions $C_b^2(I_X)$;

C3 The bandwidth h_n converges to zero and $h = O(n^{-\frac{1}{5}})$ as n tends to infinity.

Under Conditions C, the estimators $\widehat{m}_{1nh}$ and respectively $\widehat{m}_{2nh}$ are uniformly a.s. consistent under P_0, on the subsets $\{X \leq \gamma_0\}$ and respectively $\{X > \gamma_0\}$ of I_X, and $E_0\|\widehat{m}_{knh} - m_{knh}\|_p = O((nh)^{-\frac{1}{p}})$, for $k = 1, 2$.

Let $m = (m_1, m_2)$ belong to a space $\mathcal{M}$ of two-phases regression functions with a change-point and let

$$\sigma_n^2(m, \gamma) = n^{-1} \sum_{i=1}^{n} |Y_i - (1 - \delta_{\gamma,i}) m_{1,\gamma}(X_i) + \delta_{\gamma,i} m_{2,\gamma}(X_i)|^2$$

be the variance of the sequence $(Y_i - m(X_i))_{i=1,\dots,n}$. The process σ_n^2 is estimated by $\widehat{\sigma}_{nh}^2$, the mean square error of estimation of $(Y_i)_{i=1,\dots,n}$ with the estimators $\widehat{m}_{nh}$ and $\widehat{\gamma}_{nh}$. The difference of the estimators of the variance in the estimated model and under P_0 is

$$\begin{aligned}
\widehat{l}_{nh} &= \widehat{\sigma}_{nh}^2 - \sigma_n^2(m_0, \gamma_0) \\
&= n^{-1} \sum_{i=1}^{n} [\{Y_i - (1 - \delta_{\widehat{\gamma}_{nh},i}) \widehat{m}_{1nh}(X_i) + \delta_{\widehat{\gamma}_{nh},i} \widehat{m}_{2nh}(X_i)\}^2 \\
&\quad - \{Y_i - (1 - \delta_{\gamma_0,i}) m_{01}(X_i) + \delta_{\gamma_0,i} m_{02}(X_i)\}^2].
\end{aligned}$$

The differences of the means $\widehat{m}_{1nh,\gamma} - \widehat{m}_{1nh,\gamma_0}$ is biased if $\gamma > \gamma_0$, for $x > \gamma_0$, it is asymptotically equivalent to

$$m_{2nh,\gamma}(x) + m_{1nh}(x)\{f_{2nh,\gamma}(x) - f_{1nh}(x)\} f_{2nh,\gamma}^{-1}(x)$$

where $m_{knh,\gamma} = E_0 \widehat{m}_{knh,\gamma}$ and $f_{knh,\gamma} = E_0 \widehat{f}_{knh,\gamma}$, for $k = 1, 2$. Then $m_{1nh,\gamma} - (x) m_{1nh,\gamma_0}(x)$ converges to $b_1(x, \gamma) = m_\gamma(x) m_1(x)\{f_{2,\gamma}(x) - f_1(x)\} f_{2,\gamma}^{-1}(x)$, as n tends to infinity. In the same way, the differences of the means $\widehat{m}_{2nh,\gamma} - \widehat{m}_{2nh,\gamma_0}$ is biased if $\gamma \leq \gamma_0$, for $x \leq \gamma_0$.

Lemma 8.3. *Under Conditions* C *and* P_0, *the process* $\widehat{l}_{nh,\gamma} = l_n(\widehat{m}_{nh\gamma}, \gamma)$ *has the uniform asymptotic expansion*

$$\begin{aligned}
\widehat{l}_{nh,\gamma} &= n^{-1} \sum_{i=1}^{n} [\{(1 - \delta_{\gamma,i}) \widehat{m}_{1nh\gamma}(X_i) - (1 - \delta_{\gamma_0,i}) m_{01}(X_i)\}^2 \\
&\quad + \{\delta_{\gamma,i} \widehat{m}_{2nh\gamma}(X_i) - \delta_{\gamma_0,i} m_{02}(X_i)\}^2] + o_p((nh)^{-1}).
\end{aligned}$$

Proof. This expansion is a consequence of the expression of the conditional mean of Y under P_0 which implies that

$$E_0[Y\{\widehat{m}_{nh\gamma}(X) - m_0(X)\} \mid X] = m_0(X)E_0\{\widehat{m}_{nh\gamma}(X) - m_0(X) \mid X\}$$

and $E_0\{\widehat{l}_{nh} \mid X\} = \|\widehat{m}_{nh\gamma}(X) - m_0(X)\|_n^2$. Then

$$\begin{aligned} P_0\{|\widehat{l}_{nh,\gamma} - E_0\widehat{l}_{nh\gamma}| > \varepsilon \mid \widehat{m}_{nh}, X = x\} \\ \leq 4n^{-1}\varepsilon^{-2}E_0[Var_0(Y \mid X)\{\widehat{m}_{nh}(X) - m_0(X)\}^2], \end{aligned}$$

under the conditions, the estimator $\widehat{m}_{nh\gamma}$ is uniformly bounded on I_X hence to remainder term is uniform. □

For every (m, γ) of $\mathcal{M} \times \Gamma$, the difference of the estimators of the empirical variances of Y under $P_{m,\gamma}$ and P_0 is

$$\begin{aligned} l_n(m,\gamma) = n^{-1}\sum_{i=1}^{n}[\{Y_i - (1-\delta_{\gamma,i})m_1(X_i) + \delta_{\gamma,i}m_2(X_i)\}^2 \\ -\{Y_i - (1-\delta_{\gamma_0,i})m_{01}(X_i) + \delta_{\gamma_0,i}m_{02}(X_i)\}^2], \end{aligned}$$

it converges a.s. under P_0, as n tends infinity, to

$$\begin{aligned} l(m,\gamma) &= E_0\{m_\gamma(X) - m_0(X)\}^2 \\ &= \{(1-\delta_\gamma)m_{1\gamma}(X) - (1-\delta_{\gamma_0})m_{01}(X) \\ &\qquad + \delta_\gamma m_{2\gamma}(X) - \delta_{\gamma_0}m_{02}(X)\}^2, \end{aligned} \tag{8.8}$$

the function l is minimum at m_0 and γ_0 and the process l_n is minimum at the estimators $\widehat{m}_{nh} = (\widehat{m}_{1nh}, \widehat{m}_{2nh})$ and $\widehat{\gamma}_{nh}$ of m_0 and γ_0.

Proposition 8.4. *The estimators $\widehat{\gamma}_{nh}$ and $\widehat{\sigma}^2_{nh}$ are a.s. consistent and the estimator $\widehat{m}_{nh}$ of the regression function is uniformly $(nh)^{-\frac{1}{2}}$-consistent in probability.*

Proof. At a fixed value γ, the minimum of $l_n(m,\gamma)$ according to the two sub-samples of achieved at $\widehat{m}_{nh\gamma}$ and the process $\widehat{l}_{nh}(\gamma)$ is minimal at $\widehat{\gamma}_{nh}$. The approximation of Lemma 8.3 ensures that l_n converges in probability to the quadratic limit $E_0\{\widehat{m}_{nh\gamma} - m_0(X)\}^2$ which is minimum with the value zero at change-point γ_0 under P_0. The minimum of l_n is reached as the estimators are in probability in a neighborhood of the parameters values under P_0. □

The choice of the convergence rate of the bandwidth is based on the minimization of an estimated error for the regression function m. The integrated square error for the estimation of m by a kernel estimator with bandwidth h is a measure of the adjustment of the regression function by its estimator $ISE(h) = \int \{\widehat{m}_{nh}(x) - m(x)\}^2 \, dF_X(x)$. It is estimated by its average

$$AISE(h) = n^{-1} \sum_{i=1}^{n} \{\widehat{m}_{nh} - m\}^2(X_i)$$

and its expectation is $MISE(h) = E \int \{(\widehat{m}_{nh} - m)^2(X_i)\}$. The optimal bandwidth for these distances minimize them with respect to h, they are respectively denoted h_{ISE}, h_{AISE}, h_{MISE}. These optimal bandwidths cannot be computed since they depend on m and an empirical cross-validation criterion provides an estimator

$$CV(h) = n^{-1} \sum_{i=1}^{n} \{\widehat{m}_{nh}^2(X_i) - \widehat{m}_{nh,i}^2(X_i)\}$$

where $\widehat{m}_{nh,i}$ is the nonparametric estimator of m calculated like $\widehat{m}_{nh}$ and without the i-th observation (X_i, Y_i), $CV(h)$ is minimal at h_{CV}. For a continuous regression function and under the conditions, the bandwidths $h_{AISE} = O_p(n^{-1/5})$, h_{ISE} and h_{MISE} have the same order when the support of X is compact. The approximation rate of the bandwidth h_{CV} is $(h_{CV} - h_{ISE})h_{ISE}^{-1} = O_p(n^{-1/5})$ and for the cross-validation $(CV(h_{CV}) - ISE(h_{ISE}))ISE^{-1}(h_{ISE}) = O_p(n^{-1/5})$. With the $MISE$ criterion, the ratios defined for h_{MISE} and $MISE(h_{MISE})$ satisfy the same results as h_{CV} and $CV(h_{CV})$ (Haerdle, Hall and Marron 1988). They are similar to crux of the estimations nonparametric of densities (Hall and Marron 1987). Under the conditions $h_n = o_p(n^{-1/5})$, the bias of the estimator is neglectable and the limit in Theorem 8.5 is a centered process.

For the weak convergence of the change-point estimator, we denote $\|\varphi\|_X$ the uniform norm of a function φ on I_X

$$\rho(\theta, \theta') = (|\gamma - \gamma'| + \|m - m'\|_X^2)^{\frac{1}{2}}$$

the distance between $\theta = (m^T, \gamma)^T$ and $\theta' = (m'^T, \gamma')^T$, and $V_\varepsilon(\theta_0)$ a neighborhood of θ_0 of radius ε for ρ. The quadratic function $\widehat{l}_{nh} = l_n(\widehat{m}_{nh}, \widehat{\gamma}_{nh})$ defined by (8.3) converges to $l(m, \gamma)$ defined by (8.8). Let

$$W_n(m, \gamma) = nh\{l_n(m, \gamma) - l(m, \gamma)\},$$

we assume that $E_0\{Y - m(X)\}^4$ is finite in a neighborhood of θ_0.

Proposition 8.5. *For $\varepsilon > 0$ sufficiently small, there exists a constant κ_0 such that*

$$\inf_{\rho(\theta,\theta_0) \le \varepsilon} l(m, \gamma) \ge \kappa_0 \rho^2(\theta, \theta_0).$$

For every $\varepsilon > 0$, there exists a constant κ_1 such that for n sufficiently large

$$E_0 \sup_{\rho(\theta,\theta_0)\le\varepsilon} |W_n(\gamma)| \le \kappa_1 \rho(\theta,\theta_0).$$

The proof of these inequalities are the same as in Chapter 4, they imply

$$\inf_{|\gamma-\gamma_0|\le\varepsilon} l(\widehat{m}_{nh},\gamma) \ge \kappa_0(|\gamma-\gamma_0| + |\widehat{m}_{nh} - m_0|^2),$$

$$E_0 \sup_{|\gamma-\gamma_0|\le\varepsilon} |W_n(\widehat{m}_{nh},\gamma)| \le \kappa_1(|\gamma-\gamma_0| + |\widehat{m}_{nh} - m_0|^2)^{\frac{1}{2}}$$

and the convergence rate $(nh)^{-\frac{1}{2}}$ of the nonparametric estimator $\widehat{m}_{nh}$ enables to determine the convergence rate of the change-point estimator, using the arguments of Theorem 2.2 for the minimum of variance.

Theorem 8.4. *Under Conditions* C *and if* $E_0 \sup_{|\gamma-\gamma_0|\le\varepsilon}\{Y - m_\gamma(X)\}^4$ *is finite, for* $\varepsilon > 0$, *then*

$$\overline{\lim}_{n,A\to\infty} P_0(nh|\widehat{\gamma}_{nh} - \gamma_0| > A) = 0.$$

The asymptotic behaviour of the estimated regression function is a consequence of the convergence rates.

Theorem 8.5. *Under Conditions* **C**, *the estimator*

$$\widehat{m}_{nh}(x) = \widehat{m}_{1nh}(x)1_{\{x\le\widehat{\gamma}_{nh}\}} + \widehat{m}_{2nh}(x)1_{\{x>\widehat{\gamma}_{nh}\}}$$

of the regression function $m_0(x) = m_{01}(x)1_{\{x\le\gamma_0\}} + m_{02}(x)1_{\{x>\gamma_0\}}$ *is such that the process* $(nh)^{\frac{1}{2}}(\widehat{m}_{nh} - m_0)$ *converges weakly under* P_0 *to a centred Gaussian process*

$$G_m(x) = G_{1m}(x)1_{\{x\le\gamma_0\}} + G_{2m}(x)1_{\{x>\gamma_0\}}$$

on I_X, *where the variance of* G_{1m} *is* $V_1(x) = k_2 Var(Y|X = x \le \gamma_0)$ *and the variance of* G_{2m} *is* $V_2(x) = k_2 Var(Y|X = x > \gamma_0)$.

Proof. At any fixed γ, the estimators of the two regression functions are independent, by plugging the consistent estimator $\widehat{\gamma}_{nh}$ in their expression they are therefore asymptotically independent. Let $G_{nh} = (nh)^{\frac{1}{2}}(\widehat{m}_{nh} - m_0)$ on I_X, it develops as

$$G_{nh}(x) = (nh)^{\frac{1}{2}}(\widehat{m}_{nh}(x,\gamma_0) - m_0(x)) \\ +(nh)^{\frac{1}{2}}(\widehat{m}_{nh}(x,\widehat{\gamma}_{nh}) - \widehat{m}_{nh}(x,\gamma_0))$$

which is also denoted $G_{1nh}(x) + G_{2nh}(x)$. The first term is a sum of two processes on $]-\infty,\gamma_0]$ and $]\gamma_0,+\infty[$ respectively, they converge weakly to independent centred Gaussian processes with covariances zero and variances

$$C_{1m}(x) = k_2 1_{\{x\le\gamma_0\}} Var(Y|X, X \le \gamma_0),$$

$$C_{2m}(x) = k_2 1_{\{x>\gamma_0\}} Var(Y|X, X > \gamma_0).$$

The second term splits according to several sub-samples. Let

$$\eta_{ni} = 1_{\{\gamma_0 < X_i \leq \widehat{\gamma}_{nh}\}} - 1_{\{\widehat{\gamma}_{nh} < X_i \leq \gamma_0\}},$$

for every x of I_X

$$\begin{aligned} G_{2nh}(x) = (nh)^{\frac{1}{2}} \{ \widehat{m}_{1nh}(x, \widehat{\gamma}_{nh}) 1_{\{x \leq \widehat{\gamma}_{nh}\}} - \widehat{m}_{1nh}(x, \gamma_0) 1_{\{x \leq \gamma_0\}} \} \\ + (nh)^{\frac{1}{2}} \{ \widehat{m}_{2nh}(x, \widehat{\gamma}_{nh}) 1_{\{x > \widehat{\gamma}_{nh}\}} - \widehat{m}_{2nh}(x, \gamma_0) 1_{\{x > \gamma_0\}} \}, \end{aligned}$$

where the difference of the estimators for the function m_1 is

$$\begin{aligned} \widetilde{G}_{1nh}(x) = (nh)^{\frac{1}{2}} \frac{\sum_{i=1}^n K_h(x - X_i) \eta_{ni} Y_i}{\sum_{i=1}^n K_h(x - X_i)(1 - \delta_{\widehat{\gamma}_{nh} i})} \\ + \widehat{m}_{1nh}(x, \gamma_0) (nh)^{\frac{1}{2}} \frac{\sum_{i=1}^n K_h(x - X_i) \eta_{ni}}{\sum_{i=1}^n K_h(x - X_i)(1 - \delta_{\widehat{\gamma}_{nh} i})}. \end{aligned}$$

The almost sure convergence to zero of $|\widehat{\gamma}_{nh} - \gamma_0|$ implies that the empirical mean of the indicators η_{ni}, $\nu_{nh} = n^{-1} \sum_i \eta_{ni}$ is such that $(nh)^{\frac{1}{2}} \nu_{nh}$ converges a.s. to zero and $\widetilde{G}_{1nh}$ converges a.s. to zero uniformly on I_X, as n tends to infinity, under the conditions for the bandwidth.

In the same way, the difference of the estimators for the function m_2, in the expression of G_{2nh}, is

$$\begin{aligned} \widetilde{G}_{2nh}(x) = -(nh)^{\frac{1}{2}} \frac{\sum_{i=1}^n K_h(x - X_i) \eta_{ni} Y_i}{\sum_{i=1}^n K_h(x - X_i)(1 - \delta_{\widehat{\gamma}_{nh} i})} \\ - \widehat{m}_{2nh}(x, \gamma_0) (nh)^{\frac{1}{2}} \frac{\sum_{i=1}^n K_h(x - X_i) \eta_{ni}}{\sum_{i=1}^n K_h(x - X_i)(1 - \delta_{\widehat{\gamma}_{nh} i})} \end{aligned}$$

and $\widetilde{G}_{2nh}$ converges a.s. to zero uniformly on I_X. □

For the weak convergence of $nh(\widehat{\gamma}_{nh} - \gamma_0)$, we denote

$$\begin{aligned} \mathcal{U}_{nh} = \{ u = (u_m^T, u_\gamma)^T : u_m = (nh)^{\frac{1}{2}} (m - m_0), \\ u_\gamma = nh(\gamma - \gamma_0), m \in \mathcal{M}, \gamma \in \Gamma \} \end{aligned} \tag{8.9}$$

and for every $A > 0$, let

$$\mathcal{U}_{nh}^A = \{ u \in \mathcal{U}_n; \|u\|^2 \leq A \}. \tag{8.10}$$

For $u = (u_m^T, u_\gamma)^T$ belonging to $\mathcal{U}_{nh}^A$ and $\theta_{nh,u} = (m_{nh,u}^T, \gamma_{nh,u})^T$ with

$$\gamma_{nh,u} = \gamma_0 + (nh)^{-1} u_\gamma, \quad m_{k,nh,u} = m_{0k} + (nh)^{-\frac{1}{2}} u_{mk}, \ k = 1, 2,$$

and $m_{nh,u,\gamma_{nh,u}} = (1-\delta_{\gamma_{nh,u}})m_{1,nh,u} + \delta_{\gamma_{nh,u}} m_{2,nh,u}$. The process nhl_n is the sum of the processes

$$\widetilde{l}_n(u_m) = h\sum_{i=1}^{n}[\{Y_i - m_{nh,u,\gamma_0}(X_i)\}^2 - \{Y_i - m_0(X_i)\}^2] \tag{8.11}$$

$$Q_n(u) = h\sum_{i=1}^{n}\{Y_i - m_{nh,u,\gamma_{nh,u}}(X_i)\}^2 - \{Y_i - m_{nh,u,\gamma_0}(X_i)\}^2]$$

where the process $\widetilde{l}_n$ does not depend on u_γ. The expectation of $Q_n(u)$ converges to

$$\mu_Q(u) = f_X(\gamma_0)[u_{\gamma_0}\{m_{01}(\gamma_0) - m_{02}(\gamma_0)\}]^2$$

and its variance function has a finite limit v_Q, as n tends to infinity. Theorem 4.3 extends straightforwardly to the model with nonparametric regression functions.

Theorem 8.6. *Under the conditions of Theorem 8.4, for A sufficiently large and under P_0, the process Q_n converges weakly in $\mathcal{U}_n^A$ to a process Q with finite mean μ_Q and variance v_Q, as n and A tend to infinity.*

The process $Q_n(u_{\widehat{m}_n}, u_\gamma)$ reaches it minimum at $u_{\widehat{\gamma}_n} = nh(\widehat{\gamma}_{nh} - \gamma_0)$ which is bounded in probability, by Theorem 8.4. The asymptotic distribution of the estimator $\widehat{\gamma}_{nh}$ is deduced from the limit of the process Q_n, like in Theorem 4.4 for the mean square estimator of linear regressions.

Theorem 8.7. *Under the previous conditions, the variable $nh(\widehat{\gamma}_{nh} - \gamma_0)$ converges weakly to a stopping time u_0 where the Gaussian process Q achieves its minimum and u_0 is bounded in probability.*

The sum of the jumps of nhl_n in $\mathcal{U}_n^A$ defined by (8.11) is approximated by the process

$$\widetilde{Q}_n(u_\gamma) = h\sum_{i=1}^{n}\{(1_{\{\gamma_0 < X_i \le \gamma_{n,u}\}} - 1_{\{\gamma_{n,u} < X_i \le \gamma_0\}})\}\{m_{01}^2(X_i) - m_{02}^2(X_i)\}, \tag{8.12}$$

they have the same asymptotic mean and variance under P_0 and they converge weakly to the same uncentered Gaussian process, as sums of independent variables and by the tightness criterion (1.2).

By Lemma 8.3, $\widetilde{l}_n$ has the uniform asymptotic expansion

$$\widetilde{l}_n(u) = h\sum_{i=1}^{n}[\{(1-\delta_{\gamma_0,i})(m_{1nh,u} - m_{01})(X_i)\}^2 + \{\delta_{\gamma_0,i}(m_{2nh,u} - m_{02})(X_i)\}^2] + o_p(1).$$

Let $\widetilde{W}_n = \widetilde{l}_n - E_0\widetilde{l}_n$.

Theorem 8.8. *Under the conditions of Theorem 8.4, for every A sufficiently large and under P_0, the process $W_n(\theta_{n,u})$ develops as*

$$W_n(\theta_{n,u}) = \widetilde{Q}_n(u_\gamma) + \widetilde{W}_n(u_m) + o_p(1),$$

where the $o_p(1)$ is uniform on $\mathcal{U}_n^A$, as n tends to infinity.

Proof. On $\mathcal{U}_n^A$, the remainder term of the approximation of the process nhl_n by $\widetilde{l}_n$ is

$$h\sum_{i=1}^{n}[\{(\delta_{i,\gamma_{n,u}} - \delta_{i,\gamma_0})m_{1n,u}(X_i)\}^2 + \{(\delta_{i,\gamma_0} - \delta_{i,\gamma_{n,u}})m_{2n,u}(X_i)\}^2]$$

and it is uniformly approximated by $\widetilde{Q}_n(u_\gamma)+o_p(1)$ due to the convergence rates of Theorem 8.4. □

With the estimator of the regression function, the process $\widetilde{W}_n(u_{\widehat{m}_{nh}})$ is asymptotically equivalent to

$$n^{-1}\sum_{i=1}^{n}[\{(1-\delta_{\gamma_0,i})G_{1nh}(X_i)\}^2 + \{\delta_{\gamma_0,i}(nh)^{\frac{1}{2}}G_{2nh}(X_i)\}^2]$$

where the processes $G_{1nh}(x) = (nh)^{\frac{1}{2}}\{\widehat{m}_{1nh}(x,\gamma_0) - m_{01}(x)\}$ and, respectively $G_{2nh}(x) = (nh)^{\frac{1}{2}}\{\widehat{m}_{2nh}(x,\gamma_0) - m_{02}(x)\}$, converge to Gaussian distributions G_1 and, respectively G_2, defined by Theorem 8.5.

8.6 Test of models without changes

We assume that the support of observations of the variable X is a bounded compact I_X. The mean square test for the hypothesis H_0 of a model without change-point against the alternative of model 8.7, relies on the statistic

$$T_n = \inf_{\gamma\in I_X} nh\{\widehat{\sigma}_n^2(\gamma) - \widehat{\sigma}_{0n}^2\}$$

where $\widehat{\sigma}_{0n}^2 = \|Y - \widehat{m}_{0,nh}(X)\|_n^2$ is the estimator of the variance in the model with the regression function m_0 of the sample under H_0. Let $\mathcal{M}_0$ be the class of regression functions of $C^2(I_X)$ without change-point. Under H_0, m_{γ_0} is a function of $\mathcal{M}_0$, with γ_0 is at the end point of I_X, and $\inf_{m\in\mathcal{M}_0}\sigma_n^2(m) - \sigma_n^2(m_0) = 0$ and $\inf_{\rho(\theta,\theta_0)\leq\varepsilon} l(m,\gamma) \geq \kappa_0\rho^2(\theta,\theta_0)$.

Let $\widehat{m}_{0nh}$ be the nonparametric estimator of the function m_0 and let $G_{0nh} = (nh)^{\frac{1}{2}}(\widehat{m}_{0nh} - m_0)$, it converges weakly to a Gaussian process G_{0m}

with a finite variance under P_0 and Conditions C. For γ in a neighborhood of γ_0, let $G_{nh,\gamma} = (nh)^{\frac{1}{2}}(\widehat{m}_{nh,\gamma} - m_{0,\gamma})$.

Proposition 8.6. *Under H_0, Conditions C and $E_0\{Y - m(X)\}^4$ finite, the statistic T_n converges in probability under P_0 to $T_0 = k_2 u_0 Var(Y|X = \gamma_0)$.*

Proof. Under H_0, the inequalities of Proposition 8.5 are satisfied on a compact interval I_X and the convergence rate of $\widehat{\gamma}_{nh}$ to γ_0 is still $(nh)^{-1}$. The process $G_{nh,\gamma} - G_{0nh}$ is the difference of $G_{nh,\gamma} - G_{0nh} = G_{2nh,\gamma} - G_{1nh,\gamma}$ of the processes

$$G_{1nh,\gamma} = (nh)^{\frac{1}{2}}(\widehat{m}_{1nh,\gamma} - m_{01,\gamma} - \widehat{m}_{0nh} + m_0)1_{]\widehat{\gamma}_{nh},\gamma_0]}$$
$$G_{2nh,\gamma} = (nh)^{\frac{1}{2}}(\widehat{m}_{2nh,\gamma} - m_{02,\gamma})1_{]\widehat{\gamma}_{nh},\gamma_0]}.$$

Under H_0, $(nh)^{\frac{1}{2}}(m_{01,\gamma} - m_0)$ and the process $G_{1nh,\widehat{\gamma}_{nh}}$ converges in probability to zero due to the convergence rate of $\widehat{\gamma}_{nh}$. The statistic T_n is the value at $\widehat{\gamma}_{nh}$ of the process

$$nh(\widehat{l}_{nh\gamma} - \widehat{l}_{0nh}) = h\sum_{i=1}^{n}[\{\widehat{m}_{0nh}(X_i) - \widehat{m}_{nh\gamma}(X_i)\}^2 + \{\widehat{m}_{0nh}(X_i) - \widehat{m}_{nh\gamma}(X_i)\}\{Y_i - \widehat{m}_{0nh}(X_i)\}].$$

At $\widehat{\gamma}_{nh}$, the first term is asymptotically equivalent to

$$\|G_{nh,\widehat{\gamma}_{nh}} - G_{0nh}\|_n = \|G_{2nh,\widehat{\gamma}_{nh}}\|_n + o_p(1)$$

where the expectation of $G_{2nh,\widehat{\gamma}_{nh}}(x)$ converges to zero and its variance is asymptotically equivalent to $V_0 = k_2 u_0 Var(Y|X = \gamma_0)$, therefore $\|G_{2nh,\widehat{\gamma}_{nh}}\|_n$ converges in probability to V_0.

The expectation of $Y_i - \widehat{m}_{0nh}(X_i)$ conditionally on X_i is $(m_0 - \widehat{m}_{0nh})(X_i)$. Let $\gamma_{nh,u} = \gamma_0 - (nh)^{-1}u_{nh}$ and let $Z_i = nh\{\widehat{m}_{0nh}(X_i) - \widehat{m}_{nh\gamma_{nh,u}}(X_i)\}\{Y_i - \widehat{m}_{0nh}(X_i)\}$, the expectation of Z_i conditionally on X_i is therefore asymptotically equivalent to

$$nhE_0\{(m_0 - \widehat{m}_{0nh})(X_i)(m_{0nh} - m_{2nh\gamma_{nh,u}})(X_i)1_{\{X_i > \gamma_{nh,u}\}} \mid X_i\} + o_p(1)$$

and it converges to zero. The variance of $Y_i - \widehat{m}_{0nh}(X_i)$ conditionally on X_i converges to $\sigma_0^2(X_i)$ and the variance of Z_i conditionally on X_i is asymptotically equivalent to

$$nh^2\sigma_0^2(X_i)E_0\{(\widehat{m}_{0nh} - \widehat{m}_{nh\gamma_{nh,u}})^2(X_i) \mid X_i\}$$
$$+nh^2E_0[\{(\widehat{m}_{0nh} - m_0)^2(X_i)\}\{(\widehat{m}_{0nh} - \widehat{m}_{nh\gamma_{nh,u}})^2(X_i)\} \mid X_i]$$

the second expectation is bounded by the product of the conditional moments $[E_0\{(\widehat{m}_{0nh} - m_0)^4(X_i) \mid X_i\}]^{\frac{1}{2}}$ and $[E_0\{(\widehat{m}_{0nh} - \widehat{m}_{nh\gamma_{nh,u}})^4(X_i)\}]^{\frac{1}{2}}$, it is a $O_p((nh)^{-1})$. It follows that T_n converges in probability to V_0. □

Under an alternative with a change-point at γ_0 and a regression function $m_\gamma = (1-\delta_0)m_1 + \delta_0 m_2$ with m_1 or m_2 distinct from m_0, one of the processes $G_{knh} = (nh)^{\frac{1}{2}}\{(\widehat{m}_{knh} - m_1) + (nh)^{\frac{1}{2}}(m_k - m_0)$, $k = 1, 2$, diverges and the test statistic T_n diverges. Under an alternative with a change-point at a fixed point γ distinct from γ_0 and a regression function $m_\gamma = (1-\delta_\gamma)m_1 + \delta_\gamma m_2$, with distinct functions m_1 and m_2, the test statistic T_n diverges in probability.

Under local alternatives K_n defined by sequences $(m_n)_n$ converging to m_0 with the rate $(nh)^{-\frac{1}{2}}$ and $(\gamma_n)_n$ converging to γ_0 with the rate $(nh)^{-1}$, let $\gamma_{nh,u} = \gamma_0 - (nh)^{-1}u_n$, where u_n converges to u in $\mathcal{U}^A_{nh}$, and let

$$m_{1nh,\gamma_{nh,u}} = m_0 + (nh)^{-\frac{1}{2}} g_{1n},\ y \le \gamma_n,$$
$$m_{2nh,\gamma_{nh,u}} = m_0 + (nh)^{-\frac{1}{2}} g_{2n},\ y > \gamma_n,$$

where the functions g_{kn} converges uniformly to g_k in I_X, for $k = 1, 2$. We assume that the conditions of Section 8.5 are satisfied.

Proposition 8.7. *Under the local alternatives K_n, the statistic T_n converges weakly to $T_0 + \int g_1^2(x)\, dF_{0X}(x)$.*

Proof. Under K_n, the process $(nh)^{\frac{1}{2}}(\widehat{m}_{1nh} - \widehat{m}_{0nh})$ is the sum $(nh)^{\frac{1}{2}}(\widehat{m}_{1nh} - m_{1nh}) + g_{1n} - (nh)^{\frac{1}{2}}(\widehat{m}_{0nh} - m_0)$ where $(nh)^{\frac{1}{2}}(\widehat{m}_{1nh} - m_{1nh})$ and $(nh)^{\frac{1}{2}}(\widehat{m}_{0nh} - m_0)$ converge weakly to the same distribution and the integral $nh\mathcal{I}_{1nh}(\widehat{\gamma}_{nh})$ converges in probability to $\frac{1}{2}\int g_1^2(y)\, dy$. The process $\widehat{l}_{1n}(u)$ converges to $E_0 g_1^2(X)$ under H_0. The variable $\widehat{\gamma}_{nh}$ converges weakly to γ_0 and the integral $nh\mathcal{I}_{2nh}(\widehat{\gamma}_{nh})$ converges to T_0. □

8.7 Maximum likelihood for nonparametric regressions

Let (X, Y) be a variable defined on a probability space (Ω, A, P_0), with values in $(\mathbb{R}^2, \mathcal{B})$ and satisfying the model (8.7) with distinct functions m_{01} and m_{02} and a change-point at γ_0, under P_0. The density of Y conditionally on X under a probability measure $P_{m,\gamma}$ for the model (8.7) is written as

$$f(Y - m_\gamma(X)) = (1-\delta_\gamma) f(Y - m_1(X)) + \delta_\gamma f(Y - m_2(X)).$$

We assume that the density f belongs to $C^2(\mathbb{R})$ and the regression functions m_{01} and m_{01} belong to $C^2(\mathcal{I}_X)$. The log-likelihood of the sample under the probability $P_{m,\gamma}$ is denoted $l_n(m,\gamma) = \sum_{i=1}^n \log f(Y_i - m_\gamma(X_i))$.

The nonparametric estimators of the regression functions $\widehat{m}_{k,nh,\gamma}$, for $k = 1, 2$, define the estimators of the density

$$f(Y - \widehat{m}_{nh,\gamma}(X)) = (1-\delta_\gamma) f(Y - \widehat{m}_{1,nh,\gamma}(X)) + \delta_\gamma f(Y - \widehat{m}_{2,nh,\gamma}(X)),$$

the maximum likelihood estimator $\widehat{\gamma}_n$ of γ_0 maximizes the estimated log-likelihood of the sample

$$\widehat{l}_{nh}(\gamma) = \sum_{i=1}^{n} \log f(Y_i - \widehat{m}_{nh,\gamma}(X_i))$$

and $\widehat{m}_{nh} = \widehat{m}_{nh,\widehat{\gamma}_n}$. The process $\widehat{l}_{nh}$ splits as

$$\widehat{l}_{nh}(\gamma) = \sum_{i=1}^{n} \log f(Y_i - \widehat{m}_{1,nh,\gamma}(X_i)) + \sum_{i=1}^{n} \delta_{\gamma,i} \log \frac{f(Y_i - \widehat{m}_{2,nh,\gamma}(X_i))}{f(Y_i - \widehat{m}_{1,nh,\gamma}(X_i))}$$

and the log-likelihood of the sample under $P_{m,\gamma}$ and P_0 defines the process $X_n(m,\gamma) = n^{-1}\{l_n(m,\gamma) - l_n(m_0,\gamma_0)\}$, it is the sum

$$\begin{aligned}X_n(m,\gamma) &= n^{-1}\sum_{i=1}^{n}(1-\delta_{0,i}) \log \frac{f(Y_i - m_1(X_i))}{f(Y_i - m_{01}(X_i))} \\ &\quad + n^{-1}\sum_{i=1}^{n} \delta_{0,i} \log \frac{f(Y_i - m_2(X_i))}{f(Y_i - m_{02}(X_i))} \\ &\quad + n^{-1}\sum_{i=1}^{n}(\delta_{\gamma,i} - \delta_{0,i}) \log \frac{f(Y_i - m_2(X_i))}{f(Y_i - m_1(X_i))}.\end{aligned}$$

The parameter $\theta = (m,\gamma)$ has the value θ_0 under P_0 and the norm on the parameter space is

$$\rho(\theta,\theta_0) = (|\gamma - \gamma_0| + \|m_1 - m_{01}\|_{I_X}^2 \|m_2 - m_{02}\|_{I_X}^2)^{\frac{1}{2}}.$$

The process X_n converges a.s. uniformly under P_0 to its expectation $X = E_0 X_n$, minimum at θ_0, it follows that the maximum likelihood estimator $\widehat{\gamma}_n$ of the change-point γ_0 is a.s. consistent.

The arguments of Section 4.5 imply that for $\varepsilon > 0$ sufficiently small, there exists a constant κ_0 and such that for every θ in $V_\varepsilon(\theta_0)$

$$X(\theta) \geq -\kappa_0 \rho^2(\theta,\theta_0). \tag{8.13}$$

Let $W_n(\theta) = n^{\frac{1}{2}}\{X_n(\theta) - X(\theta)\}$, its variance is finite in an ε-neighborhood $V_\varepsilon(\theta_0)$ of θ_0, under the next condition.

Condition C2:

$$E_0\Big\{\sup_{\theta \in V_\varepsilon(\theta_0)} \log^2 f(Y - m_\gamma(X))\Big\} \tag{8.14}$$

is finite.

Lemma 8.4. *Under the conditions* C *and* C2, *for* $\varepsilon > 0$ *sufficiently small, there exists a constant* κ_1 *and such that as* n *tends to infinity*

$$E_0 \sup_{\theta \in V_\varepsilon(\theta_0)} |W_n(\theta)| \leq \kappa_1 \varepsilon. \tag{8.15}$$

Proof. The process X_n is the sum $X_{1n} + X_{2n}$ with

$$X_{1n}(\theta) = n^{-1} \sum_{i=1}^n (1 - \delta_{0,i}) \log \frac{f(Y_i - m_1(X_i))}{f(Y_i - m_{01}(X_i))} + n^{-1} \sum_{i=1}^n \delta_{0,i} \log \frac{f(Y_i - m_2(X_i))}{f(Y_i - m_{02}(X_i))}$$

$$X_{2n}(\theta) = n^{-1} \sum_{i=1}^n (\delta_{\gamma,i} - \delta_{0,i}) \log \frac{f(Y_i - m_2(X_i))}{f(Y_i - m_1(X_i))}$$

and W_n is the sum of the processes $W_{kn}(\theta) = n^{\frac{1}{2}}\{X_{kn}(\theta) - E_0 X_{kn}(\theta)\}$, for $k = 1, 2$. As $\delta_{\gamma,i} - \delta_{0,i} = O_p(|\gamma - \gamma_0|)$, $E_0\{\sup_{\theta \in V_\varepsilon(\theta_0)} W_{2n}^2(\theta)\} = O_p(\varepsilon^2)$. For the first term, a second order expansion of $\log f(Y_i - m_k) - \log f(Y_i - m_{0k})$ as $\|m_k - m_{0k}\|_{I_X} \leq \varepsilon$ implies that for every x in I_X

$$\begin{aligned}
\log \frac{f(Y_i - m_k)}{f(Y_i - m_{0k})}(x) &= \frac{f(Y_i - m_k) - f(Y_i - m_{0k})}{f(Y_i - m_{0k})}(x) \\
&\quad - \frac{\{f(Y_i - m_k) - f(Y_i - m_{0k})\}^2}{2f^2(Y_i - m_{0k})}(x) \\
&\quad + o_p(\|m_k - m_{0k}\|_{I_X}^2) \\
&= -\|m_k - m_{0k}\|_{I_X} \frac{f'(Y_i - m_{0k})}{f(Y_i - m_{0k})}(x) \\
&\quad + o_p(\|m_k - m_{0k}\|_{I_X}),
\end{aligned}$$

the first term of this expansion has the conditional expectation

$$\int \{f(y - m_k(X)) - f(Y - m_{0k}(X))\}\, dF_{Y|X}(y) = O_p(\|m_k - m_{0k}\|_{I_X})$$

and $E_0\{\sup_{\theta \in V_\varepsilon(\theta_0)} W_{1n}^2(\theta)\} = O_p(\varepsilon^2)$, the result is a consequence of the Cauchy–Schwarz inequality. □

Theorem 8.9. *Under Conditions* C *and* C2, *for* $\varepsilon > 0$ *sufficiently small*

$$\overline{\lim}_{n,A\to\infty} P_0(nh|\widehat{\gamma}_{nh} - \gamma_0| > A) = 0.$$

This is a consequence of the inequalities (8.13) and (8.15), of the weak convergence of the processes $(nh)^{\frac{1}{2}}(\widehat{m}_{k,nh} - m_{0k})$ under the conditions C.

For $A > 0$, let $u = (u_m^T, u_\gamma)^T$ belonging to $\mathcal{U}_{nh}^A$ defined by (8.10) and let $\theta_{nh,u} = (m_{nh,u}^T, \gamma_{nh,u})^T$ with $\gamma_{nh,u} = \gamma_0 + (nh)^{-1} u_\gamma$ and $m_{k,nh,u} = m_{0k} + (nh)^{-\frac{1}{2}} u_{mk}$, for $k = 1, 2$. On $\mathcal{U}_{nh}^A$, let

$$\widetilde{X}_{2n}(u_\gamma) = h \sum_{i=1}^n (\delta_{\gamma_{nh,u},i} - \delta_{0,i}) \log \frac{f(Y_i - m_{02}(X_i))}{f(Y_i - m_{01}(X_i))}, \tag{8.16}$$

the process X_{1n} is also defined on $\mathcal{U}_{nh}^A$ by the map $\theta \mapsto \theta_{nh,u}$.

Theorem 8.10. *For A sufficiently large and under P_0, the process nhl_n has the uniform asymptotic expansion on $\mathcal{U}_n^A$*

$$nhX_n(\theta_{n,u}) = X_{1n}(u_m) + \widetilde{X}_{2n}(u_\gamma) + o_p(1).$$

Proof. On $\mathcal{U}_n^A$, the remainder term of the approximation of the process nhX_n by $X_{1n} + \widetilde{X}_{2n}$ is

$$h\sum_{i=1}^n (\delta_{\gamma_{nh,u},i} - \delta_{\gamma_0}, i)\Big\{\log \frac{f(Y_i - m_{nh,2}(X_i))}{f(Y_i - m_{02}(X_i))} - \log \frac{f(Y_i - m_{nh,1}(X_i))}{f(Y_i - m_{01}(X_i))}\Big\},$$

and it converges uniformly to zero in probability due to the convergence rates of Theorem 8.9. □

Theorem 8.11. *Under Conditions* C *and* C2, *the process $Q_n = nhX_{2n}$ converges weakly in $\mathcal{U}_n^A$, under P_0, to an uncentered Gaussian process Q with a finite variance function. The variable $nh(\widehat{\gamma}_{nh} - \gamma_0)$ converges weakly to a stopping time u_0 where the Gaussian process Q achieves its maximum and u_0 is bounded in probability.*

Proof. In the approximation of Theorem 8.10, the process X_{1n} does not depend on the parameter γ and the estimator $\widehat{u}_{nh,\gamma}$ of u_γ maximizes the process $\widetilde{X}_{2n}$. Its expectation is

$$nhE_0\Big\{(\delta_{\gamma_{nh,u}} - \delta_0)\log \frac{f(Y - m_{02}(X))}{f(Y - m_{01}(X))}\Big\} = O(u_\gamma)$$

and its variance has the same order $O(|u_\gamma|)$, it satisfies Billingsley's criterion (15.21) for tightness of processes of $D(I)$ defined on small intervals and it converges weakly on $\mathcal{U}_{nh}^A$ to an uncentered Gaussian process as a sum of independent variables. The weak convergence of the variable $nh(\widehat{\gamma}_{nh} - \gamma_0)$ follows and it is bounded in probability, by Theorem 8.9. □

8.8 Likelihood ratio test

The likelihood ratio test for the hypothesis H_0 of a model without change-point against the alternative of model 8.7, relies on the statistic

$$T_n = \sup_{\gamma\in\Gamma} nh\{X_n(\widehat{m}_{nh,\gamma}, \gamma) - \widehat{X}_{0n}\}$$

where X_{0n} is the process X_n under H_0, with γ_0 the end-point of I_X, and

$$X_n(m,\gamma) = n^{-1}\sum_{i=1}^n \log\frac{f(Y_i - m_1(X_i))}{f(Y_i - m_0(X_i))} + n^{-1}\sum_{i=1}^n \delta_{\gamma,i}\log\frac{f(Y_i - m_2(X_i))}{f(Y_i - m_1(X_i))},$$

$$X_{0n}(m) = n^{-1}\sum_{i=1}^n \log\frac{f(Y_i - m(X_i))}{f(Y_i - m_0(X_i))},$$

and $\widehat{X}_{0n} = X_{0n}(\widehat{m}_{0nh})$, estimated with the kernel estimator $\widehat{m}_{0nh}$, under H_0, of an unknown regression function m belonging to $C^2(I_X)$ and m_0 is the true regression function under H_0. In parametric regression models, the convergence rate of the estimators of the parameters may be modified under H_0 with the convergence rate of estimator $\widehat{\gamma}_n$ of γ_0 at the end-point of the support of the observations. With nonparametric regressions, a change in the regression model is a change in the mean of the density of the variable Y but the convergence rate of the kernel estimators of the regression functions is $(nh)^{\frac{1}{2}}$. Let u_0 be the variable defined by Theorem 8.11.

Proposition 8.8. *Under the hypothesis H_0, Conditions C and C2, if the interval $\mathcal{I}_X$ is bounded, then the test statistic T_n converges in probability to $T_0 = -u_0 \int Var_0\widehat{m}_{0nh}(x)f'^2(y - m_0(x))f^{-1}(y - m_0(x))\, F_{XY}(dx, dy)$.*

Proof. Under H_0, the test statistic $T_n = nhX_n(\widehat{m}_{nh}, \widehat{\gamma}_{nh}, \widehat{m}_{0,nh})$ is the sum of nhX_{1n} and nhX_{2n} which reduce to

$$nhX_{1n}(\theta, m_0) = h\sum_{i=1}^n (1-\delta_{\gamma,i})\log\frac{f(Y_i - m_1(X_i))}{f(Y_i - m_0(X_i))}$$

$$nhX_{2n}(\theta, m_0) = h\sum_{i=1}^n \delta_{\gamma,i}\log\frac{f(Y_i - m_2(X_i))}{f(Y_i - m_0(X_i))}.$$

The process X_n converges a.s. uniformly under H_0 to its expectation

$$X(\theta) = E_0\left\{1_{\{X\le\gamma\}}\log\frac{f(Y - m_1(X))}{f(Y - m_0(X))}\right\} + E_0\left\{1_{\{X>\gamma\}}\log\frac{f(Y - m_2(X))}{f(Y - m_0(X))}\right\},$$

with distinct functions m_1 and m_2, the function X is maximum for every function m_2 as $m_1 = m_0$ and $\gamma = \gamma_0$. The a.s. convergence of the process X_n implies the a.s. consistency of the estimators under H_0. Under H_0, the function X satisfies the inequality (8.13) and the process W_n has the uniform bound (8.15), then by the same arguments as in Section 8.7, the convergence rate of $\widehat{\gamma}_{nh}$ is nh.

By a second order expansion of $\log f(Y_i - m_{1nh})(x) - \log f(Y_i - m_{0nh})(X_i)$ as n tends to infinity, it is asymptotically equivalent to

$$-(m_{1nh} - m_{0nh})(X_i)\frac{f'(Y_i - m_0(X_i))}{f(Y_i - m_0(X_i))}(X_i)$$
$$+(m_{1nh} - m_{0nh})^2(X_i)\frac{ff''(Y_i - m_0(X_i)) - f'^2(Y_i - m_0(X_i))}{f^2(Y_i - m_0(X_i))}.$$

Let $m_{0nh} = m_0 + (nh)^{-\frac{1}{2}}u_{0n}$, and respectively $m_{1nh} = m_0 + (nh)^{-\frac{1}{2}}u_{1n}$, such that u_{0n} converges to a limit u_0, and respectively u_{1n} converges to u_1. At $\gamma_{nh,u} = \gamma_0 + (nh)^{-\frac{1}{2}}v_n$ with a sequence $(v_n)_n$ converging to a non-null limit v, the variable $(nh)^{\frac{1}{2}}\int 1_{\{x\leq\widehat{\gamma}_{nh}\}}f'(y - m_0(x))\,dy$ converges to zero for every x and

$$h\sum_{i=1}^n 1_{\{X_i\leq\gamma\gamma_{nh,u}\}}\{\log f(Y_i - m_{1nh,u_1}(X_i)) - \log f(Y_i - m_{0nh,u_0}(X_i))\}$$
$$= -h\sum_{i=1}^n (u_1 - u_0)^2(X_i)1_{\{X_i\leq\gamma\gamma_{nh,u}\}}\frac{f'^2(Y_i - m_0(X_i))}{f(Y_i - m_0(X_i))} + o_p(1).$$

The function $I_0 = f'^2(y - m_0(x))f^{-1}(y - m_0(x))$ is strictly positive and $nhX_{1n}(\theta_{nh,u}, m_0)$ is approximated by

$$-\int (u_1 - u_0)^2(x)1_{\{x\leq\gamma_{nh,u}\}}I_0(x,y)\,dF_{Y|X}(y,x)\,dx + o_p(1).$$

Under H_0, the processes $G_{1n} = (nh)^{\frac{1}{2}}(\widehat{m}_{1nh} - m_0)$ and $G_{0n} = (nh)^{\frac{1}{2}}(\widehat{m}_{0nh} - m_0)$ converge weakly to the same Gaussian process G_0 and $G_{0n} - G_{1n}$ converges in probability to zero. The variable $nhX_{1n}(\widehat{\theta}_{nh}, \widehat{m}_{0nh})$ is therefore asymptotically equivalent to

$$-\int 1_{\{x\leq\gamma\}}(G_{0n} - G_{1n})^2(x)I_0(x,y)\,dF_{Y|X}(y,x)\,dx$$

and it converges in probability to zero.

Under H_0, the estimators $\widehat{m}_{2nh,\gamma}(x)$ and $\widehat{m}_{0nh}(x)1_{\{x>\gamma\}}$ converge a.s. to the same limit $m_0(x)1_{\{x>\gamma\}}$ and, by the convergence rate of $\widehat{\gamma}_{nh}$, the expectation of the variable $Z_i = (nh)^{\frac{1}{2}}\{\widehat{m}_{0nh}(X_i) - \widehat{m}_{2nh}(X_i)\}1_{\{X_i>\widehat{\gamma}_{nh}\}}$ converges to zero, its variance converges to a $u_0 Var_0\widehat{m}_{0nh}(X_i)$ which is strictly positive, it converges to a centered Gaussian variable. A second order expansion of the process $nh\widehat{X}_{2n} = nhX_{2n}(\widehat{\theta}_{nh}, \widehat{m}_{0nh})$ has the form $n^{-1}\sum_{i=1}^n \xi_{inh} + o_p(1)$ with

$$\xi_{inh} = -nh\{\widehat{m}_{0nh}(X_i) - \widehat{m}_{2nh}(X_i)\}^2\frac{f'^2}{f}(Y_i - m_0(X_i))1_{\{X_i>\widehat{\gamma}_{nh}\}}$$

and it converges in probability to a T_0. □

Under an alternative with a change-point at γ_0 and a regression function

$$m_\gamma = (1-\delta_0)m_1 + \delta_0 m_2,$$

with m_1 or m_2 distinct from m_0, the process $G_{nh} = (nh)^{\frac{1}{2}}\{(1-\delta_0)(\widehat{m}_{1nh} - m_1) + \delta_0(\widehat{m}_{2nh} - m_2) + (nh)^{\frac{1}{2}}\{(1-\delta_0)(m_1 - m_0) + \delta_0(m_2 - m_0)$ is an uncentered asymptotically Gaussian process with a mean that diverges and the process Q_n is zero. It follows that the test statistic diverges. Under an alternative with a change-point at γ distinct from γ_0 and a regression function

$$m_\gamma = (1-\delta_\gamma)m_1 + \delta_\gamma m_2,$$

with distinct functions m_1 and m_2, the maximum likelihood estimator of the change point converges in probability to γ under the alternative and the process Q_n diverges in probability.

Let P_n be the probability measure of the observations under local alternatives defined by sequences $(m_n)_n$ converging to m_0 with the rate $(nh)^{-\frac{1}{2}}$, with distinct functions m_{1n} and m_{2n}, and $(\gamma_n)_n$ converging to γ_0 with the rate $(nh)^{-1}$. Let $m_n = m_0 + (nh)^{-\frac{1}{2}} v_n$ and $\gamma_n = \gamma_0 + (nh)^{-1} u_n$ where v_n, and respectively u_n, converge to non-zero limits v, and respectively u.

Proposition 8.9. *Under local alternatives and the conditions, the statistic T_n converges in probability to $T_0 - T$, where T is strictly positive.*

Proof. The local alternatives are defined by distinct functions m_{1n} and m_{2n}, and γ_n different from γ_0, such that the sequence $(nh)^{\frac{1}{2}}\{(1-\delta_0)(m_{1n} - m_0) + \delta_0(m_{2n} - m_0)\}$ converges to the non-zero limit $\mu_{0u} = (1-\delta_0)u_1 + \delta_0 u_2$ and, under P_n, the process $G_{nh} = (nh)^{\frac{1}{2}}\{(1-\delta_0)(\widehat{m}_{1nh} - m_{1n}) + \delta_0(\widehat{m}_{2nh} - m_{2n})\}$ converges weakly to a Gaussian process, by Theorem 8.5. The variable $nh(\gamma_{nh} - \gamma_0)$ converges weakly to $u_0 + u$, by Theorem 8.11.

The variable $\widehat{X}_{1n}$ converges in probability to $-u_1^2 \int I_0(x,y)\, F_{XY}(dx,dy)$, as the expectation of the process $G_{0n} - G_{1n}$ converges to u_1 and its variance converges to zero.

The variable $\widetilde{X}_{2n}(\widehat{\theta}_n, \widehat{m}_{0nh})$ defined by (8.16) converges in probability to zero. The expectation of the variable $Z_i = (nh)^{\frac{1}{2}}\{\widehat{m}_{0nh}(X_i) - \widehat{m}_{2nh}(X_i)\}1_{\{X_i > \widehat{\gamma}_{nh}\}}$ converges to $-v_2$ and its variance converges to a strictly positive limit $V_1 = (u_0 + u) Var_0 \widehat{m}_{0nh}(X_i)$, it converges weakly to an uncentered Gaussian variable. A second order expansion of the process $nh\widehat{X}_{2n} = nhX_{2n}(\widehat{\theta}_{nh}, \widehat{m}_{0nh})$ has the form $n^{-1}\sum_{i=1}^n \xi_{inh} + o_p(1)$ with

$$\xi_{inh} = -nh\{\widehat{m}_{0nh}(X_i) - \widehat{m}_{2nh}(X_i)\}^2 \frac{f'^2}{f}(Y_i - m_0(X_i))1_{\{X_i > \widehat{\gamma}_{nh}\}}$$

and it converges in probability to a $T_0 - T$, where T is strictly positive. □

In a model including K consecutive change-points at γ_k, for $1 \leq k \leq K$, the expectation of Y conditionally on $X = x$ is

$$m_\gamma(x) = \sum_{k=1}^{K} \delta_{\gamma_k} m_k(x)$$

where $\gamma = (\gamma_k)_{1 \leq k \leq K}$ is an ordered vector, and the indicator variables

$$\delta_{\gamma_k} = 1_{\{\gamma_{k-1} < X \leq \gamma_k\}}$$

form a partition of I_X. At fixed change-point values, the estimation of the regression functions m_k on each sub-interval is similar to the estimation of m_1 and m_2 in the model where $K = 1$. The thresholds are then consecutively estimated by

$$\widehat{\gamma}_{1,nh} = \arg \inf_{\gamma_1 \in I_X} \widehat{l}_{nh}(\gamma_1, \ldots, \gamma_K),$$

$$\widehat{\gamma}_{k,nh} = \arg \inf_{\gamma_k \in I_X, \gamma_k \geq \widehat{\gamma}_{k-1,nh}} \widehat{l}_{nh}(\widehat{\gamma}_{1nh}, \ldots, \widehat{\gamma}_{k-1,nh}, \gamma_k, \ldots, \gamma_{K,nh}).$$

The limits of the estimators are similar to the previous ones, the K change-point values define K partial log-likelihood processes $(X_{1n}, \ldots, X_{Kn})$ such that X_{kn} is similar to X_n but restricted to an interval starting at $\widehat{\gamma}_{k-1,nh}$ for the regression variable. They fulfill Theorem 8.10 and they converge to K processes $(X_1, \ldots, X_K)$. The variables $nh(\widehat{\gamma}_{k,nh} - \gamma_k)_{1 \leq k \leq K}$ converges weakly to the vector of the K independent stopping times where uncentered Gaussian processes achieve their maximum.

A test about the number of change-points can be performed successively with hypotheses of $k - 1$ change-points against alternatives of k change-points, starting from a maximal number K, or directly against an alternative of K change-points.

8.9 Nonparametric series with change-points

Consider a real nonparametric auto-regressive regression model

$$X_t = m(X_{t-1}) + \varepsilon_t \tag{8.17}$$

with an error ε_t such that $E\varepsilon_t = 0$ and $E\varepsilon_t^2 = \sigma^2$, the ε_k's being mutually independent and identically distributed, $k \geq 1$, with a continuous distribution function F, and ε_t is independent of X_{t-1}. The covariance between the variables X_k and X_{k-1} is therefore $E\{X_{k-1} m(X_{k-1})\} - EX_{k-1}\, E\{m(X_{k-1})\}$, for $k \geq 0$.

The series $(X_k)_{k\geq 0}$ is a discrete Markov chain with initial value X_0. Let Π_0 be the distribution function of X_0, the distribution function of X_k conditionally on X_{k-1} is

$$\Pi(X_{k-1}, x) = P(X_k \leq x \mid X_{k-1}) = \int 1_{\{s\leq x\}} \, dF(s - m(X_{k-1})),$$

for every $k \geq 1$, then X_k has the distribution function

$$\Pi_k(x) = \Pi_0 \otimes \Pi^{\otimes k}(x) = \int_{\mathbb{R}^{k+1}} 1_{\{x_k\leq x\}} \Pi(x_{k-1}, dx_k) \cdots \Pi(x_0, dx_1) \, d\Pi_0(x_0)$$

with $\Pi(x_{n-1}, dx_n) = F(dx_n - m(x_{n-1}))$ for $n = 1, \ldots, k$. The expectation of X_k conditionally on X_0 is

$$E(X_k \mid X_0) = \int x \, \Pi^{\otimes k}(X_0, dx).$$

If the error ε has a density f on $\mathbb{R}$, the distribution functions Π_k have the densities

$$\pi_k(x_k) = \int_{\mathbb{R}^k} f(x_k - m(x_{k-1}))\Pi(x_{k-2}, dx_{k-1}) \cdots \Pi(x_0, dx_1) \, d\Pi_0(x_0).$$

For every real x, the stopping time $T_x = \min\{k \geq 1 : X_k = x\}$ has the conditional probability distribution and density given $X_0 = x_0$

$$H(\{x\}, x_0) = P_{x_0}(T_x \leq k) = \sum_{n=1}^{k} \Pi^{\otimes n}(\{x\}, x_0),$$

$$h(x, x_0) = \sum_{n=1}^{k} \pi^{\otimes n}(x, x_0),$$

and x is transient point with $P_{x_0}(T_x < \infty) < 1$. The conditional expectation of T_x given $X_0 = x_0$ is $h(x, x_0) = \sum_{n=1}^{k} n\pi^{\otimes n}(x, x_0)$.

For every interval I having a strictly positive Lebesgue's measure, the stopping time T_I has the conditional probability distribution

$$P_{x_0}(T_I \leq k) = \sum_{n=1}^{k} \Pi^{\otimes n}(x_0, I)$$

and the interval I is recurrent with $P_{x_0}(T_I < \infty) = \sum_{k\geq 1} P_{x_0}(X_k \in I) = 1$. The conditional expectation of T_I given $X_0 = x_0$ is

$$E_{X_0=x_0} T_I = \sum_{k\geq 1} k\Pi^{\otimes k}(x_0, I)$$

and it is finite. The probability measure

$$\mu(I) = \frac{E_{X_0=x}(\sum_{k=1}^{T_x} 1_{\{X_k\in I\}})}{E_{X_0=x} T_x}$$

is the mean duration time of the chain in I starting from x_0 in I and it is strictly positive, this is the ergodic measure of the Markov chain. It has the invariant property

$$\int \Pi(s, A)\, d\mu(s) = \mu(A)$$

for every real set A.

For every continuous and bounded function φ on $\mathbb{R}^2$, the series satisfies the ergodic property of convergence in probability

$$\frac{1}{t}\sum_{k=1}^{t} \varphi(X_k, X_{k-1}) \to \int\int \varphi(x,y) F(dx - m(y))\, d\mu(y) \tag{8.18}$$

for the ergodic measure μ of the chain and the ergodic density of X_k is $g(x) = \int f(x - m(y))\, d\mu(y)$.

A nonparametric kernel estimator of the function m is defined under the following conditions.

Conditions C'

C'1 The function K is a symmetric density of $C^2([-1;1])$, such that the integrals $m_{2K} = \int v^2 K(v) dv$, $k_j = \int K^j(v) dv$ and $\int |K'(v)|^j dv$, for $j = 1, 2$ are finite;

C'2 The probability densities f, $\pi_k(x_0, \cdot)$, $k \geq 1$, and the regression function m belong to the class of functions $C_b^2(I_X)$;

C'3 The bandwidth h_t converges to zero as t tends to infinity and $h = O(t^{-\frac{1}{5}})$.

They imply that the ergodic density g of X_k belongs to $C_b^2(I_X)$. The ergodic density g and the regression function are estimated by

$$\widehat{g}_{ht}(x) = t^{-1} \sum_{k=1}^{t} K_h(x - X_{k-1}),$$

$$\widehat{m}_{ht}(x) = \frac{\sum_{k=1}^{t} K_h(x - X_{k-1}) X_k}{\sum_{k=1}^{t} K_h(x - X_{k-1})}.$$

By the ergodicity of the series, the kernel estimator of the density function $\widehat{g}_{ht}$ has the expectation

$$g_{ht}(x) = \int K_h(x - y) g(y)\, dy = g(x) - \frac{h^2}{2} m_{2K} g''(x) + o(h^2),$$

the kernel estimator

$$\widehat{E}_{ht}(x) = t^{-1} \sum_{k=1}^{t} K_h(x - X_{k-1}) X_k$$

of $E(x) = E(X_k 1_{\{X_{k-1}=x\}})$ has the expectation

$$E_{ht}(x) = \int K_h(x-y)m(y)g(y)\,dy = m(x)g(x) - \frac{h^2}{2}m_{2K}(mg)''(x) + o(h^2),$$

and they are uniformly consistent on finite intervals, hence $\widehat{m}_{ht}$ is uniformly consistent on finite intervals. Their variances have the order $(ht)^{-1}$ and the process $(ht)^{\frac{1}{2}}(\widehat{m}_{ht} - m)$ converges weakly to a Gaussian process with variance $v(x)k_2$ and covariances zero, where $v(x)$ is the conditional variance of X_k given $X_{k-1} = x$ under the ergodic density g. The variance σ^2 of the errors ε_k is estimated by

$$\widehat{\sigma}^2_{ht} = t^{-1}\sum_{k=1}^{t}\{X_k - \widehat{m}_{ht}(X_{k-1})\}^2,$$

its expectation is $\sigma^2 + E\{\widehat{m}_{ht}(X_{k-1}) - m(X_{k-1})\}^2$ and it is consistent.

In a nonparametric model with a change at a threshold of the series on a bounded interval $\mathcal{I}$, the regression function is written as

$$E(X_t \mid X_{t-1}) = (1-\delta_{\gamma,t-1})m_1(X_{t-1}) + \delta_{\gamma,t-1}m_2(X_{t-1}) + \varepsilon_t,$$

where $\delta_{\gamma,t} = 1_{\{X_t>\gamma\}}$, with an unknown change-point parameter γ and regression functions m_1 and m_2. When the threshold γ is known, independent estimators of m_1 and m_2 are

$$\widehat{m}_{1ht}(x,\gamma) = \frac{\sum_{k=1}^{t} K_h(x-X_{k-1})(1-\delta_{\gamma,k-1})X_k}{\sum_{k=1}^{t} K_h(x-X_{k-1})(1-\delta_{\gamma,k-1})},$$

$$\widehat{m}_{2ht}(x,\gamma) = \frac{\sum_{k=1}^{t} K_h(x-X_{k-1})\delta_{\gamma,k-1}X_k}{\sum_{k=1}^{t} K_h(x-X_{k-1})\delta_{\gamma,k-1}}.$$

At fixed γ, the variance σ^2 of ε_k is estimated by

$$\widehat{\sigma}^2_{ht,\gamma} = t^{-1}\sum_{k=1}^{t}\{X_k - (1-\delta_{\gamma,k-1})\widehat{m}_{1ht}(X_{k-1},\gamma) - \delta_{\gamma,k-1}\widehat{m}_{2ht}(X_{k-1},\gamma)\}^2.$$

The change-point parameter γ is estimated by minimization of the empirical variance

$$\widehat{\gamma}_{ht} = \arg\inf_{\gamma\in\mathcal{I}} \widehat{\sigma}^2_{ht,\gamma}$$

and the functions m_1 and m_2 by $\widehat{m}_{kht}(x) = \widehat{m}_{kt}(x,\widehat{\gamma}_t)$ for $k = 1, 2$. The consistency of the estimators of the change-point and the regression functions m_1 and m_2 is proved like in Section 8.5, with the ergodic property 8.18 for the convergences in probability. The inequalities of Proposition

8.5 and Theorem 8.4 are still satisfied, then the convergence rate of the change-point parameter is $(ht)^{-1}$. We assume the following condition.

Condition C"

$$E_0\Big[\sup_{\rho(\theta,\theta_0)<A(ht)^{-\frac{1}{2}}}\{X_k - m_\gamma(X_{k-1})\}^4\Big] < \infty,$$

for $A > 0$ and n sufficiently large.

Theorem 8.12. *Under P_0 and Conditions* **C'** *and* **C"**, *the variable $h(\widehat{\gamma}_{ht} - \gamma_0)$ converges weakly to a stopping time U_0 where an uncentered Gaussian process achieves its minimum and U_0 is bounded in probability.*

8.10 Maximum likelihood for nonparametric series

In the time series model with a change-point of Section 8.9, the density of the independent error variables ε_k is

$$f(\varepsilon_k) = (1-\delta_{\gamma,k})f(X_k - m_1(X_{k-1})) + \delta_{\gamma,k}f(X_k - m_2(X_{k-1})),$$

with $\delta_{\gamma,k} = 1_{\{X_{k-1}>\gamma\}}$, for $k = 1,\ldots,t$. The log-likelihood of the sample under a probability measure $P_{m,\gamma}$ is denoted

$$l_t(m,\gamma) = \sum_{k=1}^{t} \log f(X_k - m_\gamma(X_{k-1}))$$

and the log-likelihood of the sample under $P_{m,\gamma}$ and P_0 defines the process $Z_t(m,\gamma) = t^{-1}\{l_t(m,\gamma) - l_t(m_0,\gamma_0)\}$

$$\begin{aligned}
Z_t(m,\gamma) &= t^{-1}\sum_{k=1}^{t}(1-\delta_{0,k})\log\frac{f(X_k - m_1(X_{k-1}))}{f(X_k - m_{01}(X_{k-1}))}\\
&\quad + t^{-1}\sum_{k=1}^{t}\delta_{0,k}\log\frac{f(X_k - m_2(X_{k-1}))}{f(X_k - m_{02}(X_{k-1}))}\\
&\quad + t^{-1}\sum_{k=1}^{t}(\delta_{\gamma,k} - \delta_{0,k})\log\frac{f(X_k - m_2(X_{k-1}))}{f(X_k - m_1(X_{k-1}))}.
\end{aligned}$$

By the ergodic property (8.18), the process Z_t converges uniformly in probability to the functional Z expectation of Z_t under the ergodic probability measure. It is minimum at m_{01}, m_{02} and γ_0 and the maximum likelihood estimators are consistent.

Conditions C'

C'1 The function K is a symmetric density of $C^2([-1;1])$, such that the integrals $m_{2K} = \int v^2 K(v)dv$, $k_j = \int K^j(v)dv$ and $\int |K'(v)|^j dv$, for $j = 1, 2$ are finite;

C'2 The probability densities f, $\pi_k(x_0, \cdot)$, $k \geq 1$, and the regression function m belong to the class of functions $C_b^2(I_X)$;

C'3 The bandwidth h_t converges to zero as t tends to infinity and $h = O(t^{-\frac{1}{5}})$.

C'4 For every $\varepsilon > 0$, $E_0 \sup_{|\gamma-\gamma_0|\leq\varepsilon} \log^2 f_\gamma(X_k \mid X_{k-1})$ is finite.

Under Conditions C', the convergence rate of the estimator $\widehat{\gamma}_{ht}$ is established like in Theorem 8.9 and the variable $ht(\widehat{\gamma}_{ht} - \gamma_0)$ converges weakly to the variable u_0 where the Gaussian process limit of the process

$$Q_t(u_\gamma) = h \sum_{k=1}^{t} (\delta_{\gamma_{ht,u},k} - \delta_{0,k}) \log \frac{f(X_k - m_{02}(X_{k-1}))}{f(X_k - m_{01}(X_{k-1}))}$$

achieves its maximum (cf. Theorem 8.11). The asymptotic behaviour of the estimated regression function does not depend on the estimation of the change-point, the process $(ht)^{\frac{1}{2}}(\widehat{m}_{ht} - m_0)$ converges weakly under P_0 to a Gaussian process

$$G_m(x) = G_{1m}(x) 1_{\{x \leq \gamma_0\}} + G_{2m}(x) 1_{\{x > \gamma_0\}}$$

on $\mathcal{I}$, the variance of $G_{1m}(x)$ is $V_1(x) = k_2 Var(X_k | X_{k-1} = x \leq \gamma_0)$ and the variance of $G_{2m}(x)$ is $V_2(x) = k_2 Var(X_k Y | X_{k-1} = x > \gamma_0)$.

The likelihood ratio test for the hypothesis H_0 of a nonparametric autoregressive model without changes is performed like in Section 8.8 with the statistic

$$S_{ht} = \sup_{\gamma \in \mathcal{I}} ht\{Z_t(\widehat{m}_{nh,\gamma}, \gamma) - \widehat{Z}_{0t}\}$$

using the nonparametric estimator of the regression function in the model under H_0, $\widehat{Z}_{0t} = t^{-1} \sum_{k=1}^{t} \{l_{0t}(\widehat{m}_{0t}) - l_{0t}(m_0)\}$, with γ_0 at the end-point of the interval $\mathcal{I}$.

Proposition 8.10. *Under the hypothesis H_0 and Conditions C', the test statistic S_t converges weakly to T_0.*

Under an alternative with a fixed change-point different from γ_0 and distinct regression functions, the test statistic S_t diverges. Under local alternatives

K_t with parameter γ_t converging to γ_0 with the rate $(ht)^{-1}$ and with m_{1t} converging to m_0 with the rate $(ht)^{-\frac{1}{2}}$, let u be the limit of $u_t = ht(\gamma_t - \gamma_0)$ and let v_1 be the limit of $v_{1t} = (ht)^{\frac{1}{2}}(m_{1t} - m_0)$, as t tends to infinity.

Proposition 8.11. *Under the local alternatives K_t and Conditions C', the test statistic S_t converges weakly to*

$$T_0 - v_1^2 \int \frac{f'^2}{f}(y - m_0(x))\, dy\, d\mu(x).$$

The limit of S_t under K_n is negative so the test statistic is consistent under K_t.

The first order nonparametric auto-regressive model (8.17) extends to an additive model of higher order

$$X_t = m_1(X_{t-1}) + \cdots + m_p(X_{t-p}) + \varepsilon_t \tag{8.19}$$

under the same assumptions for the error variables. The function $m_j(X_{t-j})$ is now the conditional expectation of X_t given X_{t-j}. The series $(X_k)_{k\geq 0}$ is p-order Markov chain with initial values $X_0, \ldots, X_{p-1}$. The distribution function of X_k conditionally on $X_{k-1}, \ldots, X_{k-p}$ is

$$\begin{aligned}\Pi(X_{k-1}, \ldots, X_{k-p}, x) &= P(X_k \leq x \mid X_{k-1}, \ldots, X_{k-p}) \\ &= F\Big(x - \sum_{j=1}^{p} m_j(X_{k-j})\Big)\end{aligned}$$

and X_k has the distribution function

$$\Pi_k(x) = \Pi_0 \otimes \Pi^{\otimes k}(x)$$

where Π_0 is the initial distribution function of $X_0, \ldots, X_{p-1}$. Like in Model (8.17), there exists an invariant measure μ since that the empirical mean

$$\frac{1}{t}\sum_{k=p}^{t} \varphi(X_k, X_{k-1}, \ldots, X_{k-p})$$

converges in probability to its expectation under the measure μ

$$\int\int \varphi(x, y_1, \ldots, y_p) F\Big(dx - \sum_{j=1}^{p} m_j(y_j)\Big)\, d\mu(y_1, \ldots, y_p) \tag{8.20}$$

Under Conditions C', the nonparametric kernel estimators of the functions m_j are

$$\widehat{m}_{j,ht}(x) = \frac{\sum_{k=j}^{t} K_h(x - X_{k-j}) X_k}{\sum_{k=j}^{t} K_h(x - X_{k-j})}$$

and they have the same properties as the estimator $\widehat{m}_{ht}$. In Model (8.19) with change-points according to thresholds $\gamma_1, \ldots, \gamma_p$ of the regressors $X_{t-1}, \cdots, X_{t-p}$, the conditional expectation of X_t is written as

$$E(X_t \mid X_{t-1}) = \sum_{j=1}^{p} \{(1 - \delta_{\gamma_j,t-j}) m_{1j}(X_{t-j}) + \delta_{\gamma_j,t-j} m_{2j}(X_{t-j})\} + \varepsilon_t,$$

where $\delta_{\gamma_j,t-j} = 1_{\{X_{k-j} > \gamma_j\}}$. With thresholds $\gamma_1, \ldots, \gamma_p$, the estimators of m_{1j} and m_{2j} are

$$\widehat{m}_{1j,ht}(x, \gamma_j) = \frac{\sum_{k=j}^{t} K_h(x - X_{k-j})(1 - \delta_{\gamma_j,k-j}) X_k}{\sum_{k=j}^{t} K_h(x - X_{k-j})(1 - \delta_{\gamma_j,k-j})},$$

$$\widehat{m}_{2j,ht}(x, \gamma) = \frac{\sum_{k=j}^{t} K_h(x - X_{k-j}) \delta_{\gamma_j,k-j} X_k}{\sum_{k=j}^{t} K_h(x - X_{k-j}) \delta_{\gamma_j,k-j}}.$$

They have the same properties as the kernel estimators with a change-point at a single regressor. The log-likelihood of the sample is

$$l_t(m, \gamma_1, \ldots, \gamma_p) = \sum_{k=j}^{t} \log f(X_k - m_\gamma(X_{k-1}, \ldots, X_{k-p}))$$

and the maximum likelihood estimators of the thresholds are defined by maximization of $l_t(\widehat{m}_{th}, \gamma_1, \ldots, \gamma_p)$, they are consistent and have the same asymptotic properties as the estimator $\widehat{m}_{th}$ in Theorem 8.15.

The process $Z_{th}(\gamma) = t^{-1}\{l_{th}(\gamma) - l_{th}(\gamma_0)\}$ is now the sum of $Z_{th}(\gamma_0)$ and $Z_{2th}(\gamma) = Z_{th}(\gamma) - Z_{th}(\gamma_0)$, it converges uniformly to the function $Z(\gamma) = E_\mu Z_t$, the expectation of Z_t under the ergodic probability measure, by the property 8.20. Under the conditions, the function Z is a $O_p(|\gamma - \gamma_0|)$ in a neighborhood of γ_0 and the process $W_{th} = (th)^{\frac{1}{2}}\{Z_{th} - Z\}$ satisfies

$$E_\mu \sup_{|\gamma - \gamma_0| \leq \varepsilon} W_{th} = O(\varepsilon^{\frac{1}{2}})$$

as t tends to infinity, the convergence rate of $\widehat{\gamma}_t$ is therefore $(th)^{-1}$ and the regression estimators $\widehat{m}_{th,j}$ have the convergence rate $(th)^{-\frac{1}{2}}$. The variables $u_{\widehat{\gamma}_{j,th}} = th(\widehat{\gamma}_{j,th} - \gamma_{0j})$ converge weakly to a variable u_j, where a Gaussian process reaches its maximum, for $j = 1, \ldots, p$.

The likelihood ratio test for the hypothesis H_0 of a nonparametric model without changes relies on the statistic $S_{ht} = \sup_{\gamma_1, \ldots, \gamma_p \in \mathcal{I}} h\{Z_{th}(\gamma) - \widehat{\gamma}_{0ht}\}$, its asymptotic behaviour is like in Propositions 8.10 and 8.11.

8.11 Chronological change in series

In the nonparametric model with a chronological change of regression function at an unknown index τ, we assume that $\gamma_t = t^{-1}\tau$ converges to a limit γ in $]0,1[$ as t tends to infinity and we denote $\tau_\gamma = [\gamma t]$. Let $\delta_{\gamma,k} = 1\{k > \tau_\gamma\}$, for $k = 1, 2$, and let

$$X_{1,k} = X_k(1 - \delta_{k,\tau}), \text{ and } X_{2,k} = X_k \delta_{k,\tau},$$

then the series has a two-phases regression model with a change at the integer part of γt, $\tau_\gamma = [\gamma t]$

$$X_{j,k} = E(X_{j,k} \mid X_{j,k-1} = x) + \varepsilon_k, \tag{8.21}$$

for $j = 1, 2$, with independent and identically distributed error variables ε_k with mean zero and variance σ^2 and with

$$\begin{aligned} E(X_{1,k} \mid X_{1,k-1} = x) &= m_{1,\gamma}(x), \\ E(X_{2,k} \mid X_{1,k-1} = x) &= m_{2,\gamma}(x). \end{aligned} \tag{8.22}$$

The kernel estimators of the regression functions $m_{1,\gamma}$ and $m_{2,\gamma}$ are

$$\begin{aligned} \widehat{m}_{1th}(x,\gamma) &= \frac{\sum_{i=1}^t K_h(x - X_i)(1 - \delta_{\gamma,i})Y_i}{\sum_{i=1}^t K_h(x - X_i)(1 - \delta_{\gamma,i})}, \ x \le \gamma, \\ \widehat{m}_{2th}(x,\gamma) &= \frac{\sum_{i=1}^t K_h(x - X_i)\delta_{\gamma,i}Y_i}{\sum_{i=1}^t K_h(x - X_i)\delta_{\gamma,i}}, \ x > \gamma, \end{aligned}$$

under P_0 their expectations are asymptotically equivalent to

$$\begin{aligned} m_{1th}(x,\gamma) &= m_{01}(x)1_{\{\gamma\wedge\gamma_0\}} + m_{02}(x)1_{\gamma_0>\gamma}, \ x \le \gamma, \\ m_{2th}(x,\gamma) &= m_{02}(x)1_{\{\gamma\vee\gamma_0\}} + m_{01}(x)1_{\gamma\le\gamma_0}, \ x > \gamma. \end{aligned}$$

The estimated log-likelihood of the sample with a change of parameter at γ in $]0,1[$ is

$$l_{th}(\gamma) = \sum_{k=1}^{\tau_\gamma} \log f(X_k - \widehat{m}_{1,th}(X_{k-1},\gamma)) + \sum_{k=\tau_\gamma+1}^{t} \log f(X_k - \widehat{m}_{2,th}(X_{k-1},\gamma)),$$

it is maximum at $\widehat{\gamma}_{th}$ and the estimators of the functions m_1 and m_2 are defined as $\widehat{m}_{k,th}(x) = \widehat{m}_{k,th}(x, \widehat{\gamma}_{th})$, for $k = 1, 2$.

Under P_0, the process $t^{-1}l_{th}$ converges uniformly on $\mathcal{I}$ to the function

$$l(\gamma) = (1-\gamma)E_0 \log f(X_k - m_1(X_{k-1},\gamma)) + \gamma E_0 \log f(X_k - m_2(X_{k-1},\gamma))$$

which is maximum at γ_0, the maximum likelihood estimator $\widehat{\gamma}_{th}$ and the estimators of the regression functions m_1 and m_2 are therefore consistent. Let $\tau_0 = \tau_{\gamma_0}$, the process $Z_{th} = l_{th}(\gamma) - l_{th}(\gamma_0)$ is the sum

$$\begin{aligned}
Z_{th}(\gamma) = \sum_{k=1}^{\tau_0} \log \frac{f(X_k - \widehat{m}_{1,th}(X_{k-1},\gamma))}{f(X_k - \widehat{m}_{1,th}(X_{k-1},\gamma_0))} \\
+ \sum_{k=\tau_0+1}^{t} \log \frac{f(X_k - \widehat{m}_{2,th}(X_{k-1},\gamma))}{f(X_k - \widehat{m}_{2,th}(X_{k-1},\gamma_0))} \\
+ \sum_{k=\tau_\gamma+1}^{\tau_0} \log \frac{f(X_k - \widehat{m}_{2,th}(X_{k-1},\gamma))}{f(X_k - \widehat{m}_{1,th}(X_{k-1},\gamma_0))} \\
+ \sum_{k=\tau_0+1}^{\tau} \log \frac{f(X_k - \widehat{m}_{1,th}(X_{k-1},\gamma))}{f(X_k - \widehat{m}_{2,th}(X_{k-1},\gamma_0))},
\end{aligned}$$

$t^{-1} Z_{th}$ converges in probability to the function

$$\begin{aligned}
Z(\gamma) = \gamma_0 \int \log \frac{f(y - m_1(x,\gamma))}{f(y - m_1(x,\gamma_0))} F(dy - m_1(x,\gamma_0))\, d\mu(x) \\
+(1-\gamma_0) \int \log \frac{f(y - m_2(x,\gamma))}{f(y - m_2(x,\gamma_0))} F(dy - m_2(x,\gamma_0))\, d\mu(x) \\
+(\gamma_0-\gamma) \int \log \frac{f(y - m_2(x,\gamma))}{f(y - m_1(x,\gamma_0))} F(dy - m_1(x,\gamma_0))\, d\mu(x) \\
+(\gamma-\gamma_0) \int \log \frac{f(y - m_1(x,\gamma))}{f(y - m_2(x,\gamma_0))} F(dy - m_2(x,\gamma_0))\, d\mu(x).
\end{aligned}$$

Lemma 8.5. *For $\varepsilon > 0$ sufficiently small, there exists a constant $\kappa_0 > 0$ such that for every γ in an ε-neighborhood of γ_0*

$$Z(\gamma) \geq -\kappa_0 |\gamma - \gamma_0|.$$

Proof. Let $\varepsilon > 0$ and let $|\gamma - \gamma_0| < \varepsilon$, for $x \leq \gamma \vee \gamma_0$ we have

$$|m_1(x,\gamma) - m_1(x,\gamma_0)| = 1_{]\gamma_0,\gamma]}(x) |m_{02}(x)|$$

and the measure of the interval $]\gamma_0, \gamma]$ tends to zero as ε tends to zero. Let

$$U_1(x,\gamma) = \frac{f(y - m_1(x,\gamma)) - f(y - m_1(x,\gamma_0))}{f(y - m_1(x,\gamma_0))}, \tag{8.23}$$

by second order expansion of the logarithm we have

$$\log \frac{f(y - m_1(x,\gamma))}{f(y - m_1(x,\gamma_0))} = U_1(x,\gamma) - \frac{U_1^2(x,\gamma)}{2} + o(U_1(x,\gamma)),$$

and

$$U_1(x,\gamma) = -\{m_1(x,\gamma) - m_1(x,\gamma_0)\}\frac{f'(y - m_1(x,\gamma_0))}{f(y - m_1(x,\gamma_0))}$$
$$+\{m_1(x,\gamma) - m_1(x,\gamma_0)\}^2\frac{ff'' - f'^2}{2f^2}(y - m_1(x,\gamma_0))\{1 + o(1)\}.$$

Since $\int f'(y - m_1(x,\gamma_0))\,dy$ and $\int f''(y - m_1(x,\gamma_0))\,dy$ are zero for every x, there exists a constant κ such that the first integral is

$$-\Big\{\int\{m_1(x,\gamma)-m_1(x,\gamma_0)\}^2\frac{f'^2}{f}(y-m_1(x,\gamma_0))\,dy\,d\mu(x)\Big\}\{1+o(1)\} > -\kappa\varepsilon.$$

The second integral has a similar lower bound with

$$|m_2(x,\gamma) - m_2(x,\gamma_0)| = 1_{]\gamma,\gamma_0]}(x)|m_{01}(x)|$$

and by an expansion of $\log f(y - m_2(x,\gamma)) - \log f(y - m_2(x,\gamma_0))$. The last two integrals have negative logarithms and the order $|\gamma - \gamma_0|$. □

Let $\nu_{0t} = t^{\frac{1}{2}}(\widehat{G}_t - G_0)$ be the empirical process of the sample under P_0, where $\widehat{G}_t$ is the empirical distribution function of (X_{k-1}, X_k) and $G_0(dy,dx) = F(dy - m_0(x))\,\mu(dx)$. The process

$$W_{th} = (th)^{\frac{1}{2}}\{t^{-1}Z_{th} - Z\}$$

is the sum of the four integrals of the function Z with respect to the empirical process ν_{0t}.

Lemma 8.6. *For every $\varepsilon > 0$, under Conditions C', there exists a constant $\kappa_1 > 0$ such that for n sufficiently large*

$$E_0 \sup_{|\gamma-\gamma_0|\leq\varepsilon} |W_{th}(\gamma)| \leq \kappa_1\varepsilon^{\frac{1}{2}}.$$

Proof. At the estimators of the regression functions the integral of the processes $th\{\widehat{m}_{jth}(x,\gamma) - \widehat{m}_{jth}(x,\gamma_0)\}^2$, $j = 1, 2$, with respect to the empirical process ν_{0t} is a $O(|\gamma - \gamma_0|)$ and the first two integrals of the process W_{th} have the same order, by a first order expansion of the logarithms. The last two terms of W_{th} are defined as integral on the intervals $]\gamma,\gamma_0]$ and $]\gamma_0,\gamma]$ and their variance have the order $|\gamma - \gamma_0|$. □

By the consistency of the estimator $\widehat{\gamma}_{th}$, the regressions functions $\widehat{m}_{1th}$ and $\widehat{m}_{2th}$ have the convergence rate $(th)^{\frac{1}{2}}$, according to Theorem 8.5. Like in Theorem 3.2, Lemmas 8.5 and 8.6 determine the convergence rate of the estimator $\widehat{\gamma}_{th}$ from the convergence rate of the kernel estimators of the regressions functions.

Theorem 8.13. *Under Conditions C'*

$$\overline{\lim}_{n,A\to\infty} P_0\{th|\widehat{\gamma}_{th} - \gamma_0| > A) = 0.$$

For the weak convergence of $th(\widehat{\gamma}_{th} - \gamma_0)$, let

$$\mathcal{U}_t = \{u_\gamma)(th)^{-1}(\gamma - \gamma_0), \gamma \in]0,1[\}$$

and for every $A > 0$, let $\mathcal{U}_{th}^A = \{u \in \mathcal{U}_t; |u| \leq A\}$. Then for every u belonging to $\mathcal{U}_{th}^A$, the parameter $\gamma_{t,u}$ defines a map from $]0,1[$ to $\mathcal{U}_{th}^A$ and we also denote $Z_{th}(u) = Z_{th}(\gamma_{t,u})$.

Theorem 8.14. *Under Conditions C' and if the measure μ has a density, the process Z_{th} converges weakly in $D(]0,1[)$ to an uncentered Gaussian process W_Z with a finite variance function.*

Proof. Let $\gamma_{t,u} = \gamma_0 + (th)^{-1}u_\gamma$, with u_γ in $\mathcal{U}_{th}^A$. From the approximation of Lemma 8.5 and by the ergodic property, in $\mathcal{U}_{th}^A$, the expectation of the process $Z_{th}(u)$ is asymptotically equivalent under P_0 to

$$\begin{aligned}
&-th\gamma_0 \int \{m_1(x,\gamma_{t,u}) - m_{01}(x)\}^2 \frac{f'^2}{f}(y - m_{01}(x))\,dy\,d\mu(x)\\
&-th(1-\gamma_0) \int \{m_2(x,\gamma_{t,u}) - m_{02}(x)\}^2 \frac{f'^2}{f}(y - m_{02}(x))\,dy\,d\mu(x)\\
&+th(\gamma_0 - \gamma_{t,u}) \int \log \frac{f(y - m_{02}(x))}{f(y - m_{01}(x))} F(dy - m_{01}(x))\,d\mu(x)\\
&+th(\gamma_{t,u} - \gamma_0) \int \log \frac{f(y - m_1(x,\gamma))}{f(y - m_{02}(x))} F(dy - m_{02}(x))\,d\mu(x)\\
&= -th\gamma_0 \int_{\gamma_0}^{\gamma_{t,u}} m_{02}^2(x) \frac{f'^2}{f}(y - m_{01}(x))\,dy\,d\mu(x)\\
&-th(1-\gamma_0) \int_{\gamma_{t,u}}^{\gamma_0} m_{01}^2(x) \frac{f'^2}{f}(y - m_{02}(x))\,dy\,d\mu(x)\\
&-u_\gamma \int \log \frac{f(y - m_{02}(x))}{f(y - m_{01}(x))} F(dy - m_{01}(x))\,d\mu(x)\\
&+u_\gamma \int \log \frac{f(y - m_{01}(x))}{f(y - m_{02}(x))} F(dy - m_{02}(x))\,d\mu(x),
\end{aligned}$$

as t tends to infinity, the sum of the first two integrals converges to

$$u_\gamma \mu'(\gamma_0) m_{02}^2(\gamma_0) \frac{f'^2}{f}(y - m_{01}(\gamma_0))\,dy - u_\gamma \mu'(\gamma_0) m_{01}^2(\gamma_0) \frac{f'^2}{f}(y - m_{02}(\gamma_0))\,dy.$$

In the same way as for the expectation, its variance has the order $th|\gamma_{t,u} - \gamma_0| = 0(1)$ and it is proportional to u_γ.

According to Proposition 1.3, the process Z_{th} satisfies Billingsley's criterion (15.21) for tightness in $D(]0,1[)$ and its finite dimensional distributions converge to uncentered Gaussian variables, as sums of functions of the independent error variables with converging mean and variance. It follows that the process Z_{th} converges weakly to a Gaussian process. □

The asymptotic behaviour of the estimator $u_{\widehat{\gamma}_{th}} = th(\widehat{\gamma}_{th} - \gamma_0)$ is deduced from this weak convergence of Z_{th} by continuity of the maximum and it is bounded in probability by Theorem 8.13.

Theorem 8.15. *Under Conditions C' and if the measure μ has a density, the variable $th(\widehat{\gamma}_{th} - \gamma_0)$ converges weakly to a stopping time u_0 where the Gaussian process W_Z achieves its maximum.*

The likelihood ratio test for the hypothesis H_0 of a nonparametric model without changes for autoregressive series is performed like in Section 8.8 with the statistic

$$S_{ht} = \sup_{\gamma \in \mathcal{I}} h\{l_{th}(\gamma) - \widehat{l}_{0th}\}$$

with the nonparametric estimator of the log-likelihood under H_0, as $\gamma_0 = 1$ and $\tau_0 = t$

$$\widehat{l}_{0t} = \sum_{k=1}^{t} \log f(X_k - \widehat{m}_{0,th}(X_{k-1})).$$

Under H_0, the process $Z_{0th} = l_{th}(\gamma) - \widehat{l}_{0th}$ reduces to the sum

$$\begin{aligned} Z_{0th}(\gamma) &= \sum_{k=1}^{t} \log \frac{f(X_k - \widehat{m}_{1,th}(X_{k-1}, \gamma))}{f(X_k - \widehat{m}_{0,th}(X_{k-1}))} \\ &\quad + \sum_{k=\tau_\gamma+1}^{t} \log \frac{f(X_k - \widehat{m}_{2,th}(X_{k-1}, \gamma))}{f(X_k - \widehat{m}_{0,th}(X_{k-1}))} \end{aligned}$$

and $t^{-1} Z_{0th}$ converges in probability under H_0 to the function

$$\begin{aligned} Z_0(\gamma) &= \int \log \frac{f(y - m_1(x, \gamma))}{f(y - m_0(x))} F(dy - m_0(x))\, d\mu(x) \\ &\quad + (1 - \gamma) \int \log \frac{f(y - m_2(x, \gamma))}{f(y - m_0(x))} F(dy - m_0(x))\, d\mu(x). \end{aligned}$$

The function Z_0 is maximum as $m_1 = m_0$ and $\gamma = 1$, it follows that the maximum likelihood estimator $\widehat{\gamma}_{th}$ is consistent, the kernel estimator $\widehat{m}_{1,th}$ converges in probability to m_0 uniformly on $[0, 1]$. The asymptotic distributions of the test statistic S_{th} is similar to the behaviour of the likelihood ratio test for a nonparametric regression, given in Proposition

8.8, where the limits are integrals with respect to the invariant measure of the series.

Proposition 8.12. *Under the hypothesis H_0 and Conditions C', the test statistic S_t converges in probability to*

$$T_0 = -u_0 \int Var_0 \widehat{m}_{0th}(x) \frac{f'^2(y - m_0(x))}{f(y - m_0(x))} \, dF(y - m_0(x)) \, d\mu(x).$$

Proof. The function Z_0 satisfies the inequality of Lemma 8.5 and the process W_{th} defined by $t^{-1} Z_{0th}$ and Z_0 under H_0 satisfies Lemma 8.6, the convergence rate of the estimator $\widehat{\gamma}_{th}$ is therefore th, by Theorem 8.13. Like in Section 8.8, the difference of the processes $G_{1t} = (th)^{\frac{1}{2}}(\widehat{m}_{1th} - m_0)$ and $G_{0t} = (th)^{\frac{1}{2}}(\widehat{m}_{0th} - m_0)$ converges in probability to zero and, by an expansion of the process $U_1(x, \gamma)$ defined by (8.23) according to $m_1(x, \gamma) - m_{01}(x)$, the variable $h \sum_{k=1}^{t} \{\log f(X_k - \widehat{m}_{1,th}(X_{k-1})) - \log f(X_k - \widehat{m}_{0,th}(X_{k-1}))\}$ converges in probability to zero.

The process Z_{0th} has then the approximation

$$Z_{0th}(\gamma) = \sum_{k=\tau_\gamma+1}^{t} \log \frac{f(X_k - \widehat{m}_{2,th}(X_{k-1}, \gamma))}{f(X_k - \widehat{m}_{0,th}(X_{k-1}))} + o_p(1)$$

where $th(1 - \widehat{\gamma}_{th})$ converges weakly to a stopping time u_0 of a Gaussian process by Theorem 8.15. Under H_0, the expectation of $(th)^{\frac{1}{2}}\{\widehat{m}_{0th}(X_{k-1}) - \widehat{m}_{2th}(X_{k-1})\}$, for $k > \widehat{\tau}_{th}$, converges to zero and its variance converges to a $u_0 Var_0 \widehat{m}_{0,th}(X_{k-1})$ which is strictly positive, it converges to a centered Gaussian variable. A second order expansion of the process $Z_{0th}(\widehat{\gamma}_{th})$ has the form $n^{-1} \sum_{k=\widehat{\tau}_{th}+1}^{n} \xi_{i,th} + o_p(1)$ with

$$\xi_{i,th} = -\{\widehat{m}_{0th}(X_{k-1}) - \widehat{m}_{2th}(X_{k-1})\}^2 \frac{f'^2}{f}(X_k - m_0(X_{k-1}))$$

and it converges in probability to a T_0. □

Under fixed alternatives with a change of regression function at γ, the expectation of $\widehat{m}_{j,th}(x)$ converges to $m_j(x)$, for $k = 1, 2$, with distinct functions $m_1(x)$ and $m_2(x)$. If $m_1 = m_0$, $th(\gamma - \widehat{\gamma}_{th})$ converges weakly to a stopping time u_0 of a Gaussian process and the first term of T_n converges in probability to zero, the second term diverges with $th(1 - \gamma)$. If m_1 differs from m_0, the first term diverges.

Under local alternatives K_t with γ_t converging to γ_0 with the rate $(ht)^{-1}$ and m_{1t} converging to m_0 with the rate $(ht)^{-\frac{1}{2}}$, m_{1t} distinct from m_{2t}, let $m_t = m_0 + (th)^{-\frac{1}{2}} v_t$ and let $\gamma_t = \gamma_0 + (th)^{-1} u_t$ where v_t, and respectively u_t, converge to non zero limits v, and respectively u.

Proposition 8.13. *Under local alternatives and the conditions, the test statistic S_t converges in probability to $T_0 - T$, where T is strictly positive.*

Proof. Under the local alternatives, the sequence $(th)^{\frac{1}{2}}\{(1-\delta_0)(m_{1t} - m_0)+\delta_0(m_{2t}-m_0)\}$ converges to the non-zero limit $\mu_{0u} = (1-\delta_0)u_1+\delta_0 u_2$ and the process $G_{th} = (th)^{\frac{1}{2}}\{(1-\delta_0)(\widehat{m}_{1th} - m_{1t}) + \delta_0(\widehat{m}_{2th} - m_{2t})\}$ converges weakly to a Gaussian process. The difference of the processes $G_{1t} = (th)^{\frac{1}{2}}(\widehat{m}_{1th} - m_0)$ and $G_{0t} = (th)^{\frac{1}{2}}(\widehat{m}_{0th} - m_0)$ converges in probability to u_1, then the variable $h\sum_{k=1}^{t}\{\log f(X_k - \widehat{m}_{1,th}(X_{k-1})) - \log f(X_k - \widehat{m}_{0,th}(X_{k-1}))\}$ converges in probability to $-u_1^2\int f^{-1}(y - m_0(x))f'^2(y - m_0(x))\,dy\,d\mu(x)$, from a second order of the logarithm.

The variable $th(\gamma_0 - \widehat{\gamma}_{th})$ converges weakly under the alternatives to $u_0 + u$, by Theorem 8.15. The second sum of $Z_{0th}(\widehat{\gamma}_{th})$ has a second order expansion $n^{-1}\sum_{k=\widehat{\tau}_{th}+1}^{n}\xi_{i,th} + o_p(1)$ with

$$\xi_{i,th} = -\{\widehat{m}_{0th}(X_{k-1}) - \widehat{m}_{2th}(X_{k-1})\}^2\frac{f'^2}{f}(X_k - m_0(X_{k-1}))$$

where the expectation of $(th)^{\frac{1}{2}}\{\widehat{m}_{0th}(X_{k-1}) - \widehat{m}_{2th}(X_{k-1})\}$, for $k > \widehat{\tau}_{th}$, converges to zero due to the convergence rate of $\widehat{\gamma}_{th}$ and its variance has a strictly positive limit, the limit of the test statistic follows. □

Chapter 9

Change-points in nonparametric distributions

Abstract. The models of the previous chapters extend to models for the distribution function of a variable Y and to the conditional distribution function of Y given a vector of explanatory variables X with changes at unknown points.

9.1 Distribution functions with changes

On a probability space $(\Omega, \mathcal{A}, P_0)$, let Y be a real random variable with a distribution function F_0 having two-phases with continuous sub-distribution functions F_{01} and F_{02} and a discontinuity at an unknown change-point γ_0. Under a probability P_γ, the distribution function of Y is F_γ

$$F_\gamma(y) = 1_{\{y \le \gamma\}} F_{1\gamma}(y) + 1_{\{y > \gamma\}} F_{2\gamma}(y), \tag{9.1}$$

with continuous sub-distribution functions

$$F_{1\gamma}(y) = P_\gamma(Y \le y \wedge \gamma), \quad F_{2\gamma} = P_\gamma(\gamma < Y \le y)$$

and $F_{2\gamma}(\gamma^+)$ distinct from $F_{1\gamma}(\gamma)$. The sub-distribution functions are unknown, they are estimated from a n-sample $Y_1, \ldots, Y_n$ of the variable Y by

$$\widehat{F}_{1n,\gamma}(y) = n^{-1} \sum_{i=1}^{n} 1_{\{Y_i \le y \wedge \gamma\}},$$

$$\widehat{F}_{2n,\gamma}(y) = n^{-1} \sum_{i=1}^{n} 1_{\{\gamma < Y_i \le y\}},$$

and $\widehat{F}_{n,\gamma}(y) = 1_{\{Y \le \gamma\}} \widehat{F}_{1n,\gamma}(y) + 1_{\{Y > \gamma\}} \widehat{F}_{2n,\gamma}(y)$. Under P_0, the empirical sub-distribution functions converge a.s. to

$$F_{01,\gamma}(y) = F_0(y \wedge \gamma) = F_{01}(y \wedge \gamma \wedge \gamma_0) + 1_{\{y > \gamma_0\}} \{F_{02}(y \wedge \gamma) - F_{02}(\gamma_0)\},$$

and, respectively

$$\begin{aligned}F_{02,\gamma}(y) &= 1_{\{y>\gamma\}}\{F_0(y) - F_0(\gamma)\}\\&= 1_{\{y>\gamma>\gamma_0\}}\{F_{02}(y) - F_{02}(\gamma)\}\\&\quad +1_{\{y>\gamma_0>\gamma\}}\{F_{02}(y) - F_{02}(\gamma_0) + F_{01}(\gamma_0) - F_{01}(\gamma)\}\\&\quad +1_{\{\gamma_0>y>\gamma\}}\{F_{01}(y) - F_{01}(\gamma)\},\end{aligned}$$

and by (9.1), under P_0, $\widehat{F}_{n,\gamma}$ converges a.s. uniformly on $\mathbb{R}$ to the distribution function $F_{0,\gamma}(y) = 1_{\{Y\leq\gamma\}}F_{01,\gamma}(y) + 1_{\{Y>\gamma\}}F_{02,\gamma}(y)$.

Under the model (9.1), the expectation of Y follows the model of change in the mean of Chapter 2. The empirical variance of the sample is

$$\widehat{\sigma}^2_{n\gamma} = \int y^2\, d\widehat{F}_{n,\gamma}(y) - \Big\{\int y\, d\widehat{F}_{n,\gamma}(y)\Big\}^2,$$

and the parameter γ may be estimated by minimization of $\widehat{\sigma}^2_{n\gamma}$ as

$$\widehat{\gamma}_n = \arg\inf_{\gamma} \widehat{\sigma}^2_{n\gamma}.$$

Then functions F_{01} and F_{02} are estimated by $\widehat{F}_{1n} = \widehat{F}_{1n,\widehat{\gamma}_n}$ and, respectively $\widehat{F}_{2n} = \widehat{F}_{2n,\widehat{\gamma}_n}$, then the estimator of the distribution function F_0 is

$$\widehat{F}_n(y) = 1_{\{y\leq\widehat{\gamma}_n\}}\widehat{F}_{1n}(y) + 1_{\{y>\widehat{\gamma}_n\}}\widehat{F}_{2n}(y).$$

For every γ, the empirical variance $\widehat{\sigma}^2_{n\gamma}$ converges under P_0 to the variance $\sigma^2_{0\gamma}$ of Y under the distribution function $F_{0,\gamma}$, if this limit is finite, the convergence of the estimator is deduced like in Theorem 2.1.

Theorem 9.1. *If $\sigma^2_{0\gamma}$ is finite for γ in a neighborhood of γ_0, the maximum likelihood estimator $\widehat{\gamma}_n$ of γ_0 is a.s. consistent under P_0.*

Let

$$l_n(\gamma) = \widehat{\sigma}^2_{n\gamma} - \widehat{\sigma}^2_{n\gamma_0}.$$

For every γ, the process $l_n(\gamma)$ converges a.s. uniformly on $\mathbb{R}$ to the function

$$l(\gamma) = \sigma^2_{0\gamma} - \sigma^2_0,$$

here σ^2_0 is the variance of Y under P_0.

Lemma 9.1. *If $E_0(Y^2)$ is finite, for ε small enough, there exists a constant $\kappa_0 > 0$ such that for every γ*

$$l(\gamma) \leq \kappa_0|\gamma - \gamma_0|.$$

Proof. The differences between $F_{0k,\gamma}$ and F_{0k}, $k=1,2$, are

$$\begin{aligned}F_{01,\gamma}(y)-F_{01}(y) &= 1_{\{\gamma\leq\gamma_0\}}\{F_{01}(y\wedge\gamma)-F_{01}(y\wedge\gamma_0)\}\\ &\quad +1_{\{\gamma\geq\gamma_0\}}\{F_{02}(y\wedge\gamma)-F_{02}(y\wedge\gamma_0)\}\\ &= 1_{\{\gamma\geq\gamma_0\}}\{F_{02}(y\wedge\gamma)-F_{02}(y\wedge\gamma_0)\}\\ &= 1_{\{\gamma_0<y\leq\gamma\}}\{F_{02}(y)-F_{02}(\gamma_0)\},\end{aligned}$$

for every $y\leq\gamma$, and

$$\begin{aligned}F_{02,\gamma}(y)-F_{02}(y) &= 1_{\{y>\gamma\geq\gamma_0\}}\{F_{02}(\gamma_0)-F_{02}(\gamma)\}\\ &\quad +1_{\{y>\gamma_0\geq\gamma\}}\{F_{01}(\gamma_0)-F_{01}(\gamma)\}\\ &\quad +1_{\{\gamma_0\geq y>\gamma\}}\{F_{01}(y)-F_{01}(\gamma)\}.\end{aligned}$$

The bound for the function l follows from the difference of integrals with respect to $F_{0,\gamma}$ and F_0. □

The process $W_n=n^{\frac{1}{2}}(l_n-l)$ is expressed as integrals with respect to the empirical process $\nu_{0n,\gamma}=n^{\frac{1}{2}}(\widehat{F}_{n,\gamma}-F_{0,\gamma})$, and $\nu_{0n}=n^{\frac{1}{2}}(\widehat{F}_n-F_0)$. Let $V_\varepsilon(\gamma_0)$ be an ε-neighborhood of γ_0 for the norm $\rho(\gamma,\gamma_0)=(|\gamma-\gamma_0|)^{\frac{1}{2}}$.

Lemma 9.2. *If $E_0(Y^4)$ is finite, for every $\varepsilon>0$, there exists a constant $\kappa_1>0$ such that for n large enough*

$$E_0\sup_{\gamma\in V_\varepsilon(\gamma_0)}|W_n(\gamma)|\leq\kappa_1\varepsilon^{\frac{1}{2}}.$$

Proof. The proof relies on the same arguments as Lemma 9.1. The condition $E_0(Y^4)$ is finite implies $E_0\{\sup_{|\gamma-\gamma_0|<\varepsilon}W_n^2(\gamma)\}$ is finite. The process $\nu_{0n,\gamma}$ has the variance $F_{0\gamma}\{1-F_{0\gamma}\}$ under P_0 and integrals with respect to the difference of the variances under $F_{0\gamma}$ and F_0 is a $O(|\gamma-\gamma_0|)$. This implies

$$E_0\sup_{\gamma\in V_\varepsilon(\gamma_0)}W_n^2(\gamma)\leq\kappa_1^2\varepsilon$$

and the lemma is deduced from the Cauchy–Schwarz inequality. □

Theorem 9.2. *If $E_0(Y^4)$ is finite, then*

$$\overline{\lim}_{n,A\to\infty}P_0(n|\widehat{\gamma}_n-\gamma_0|>A)=0.$$

The proof is the same as for Theorem 2.2, as a consequence of Lemmas 9.1 and 9.2. Let $\mathcal{U}_n=\{u_n=n(\gamma-\gamma_0),\gamma\in\mathbb{R}\}$ and let $\mathcal{U}_n^A=\{u\in\mathcal{U}_n:|u|<A\}$.

Theorem 9.3. *Under the conditions $E_0(Y^4)$ finite and under P_0, the process nl_n converges weakly in $\mathcal{U}_n^A$ to an uncentered Gaussian process G_0, as n and A tend to infinity. The variable $n(\widehat{\gamma}_n-\gamma_0)$ converges weakly to the location of the minimum U_0 of the process G_0.*

Proof. Let $\gamma_{n,u} = \gamma_0 + n^{-1}u_n$, for u_n in $\mathcal{U}_n^A$, for $k = 1, 2$, the variables $n \int y^k \, d(\widehat{F}_{n,\gamma_{n,u}} - \widehat{F}_{n,\gamma_0})$ are expressed as integrals with respect to F_{01} on the interval $]\gamma, \gamma_0]$ or F_{02} on the interval $]\gamma_0, \gamma]$, the expectation and the variance of the process $nl_n(\gamma_{n,u})$ are therefore $O(|u_n|)$ and they converge to non-null limits. By the weak convergence of the empirical distribution functions, the process $nl_n(\gamma_{n,u})$ converges weakly in $D([-A, A])$ to an uncentered Gaussian process G_0, as n and A tend to infinity. As the variable $n(\widehat{\gamma}_n - \gamma_0)$ maximizes this process in $[-A, A]$, its asymptotic distribution follows from its limit. □

The empirical process ν_{0n} is uniformly approximated by $n^{\frac{1}{2}}(\widehat{F}_{n,\gamma_0} - F_0)$, from Theorem 9.3, and it converges weakly under P_0 to the modified Brownian motion $B \circ F_0$.

9.2 Test of a continuous distribution function

A test of the hypothesis H_0 of a continuous distribution function F_0 against the alternative of a continuity of F_0 at an unknown threshold γ_0 belonging to the interval $]Y_{n:1}, Y_{n:n}[$ is performed with the test statistic

$$T_n = n(\widehat{\sigma}^2_{n,\widehat{\gamma}_n} - \widehat{\sigma}^2_{0n})$$

where $\widehat{\sigma}^2_{0n}$ is the empirical variance of the sample under the hypothesis, with γ_0 at the end-point of the support of F_0.

Under H_0, the test statistic T_n is the minimum value of the process

$$X_n(\gamma) = n(\widehat{\sigma}^2_{n,\gamma} - \widehat{\sigma}^2_{0n}).$$

As $\widehat{F}_n - \widehat{F}_{1n,\gamma} = \widehat{F}_{2n,\gamma}$ for every γ, the variable $\widehat{X}_n = X_n(\widehat{\gamma}_n)$ reduces under H_0 to

$$\widehat{X}_n = 2n\widehat{\mu}_{1n}\widehat{\mu}_{2n}$$

where $\widehat{\mu}_{kn} = \int y \, d\widehat{F}_{1n}$, for $k = 1, 2$.

Let U_0 be the limit of $n(\gamma_0 - \widehat{\gamma}_n)$ given by Theorem 9.3 and let ϕ_{01} be the limit of $\gamma f(\gamma)$ as γ tends to γ_0. If ϕ_{01} is finite, let G_U be the variable $U_0\phi_{01}$.

Proposition 9.1. *If $E_0(Y^4)$ is finite and if the limit of $\gamma f(\gamma)$ as γ tends to γ_0 is finite, the test statistic T_n converges weakly under H_0 to $T_0 = 2\mu_0 G_U$.*

Proof. Under H_0, by Theorem 9.1, the estimator $\widehat{\gamma}_n$ converges a.s. to γ_0, hence $\widehat{\mu}_{1n}$ converges a.s. to μ_0, the expectation of Y under F_0, and $\widehat{\mu}_{2n}$

converges a.s. to zero. By definition, the estimators $\widehat{\gamma}_n$ and $\widehat{\mu}_{2n}$ have the same convergence rate.

Replacing γ_0 by $Y_{n:n}$ in the proof of Theorem 9.3, the variable $n(\widehat{\gamma}_n - Y_{n:n})$ converges weakly to the location of the minimum U_0 of the process G_0 and the variable $n\widehat{\mu}_{2n} = n\int_{\widehat{\gamma}_n}^{Y_{n:n}} y\, d\widehat{F}_{2n}(y)$ is asymptotically equivalent to $n(Y_{n:n} - \widehat{\gamma}_n)\gamma_n f_0(\gamma_n)$, with γ_n between $\widehat{\gamma}_n$ and $Y_{n:n}$ so it converges to the variable $U_0\phi_{01}$. The limit of T_n follows. □

We assume that the conditions of Proposition 9.1 are satisfied.

Proposition 9.2. *Let the conditions of Proposition 9.1 be satisfied. Under a fixed alternative P_γ, the statistic T_n diverges. Under local alternatives P_{γ_n}, the statistic T_n converges weakly to $T_0 + c$, with a non-null constant c.*

Proof. Under an alternative P_γ with a change-point at $\gamma < \gamma_0$, the consistent estimator $\widehat{\gamma}_n$ converges a.s. to γ, the empirical mean $\widehat{\mu}_{1n}$ converges a.s. to the expectation $\mu_{1\gamma}$ of Y under $F_{1\gamma}$, and $\widehat{\mu}_{2n}$ converges a.s. to the expectation $\mu_{2\gamma}$ of Y under $F_{2\gamma}$. The estimators $\widehat{\gamma}_n$ and $\widehat{\mu}_{2n}$ have the same convergence rate n^{-1} given by Theorem 9.2, therefore the variable $\widehat{X}_n$ diverges.

Under local alternatives P_{γ_n}, with $\gamma_n = Y_{n:n} - n^{-1}u_n$, u_n converging to a non-null limit u, by consistency the estimator $\widehat{\gamma}_n$ converges a.s. to γ_0 and the variable $n(\gamma_n - \widehat{\gamma}_n)$ converges weakly to a limit U_0 like in Theorem 9.3. The variable $n\widehat{\mu}_{2n} = n\int_{\widehat{\gamma}_n}^{Y_{n:n}} y\, d\widehat{F}_{2n}(y)$ is asymptotically equivalent to $n(Y_{n:n} - \widehat{\gamma}_n)\gamma_n f_0(\gamma_n)$, with γ_n between $\widehat{\gamma}_n$ and $Y_{n:n}$, so it converges weakly under P_{γ_n} to the variable $(U_0 + u)\phi_{01}$. The limit of T_n is deduced from this convergence and the constant is $c = 2u\phi_{01}$. □

9.3 Conditional distribution functions with changes

On a probability space $(\Omega, \mathcal{A}, P_0)$, let (X, Y) be a random vector with distribution function F_0 such that Y is a real variable and X is a d-dimensional random vector of explanatory variables. Under P_0, the variable Y has the conditional distribution function $F_{0Y|X}$ and the marginal distribution function F_{0Y}, and the variable X has the marginal distribution function F_{0X}. They have discontinuities at the components of a threshold vector γ_0 for the regressor X, they determine p conditional sub-distribution functions, like in Chapter 4.

Under a probability P_γ, let δ_γ be the vector with components

$\delta_{\gamma,k} = 1_{\{X_k > \gamma_k\}}$, for $k = 1, \ldots, d$, their combination define p sets $I_{j\gamma}$ and the joint distribution function of (X, Y) is

$$F_\gamma(x, y) = \sum_{j=1}^{p} 1_{I_{j\gamma}}(x) F_{j\gamma}(x, y)$$

with $F_{j\gamma}(x, y) = P_\gamma(Y \leq y, X \leq x, X \in I_{j\gamma})$, its empirical estimator is $\widehat{F}_{n,\gamma}(x, y) = \sum_{j=1}^{p} 1_{I_{j\gamma}}(x) \widehat{F}_{nj\gamma}(x, y)$ with

$$\widehat{F}_{nj\gamma}(x, y) = n^{-1} \sum_{i=1}^{n} 1_{\{Y_i \leq y, X_i \leq x\}} 1_{I_{j\gamma}}(X_i).$$

The conditional distribution function of Y is

$$F_{\gamma,Y|X}(y; x) = \sum_{j=1}^{p} 1_{I_{j\gamma}}(x) F_{j\gamma,Y|X}(y; x) \tag{9.2}$$

with continuous and distinct conditional sub-distribution functions

$$F_{j\gamma,Y|X}(y; x) = P_\gamma(Y \leq y \mid X = x, X \in I_{j\gamma}).$$

At every γ, the sub-distribution functions $F_{j\gamma,Y|X}$ are estimated from a n-sample $(X_1, Y_1), \ldots, (X_n, Y_n)$ of the variable set (X, Y) using a kernel K with bandwidth h

$$\widehat{F}_{nh,j\gamma,Y|X}(y; x) = \frac{\sum_{i=1}^{n} 1_{\{Y_i \leq y\}} 1_{I_{j\gamma}}(X_i) K_h(x - X_i)}{\sum_{i=1}^{n} 1_{I_{j\gamma}}(X_i) K_h(x - X_i)}$$

and by (9.2), the kernel estimator of the conditional distribution function of Y, given X and thresholds at γ, is

$$\widehat{F}_{nh,\gamma,Y|X}(y; x) = \sum_{j=1}^{p} 1_{I_{j\gamma}}(x) \widehat{F}_{nh,j\gamma,Y|X}(y; x).$$

We assume the the following conditions are satisfied.

Conditions C

C1 The function K is a symmetric density of $C^2([-1;1]^d)$, such that the integrals $m_{2K} = \int v^{\otimes 2} K(v)dv$, $k_j = \int K^j(v)dv$ and $\int |K'(v)|^j dv$, for $j = 1, 2$ are finite;

C2 The conditional sub-distribution functions $F_{j\gamma,Y|X}$ belong to the class $C_b^2(\mathbb{R} \times I_{j\gamma})$, $j = 1, \ldots, p$;

C3 The bandwidth h_n converges to zero and $h = O(n^{-\frac{1}{d+4}})$, as n tends to infinity, and $K_h(x) = h^{-d} K(h^{-1}x_1, \ldots, h^{-1}x_d)$.

Under Conditions C, the estimators of the functions $F_{j\gamma,Y|X}$ converge a.s. uniformly under P_0 to functions $F_{0j\gamma,Y|X}$ on the sets $\mathbb{R} \times I_{j\gamma}$ and

$$E_0\|\widehat{F}_{nh,j\gamma,Y|X} - F_{0j\gamma,Y|X}\|_2 = O((nh^d)^{-\frac{1}{2}}),$$

for $j = 1, \ldots, p$. Under P_0 and the conditions, the estimator $\widehat{F}_{nh,j\gamma,Y|X}$ converge a.s. to the conditional sub-distribution function $F_{0j\gamma,Y|X}$, limit as n tends to infinity of

$$F_{nh,j\gamma,Y|X}(y;x) = \frac{\int 1_{\{z \le y\}} K_h(x-s)\, F_{0j\gamma}(ds,dz)}{\int K_h(x-s)\, F_{0j\gamma}(ds,dz)},$$

where the distribution function of (X, Y) with X restricted to $I_{j\gamma}$ is $F_{0j\gamma}(s,z) = P_0(Y_i \le y, X_i \in I_{j\gamma}, X_i \le s)$, for $j = 1, \ldots, p$.

Under F_γ, the expectation of Y conditionally on $X = x$ is

$$\mu_\gamma(x) = \sum_{j=1}^p 1_{I_{j\gamma}}(X)\mu_{j\gamma}(x) = \sum_{j=1}^p 1_{I_{j\gamma}}(X) \int y\, F_{j\gamma,Y|X}(dy;x)$$

and it is estimated by $\widehat{\mu}_{nh,\gamma}(x) = \sum_{j=1}^p 1_{I_{j\gamma}}(X)\widehat{\mu}_{nh,j\gamma}(x)$, with

$$\begin{aligned}\widehat{\mu}_{nh,j\gamma}(x) &= \int y\, \widehat{F}_{nh,j\gamma,Y|X}(dy;x) \\ &= \frac{\sum_{i=1}^n Y_i 1_{\{Y_i \le y\}} 1_{I_{j\gamma}}(X_i) K_h(x - X_i)}{\sum_{i=1}^n 1_{I_{j\gamma}}(X_i) K_h(x - X_i)}.\end{aligned}$$

The empirical estimator of the variance of Y conditionally on $X = x$ under $F_{\gamma,Y|X}$ is

$$\widehat{\sigma}^2_{nh,\gamma} = n^{-1} \sum_{i=1}^n \Big\{ Y_i - \sum_{j=1}^p 1_{I_{j\gamma}}(X_i)\widehat{\mu}_{nh,j\gamma}(X_i) \Big\}^2,$$

and the parameter γ is estimated by

$$\widehat{\gamma}_{nh} = \arg\inf_\gamma \widehat{\sigma}^2_{nh,\gamma}.$$

Then for $j = 1, \ldots, p$, the conditional distribution functions $F_{0j,Y|X}$ are estimated by $\widehat{F}_{nhj,Y|X} = \widehat{F}_{nh,j\widehat{\gamma}_{nh},Y|X}$ and the estimator of the conditional distribution function of Y is

$$\widehat{F}_{nh,Y|X}(y;x) = \sum_{j=1}^p 1_{I_{j\widehat{\gamma}_{nh}}}(x)\widehat{F}_{nh,j\widehat{\gamma}_{nh},Y|X}(y;x).$$

For every γ, the empirical variance $\widehat{\sigma}^2_{nh,\gamma}$ converges under P_0 to the variance of Y under $F_{0,\gamma}(x,y) = \sum_{j=1}^p 1_{I_{j\gamma}}(x) F_{0j\gamma,Y|X}(x,y)$, the convergence of the estimator $\widehat{\gamma}_n$ is deduced like in Theorem 9.4.

Theorem 9.4. *If $E_\gamma(Y^2)$ is finite for γ in a neighborhood of γ_0, then the maximum likelihood estimator $\widehat{\gamma}_n$ of γ_0 is a.s. consistent under P_0.*

Let

$$l_{nh}(\gamma) = \widehat{\sigma}^2_{nh,\gamma} - \widehat{\sigma}^2_{nh,\gamma_0}.$$

For every γ and under the conditions, the process $l_{nh}(\gamma)$ converges a.s. uniformly on $\mathcal{I}_X$ to the function

$$l(\gamma) = \sigma^2_{0\gamma} - \sigma^2_0,$$

where σ_0^2 is the variance of Y under P_0.

Lemma 9.3. *If $E_0(Y^2)$ is finite, for ε small enough, there exists a constant $\kappa_0 > 0$ such that for every γ*

$$l(\gamma) \leq \kappa_0 \|\gamma - \gamma_0\|_1.$$

Proof. For $j = 1, \ldots, p$ and for every γ, the difference between the conditional distribution functions $F_{0j,\gamma,Y|X}$ and $F_{0j,Y|X}$, is

$$\begin{aligned} F_{0j,\gamma,Y|X}(y;x) - F_{0j,Y|X}(y;x) &= P_0(Y \leq y, \mid X \leq x, X \in I_{j\gamma}) \\ &\quad - P_0(Y \leq y, \mid X \leq x, X \in I_{0j}) \end{aligned}$$

and it is a $O(\|\gamma - \gamma_0\|_1)$. It follows that the differences of first two moments with respect to $F_{0j,\gamma,Y|X}$ and $F_{0j,Y|X}$ are also $O(\|\gamma - \gamma_0\|_1)$. □

The conditional empirical processes $\nu_{0nh,\gamma,Y|X} = (nh^d)^{\frac{1}{2}}(\widehat{F}_{nh,\gamma,Y|X} - F_{0,\gamma,Y|X})$, and respectively $\nu_{0nh,Y|X} = (nh^d)^{\frac{1}{2}}(\widehat{F}_{nh,Y|X} - F_{0,Y|X})$, converge weakly under P_0 to transformed Brownian motions $B \circ F_{0,\gamma,Y|X}$, and respectively $B \circ F_{0,Y|X}$.

The process $W_{nh} = (nh^d)^{\frac{1}{2}}(l_{nh} - l)$ is expressed by the difference of integrals with respect to $\nu_{0nh,\gamma}$ and ν_{0nh}. Let $V_\varepsilon(\gamma_0)$ be an ε-neighborhood of γ_0 for the norm $\rho(\gamma, \gamma_0) = (\sum_{j=1}^p |\gamma_j - \gamma_{0j}|)^{\frac{1}{2}}$.

Lemma 9.4. *If $E_0(Y^4)$ is finite, for every $\varepsilon > 0$, there exists a constant $\kappa_1 > 0$ such that for n large enough*

$$E_0 \sup_{\gamma \in V_\varepsilon(\gamma_0)} |W_n(\gamma)| \leq \kappa_1 \varepsilon^{\frac{1}{2}}.$$

Proof. For every γ, the process $\nu_{0nh,\gamma}$ has the variance $F_{0\gamma}\{1 - F_{0\gamma}\}$ under P_0 and integrals with respect to the difference of the variances under $F_{0\gamma}$ and F_0 is a $O(\|\gamma - \gamma_0\|_1)$. The condition $E_0(Y^4)$ is finite implies $E_0\{\sup_{|\gamma-\gamma_0|<\varepsilon} W^2_{nh}(\gamma)\}$ is finite, like in Proposition 4.2, it follows that

$$E_0 \sup_{\rho(\gamma,\gamma_0)<\varepsilon} W_n^2(\gamma) \leq \kappa_1^2 \varepsilon,$$

the result is deduced from the Cauchy–Schwarz inequality. □

Theorem 9.5. *If $E_0(Y^4)$ is finite, then*

$$\overline{\lim}_{n,A\to\infty} P_0(nh^d\|\widehat{\gamma}_{nh} - \gamma_0\|_1 > A) = 0.$$

The convergence rate of the conditional expectation of Y given X is $(nh^d)^{-\frac{1}{2}}$ and the proof is the same as for Theorem 2.2, using Lemmas 9.3 and 9.4. Let $\mathcal{U}_n^A = \{u_n = n(\gamma - \gamma_0), \gamma \in \mathcal{I}_X, \|u\|_1 < A\}$.

Theorem 9.6. *Under the conditions $E_0(Y^4)$ finite and under P_0, the process $nh^d l_n$ converges weakly in $\mathcal{U}_n^A$ to an uncentered Gaussian process G_0, as n and A tend to infinity. The variable $nh^d(\widehat{\gamma}_n - \gamma_0)$ converges weakly to the location of the minimum U_0 of the process G_0.*

The proof is the same as for Theorem 9.3, with the convergence rate of the kernel estimator of the conditional distribution functions $\widehat{F}_{nh,\gamma_0,Y|X}$ and $\widehat{F}_{nh,\gamma_{nu},Y|X}$, for u in $\mathcal{U}_n^A$.

9.4 Test of a continuous conditional distribution function

A test of the hypothesis H_0 of a continuous distribution function for Y conditionally on X against the alternative of a function with a discontinuity at an unknown threshold γ_0 belonging to a bounded set I_X is performed with the test statistic

$$T_n = nh^d(\widehat{\sigma}^2_{n,\widehat{\gamma}_n} - \widehat{\sigma}^2_{0n})$$

where $\widehat{\sigma}^2_{0n}$ is the empirical variance of the sample under the hypothesis, with γ_0 at the end-point of the support of I_X. Let $I_{1\gamma}(x) = 1_{\{x\leq\gamma\}}$.

Under H_0, the test statistic T_n is the minimum value, at $\widehat{\gamma}_n$, of the process

$$\begin{aligned}X_n(\gamma) &= nh^d(\widehat{\sigma}^2_{n,\gamma} - \widehat{\sigma}^2_{0n})\\ &= h^d\sum_{i=1}^n\Big\{Y_i - \sum_{j=1}^p 1_{I_{j\gamma}}(X_i)\widehat{\mu}_{nh,j\gamma}(X_i)\Big\}^2 - h^d\sum_{i=1}^n\{Y_i - \widehat{\mu}_{0nh}(X_i)\}^2.\end{aligned}$$

Proposition 9.3. *If $E_0[\sup_{|\gamma-\gamma_0\|<\varepsilon}\{Y - m_\gamma(X)\}^4]$ is finite, the test statistic T_n converges weakly under H_0 to a $\chi^2_{d(p-1)}$ variable T_0.*

Proof. Under H_0, the estimator $\widehat{\gamma}_{nh}$ converges a.s. to γ_0 and $nh^d(\gamma_0 - \widehat{\gamma}_n)$ given by Theorem 9.6, it follows that $\widehat{\mu}_{nh,1}$ converges a.s. to μ_0 and the variable $(nh^d)^{\frac{1}{2}}(\widehat{\mu}_{nh,1} \quad \widehat{\mu}_{0n})$ converges in probability to zero with the rate nh^d. The process X_n develops as $X_{1n} + X_{2n}$ with

$$\begin{aligned}
X_{1n}(\gamma) &= h^d \sum_{i=1}^{n} \{Y_i - 1_{I_{1\gamma}}(X_i)\widehat{\mu}_{nh,1\gamma}(X_i)\}^2 - h^d \sum_{i=1}^{n} \{Y_i - \widehat{\mu}_{0nh}(X_i)\}^2 \\
&= h^d \sum_{i=1}^{n} \{1_{I_{1\gamma}}(X_i)\widehat{\mu}_{nh,1\gamma}(X_i) - \widehat{\mu}_{0nh}(X_i)\} \\
&\qquad .\{1_{I_{1\gamma}}(X_i)\widehat{\mu}_{nh,1\gamma}(X_i) + \widehat{\mu}_{0nh}(X_i) - 2Y_i\}, \\
X_{2n}(\gamma) &= h^d \widehat{\sigma}^2_{n,\gamma} - h^d \sum_{i=1}^{n} \{Y_i - 1_{I_{1\gamma}}(X_i)\widehat{\mu}_{nh,1\gamma}(X_i)\}^2 \\
&= h^d \sum_{i=1}^{n} \Big\{\sum_{j=2}^{p} 1_{I_{j\gamma}}(X_i)\widehat{\mu}_{nh,j\gamma}(X_i)\Big\}^2 \\
&\quad -2h^d \sum_{i=1}^{n} \Big\{\sum_{j=2}^{p} 1_{I_{j\gamma}}(X_i)\widehat{\mu}_{nh,j\gamma}(X_i)\Big\}\{Y_i - 1_{I_{1\gamma}}(X_i)\widehat{\mu}_{nh,1\gamma}(X_i)\}.
\end{aligned}$$

The expectation of $\widehat{\mu}_{0nh}(X_i) + 1_{I_{1\gamma}}(X_i)\widehat{\mu}_{nh,1\gamma}(X_i) - 2Y_i$ conditionally on X_i is $\mu_{0n} + 1_{I_{1\gamma}}(X_i)\mu_{nh,1\gamma}(X_i) - 2\mu_0$, at $\widehat{\gamma}_{nh} = \gamma_0 - (nh)^{-1}\widehat{u}_{nh}$ it reduces to a $O(h^2)$ and

$$1_{I_{1\gamma}}(X_i)\widehat{\mu}_{nh,1\gamma}(X_i) + \widehat{\mu}_{0nh}(X_i) - 2Y_i = O_p((nh^d)^{\frac{1}{2}}),$$

therefore $X_{1n}(\gamma)$ converges to zero in probability under H_0.

The variables $U_{nhj}(x) = 1_{I_{j\widehat{\gamma}_{nh}}}(x)\widehat{\mu}_{nh,j\widehat{\gamma}_{nh}}(x)$ have the same convergence rate to zero as $\widehat{\gamma}_n$ to γ_0 and the variables $1_{I_{1\widehat{\gamma}_{nh}}}(X_i)\widehat{\mu}_{nh,1\widehat{\gamma}_{nh}}(X_i) - Y_i$ converge in probability to zero. Then

$$X_{2n}(\widehat{\gamma}) = h^d \sum_{i=1}^{n} \Big\{\sum_{j=2}^{p} 1_{I_{j\gamma}}(X_i)\widehat{\mu}_{nh,j\gamma}(X_i)\Big\}^2 + o_p(1),$$

as the expectation and the variance of the variables $U_{nhj}(X_i)$ are $O_p(\|\gamma_0 - \widehat{\gamma}_{nh}\|)$, $X_{2n}(\widehat{\gamma})$ converges to $\chi^2_{d(p-1)}$ variable. □

Proposition 9.4. *Let the conditions of Proposition 9.3 be satisfied. Under a fixed alternative P_θ, the statistic T_n diverges. Under local alternatives P_{γ_n}, the statistic T_n converges weakly to $T_0 + c$, with a strictly positive constant c.*

Proof. Under an alternative P_γ with a change-point at $\gamma < \gamma_0$, the consistent estimator $\widehat{\gamma}_n$ converges a.s. to γ, the empirical mean $\widehat{\mu}_{1n}$ converges a.s. to the expectation $\mu_{1\gamma}$ of Y under $F_{1\gamma}$, and $\widehat{\mu}_{2n}$ converges a.s. to the expectation $\mu_{2\gamma}$ of Y under $F_{2\gamma}$. The estimators $\widehat{\gamma}_n$ and $\widehat{\mu}_{2n}$ have the same convergence rate $(nh^d)^{-1}$ given by Theorem 9.5, therefore the variable $\widehat{X}_n$ diverges.

Under local alternatives P_{γ_n}, with $\gamma_n = Y_{n:n} - (nh^d)^{-1}u_n$, u_n converging to a non-null limit u, by consistency the estimator $\widehat{\gamma}_n$ converges a.s. to γ_0 and the variable $nh^d(\gamma_n - \widehat{\gamma}_n)$ converges weakly to a limit U_0 like in Theorem 9.6. The regression functions of the alternative are $m_{jn} = m_0 + (nh^d)^{-\frac{1}{2}}\varphi_{jn}$ where the sequences $(\varphi_{jn})_n$ converge uniformly to non-null limits φ_j, for $j = 1, \ldots, p$. The means are $\mu_{jn} = \mu_0 + (nh^d)^{-\frac{1}{2}}\alpha_{jn}$, with uniformly convergent functions α_{jn}, for $j = 1, \ldots, p$. The expectation and the variance of the variables $I_{j\widehat{\gamma}_n}(X)$ are $O_p((nh^d)^{-1})$ for $j = 2, \ldots, p$, then $X_{2n}(\widehat{\gamma}_n)$ converges weakly to T_0.

Moreover $nh^d\{1_{I_{1\widehat{\gamma}_n}}(X_i)\widehat{\mu}_{nh,1\widehat{\gamma}_n}(X_i) - \widehat{\mu}_{0nh}(X_i)\}$ is asymptotically equivalent to $\alpha_1(X_i)$, it follows that $X_{1n}(\widehat{\gamma}_n)$ converges in probability to $E_0\alpha_1^2(X)$. □

9.5 Diffusion with a change of drift

Let α and β be real functions of class $C_2(I_X)$ and let B be the standard Brownian motion B on $\mathbb{R}^+$. A diffusion process $(X_t)_{t\in[0,T]}$ is defined on a probability space $(\Omega, \mathcal{A}, P_{\alpha,\beta})$ by a stochastic differential equation

$$dX_t = \alpha(X_t)dt + \beta(X_t)dB_t, t \in [0, T] \tag{9.3}$$

with an initial value X_0 such that $E(X_0^2)$ is finite. The Brownian process $(B_t)_{t\geq 0}$ is a martingale with respect to the filtration generated by the $(B_u)_{u<t}$, it satisfies $E(B_t - B_s \mid B_s) = 0$, for every $0 < s < t$, and has the moments $EB_t^{2k+1} = 0$, $EB_t^2 = t$ and $EB_t^4 = 3t^2$. The uniform norms of the functions α and β on bounded subsets of I_X are denoted $\|\alpha\|$, $\|\beta\|$, and their L^2-norms $\|\alpha^2(X)\|_{L^2}$ and $\|\beta^2(X)\|_{L^2}$ are supposed to be finite.

Equation (9.3) with locally Lipschitz drift and diffusion functions has a unique solution of $X_t - X_0 = \int_0^t \alpha(X_s)ds + \int_0^t \beta(X_s)dB_s$, it is a continuous Gaussian process, its expectation under P_0 is $E_0(X_t - X_0) = \int_0^t E\alpha(X_s)\,ds$ and its variance is

$$Var_0(X_t - X_0) = Var_0\left\{\int_0^t \alpha(X_s)\,ds\right\} + E_0\int_0^t \beta^2(X_s)\,ds.$$

With a continuous function β, the process $Y_t = \int_0^t \beta(X_s)dB_s$ is a square integrable centered martingale with variance $\int_0^t \beta^2(X_s)\,ds$ and (Pons, 2017), for every $x > 0$

$$P_0\Big(\sup_{0\le t\le T} Y_t > x\Big) \le \exp\Big\{-\frac{x^2}{2\sup_{0\le t\le T} E_0|\beta(X_t)|^2}\Big\}.$$

A diffusion is a markovian process and there exists an invariant distribution function F_X and a transition probability measure π in $\mathcal{I}_X \times \mathcal{I}_X$ such that for every bounded and continuous function ψ on $\mathcal{I}_X \times \mathcal{I}_X$

$$\lim_{T\to\infty} ET^{-1}\int_{[0,T]} \psi(X_t)\,dt = \int_{\mathcal{I}_X} \psi(x)\,dF_X(x), \tag{9.4}$$

$$\lim_{T\to\infty} ET^{-1}\int_{[0,T]^{\otimes 2}} \psi(X_s, X_t)\,ds\,dt = \int_{\mathcal{I}_X^{\otimes 2}} \psi(x,y)\pi_x(dy)\,dF_X(x),$$

let f_X be the density of F_X. Kernel estimators of the functions α and β are defined and studied by Pons (2011) from the observation of the process X on $[0,T]$, as T tends to infinity. Let K be a kernel satisfying the conditions 2.1 and let $h = h_T$ be its bandwidth such that $h_T = O(T^{-\frac{1}{5}})$.

In a model with a change-point in the drift at a threshold γ in I_X of the process X, the function α is written as

$$\alpha_\gamma(x) = \alpha_{1\gamma}(x)1_{\{x\le\gamma\}} + \alpha_{2\gamma}(x)1_{\{x>\gamma\}} \tag{9.5}$$

and γ is the first point of discontinuity of α. The parameter values under the probability measure P_0 of the observations are γ_0, $\alpha_0 = \alpha_{01}1_{\{x\le\gamma_0\}} + \alpha_{02}1_{\{x>\gamma_0\}}$ and β_0^2.

At γ, kernel estimators of the functions $\alpha_{k\gamma}$ are defined from the observation of the process X on an increasing interval $[0,T]$ as

$$\widehat{\alpha}_{1Th,\gamma}(x) = \frac{\int_0^T 1_{\{X_s\le\gamma\}}K_h(x-X_s)\,dX_s}{\int_0^T 1_{\{X_s\le\gamma\}}K_h(x-X_s)\,ds},$$

$$\widehat{\alpha}_{2Th,\gamma}(x) = \frac{\int_0^T 1_{\{X_s>\gamma\}}K_h(x-X_s)\,dX_s}{\int_0^T 1_{\{X_s>\gamma\}}K_h(x-X_s)\,ds},$$

under P_0, they converge a.s. uniformly to functions $\alpha_{01,\gamma}$ and $\alpha_{02,\gamma}$ with the rate $(Th)^{-\frac{1}{2}}$. The expectations of $\widehat{\alpha}_{1Th,\gamma}(x)$ and $\widehat{\alpha}_{2Th,\gamma}(x)$ under Conditions C are asymptotically equivalent to

$$\begin{aligned}
\alpha_{1Th,\gamma}(x) &= \frac{\int K_h(x-y)1_{\{y\le\gamma\}}\alpha_0(y)\,dF_X(y)}{f_{Th,\gamma}(x)}, \ x\le\gamma,\\
&= \alpha_{01}(x)1_{\{x\le\gamma\wedge\gamma_0\}} + \alpha_{02}(x)1_{\{\gamma_0<x\le\gamma\}} + O(h^2), \ x\le\gamma,\\
\alpha_{2Th,\gamma}(x) &= \frac{\int K_h(x-y)1_{\{y>\gamma\}}\alpha_0(y)\,dF_X(y)}{f_{Th,\gamma}(x)}, \ x>\gamma\\
&= \alpha_{02}(x)1_{\{x>\gamma\vee\gamma_0\}} + \alpha_{01}(x)1_{\{\gamma<x\le\gamma_0\}} + O(h^2), \ x>\gamma,
\end{aligned}$$

where

$$\int K_h(x-y)1_{\{y\leq\gamma\}}\alpha_0(y)\,dF_X(y) = \int K(u)1_{\{|u|<h^{-1}(\gamma-x)\}}(\alpha_0 f_X)(x+hu)$$

has the asymptotic expansion $\alpha_0(x)f_X(x) + \frac{1}{2}h^2 m_{2K}(\alpha_0 f_X)''(x)$, for $x \leq \gamma$, and the numerator of $\alpha_{2Th,\gamma}(x)$ has the same expansion for $x > \gamma$, the limits of the estimators are denoted $\alpha_{0k,\gamma}$, $k = 1, 2$. The asymptotic variance of $(Th)^{\frac{1}{2}}(\widehat{\alpha}_{Th,\gamma} - \alpha_{0\gamma})$ under P_0 is $(Th)^{-1}v_\alpha$ where $v_\alpha(x) = k_2 f_X^{-1}(x)\beta_0^2(x)$. Centering X_t under P_0 with this estimator yields the process

$$\begin{aligned} Z_{t,\gamma} &= X_t - X_0 - \int_0^t \widehat{\alpha}_{Th,\gamma}(X_s)\,ds \\ &= \int_0^t (\alpha_0 - \widehat{\alpha}_{Th,\gamma})(X_s)\,ds + \int_0^t \beta_0(X_s)\,dB_s. \end{aligned}$$

As t tends to infinity, the expectation of the process $t^{-1}Z_{t,\gamma}$ under P_0 converges to $\int_{I_X}(\alpha_0 - \alpha_{0,\gamma})(x)f_X(x)\,dx$ and Z_{t,γ_0} has the asymptotic expansion

$$Z_{t,\gamma_0} = \int_0^t \beta_0(X_s)\,dB_s + o_p(1). \tag{9.6}$$

The variance under P_0 of $t^{-1}Z_{t,\gamma}$ converges to $v_0 = \int_{I_X} E_0\beta_0^2(x)f_X(x)\,dx$ by consistency of the estimator $\widehat{\alpha}_{Th,\gamma}$. The function

$$\beta_0^2(x) = \int_{I_X} E_0\{\beta_0^2(X_s) \mid X_s = x\}\,ds$$

is estimated from the process $Z_{t,\gamma}$ as

$$\widehat{\beta}^2_{Th,\gamma}(x) = 2\frac{\int_0^T Z_{s,\gamma}K_h(X_s - x)\,dZ_{s,\gamma}}{\int_0^T K_h(X_s - x)\,ds}. \tag{9.7}$$

The estimator (9.7) is a.s. consistent at γ_0, by (9.4) and (9.6), and the variable $(Th)^{\frac{1}{2}}(\widehat{\beta}^2_{Th,\gamma_0} - \beta_0^2)$ converges weakly on I_X to a Gaussian process with variance function $v_{\beta_0} = k_2 f_X^{-1}\beta_0^4$. At γ, the estimator of the function β_0^2 has a bias depending on the asymptotic behaviour of the process $(Th)^{\frac{1}{2}}(\widehat{\alpha}_{Th,\gamma} - \alpha_0)$ which is a difference of functions at γ and γ_0 under P_0. The parameter γ_0 is estimated by least squares as

$$\widehat{\gamma}_{Th} = \arg\inf_{\gamma\in I_X} \widehat{\beta}^2_{Th,\gamma}(x),$$

then $\widehat{\alpha}^2_{Th} = \widehat{\alpha}^2_{Th,\widehat{\gamma}_{Th}}$ and $\widehat{\beta}^2_{Th} = \widehat{\beta}^2_{Th,\widehat{\gamma}_{Th}}$. By (9.4), the expectation under P_0 of the process $\widehat{\beta}^2_{Th,\gamma}(x)$ converges uniformly to

$$(\alpha_0 - \alpha_{0,\gamma})^2(x) + \beta_0^2(x)$$

which is minimum at γ_0. It follows that the estimator $\widehat{\gamma}_{Th}$ is a.s. consistent.

The process $l_{Th}(\gamma) = T^{-1}\{\widehat{\beta}^2_{Th,\gamma}(X_T) - \widehat{\beta}^2_{Th,\gamma_0}(X_T)\}$ converges a.s. uniformly to the function $l(\gamma) = \int_{I_X} (\alpha_0 - \alpha_{0,\gamma})^2(x)\, dF_X(x)$.

Lemma 9.5. *Under the conditions of the model with drift (9.5), for ε small enough, there exists a constant $\kappa_0 > 0$ such that for every γ distinct from γ_0*

$$\sup_{x \in I_X} l(x,\gamma) \geq \kappa_0 |\gamma - \gamma_0|.$$

Proof. The probabilities $P_0(\gamma_0 < X \leq \gamma)$ and $P_0(\gamma < X \leq \gamma_0)$ are $O(|\gamma - \gamma_0|)$ and the result is obtained by integration of the difference

$$(\alpha_0 - \alpha_{0\gamma})^2 = (\alpha_{01} - \alpha_{02})^2 1_{\{\gamma_0 < x \leq \gamma\}} + (\alpha_{02} - \alpha_{01})^2 1_{\{\gamma < x \gamma_0\}}$$

with respect to F_X on I_X. □

The process $W_{Th,\gamma}(x) = (Th)^{\frac{1}{2}}(l_{Th}(x,\gamma) - l(x,\gamma)$ is centered and it is expressed as variations of processes between γ and γ_0.

Lemma 9.6. *Under P_0 and the conditions of model (9.5), for every $\varepsilon > 0$, there exists a constant $\kappa_1 > 0$ such that for n large enough*

$$E_0 \sup_{\gamma \in V_\varepsilon(\gamma_0)} \|W_{Th,\gamma}\|_{I_X} \leq \kappa_1 \varepsilon^{\frac{1}{2}}.$$

Lemmas 9.5 and 9.6 imply that the convergence rate of the estimator $\widehat{\gamma}_{Th}$ is $(Th)^{-1}$. Let $\mathcal{U}^A_{Th} = \{u_{Th} = Th(\gamma_{Th} - \gamma_0), \gamma_{Th} \in \mathcal{I}_X, |u_{Th}| < A\}$.

Theorem 9.7. *Under P_0 and the conditions of model (9.5), the process Thl_{Th} converges weakly in $\mathcal{U}^A_{Th}$ to an uncentered Gaussian process L_0, as T and A tend to infinity. The variable $Th(\widehat{\gamma}_{Th} - \gamma_0)$ converges weakly to the location of the minimum U_0 of the process L_0.*

Proof. The process $l_{Th}(x,\gamma)$ develops as

$$l_{Th}(x,\gamma) = 2\frac{\int_0^T Z_{s,\gamma} K_h(X_s - x)\, dZ_{s,\gamma}}{\int_0^T K_h(X_s - x)\, ds} - 2\frac{\int_0^T Z_{s,\gamma_0} K_h(X_s - x)\, dZ_{s,\gamma_0}}{\int_0^T K_h(X_s - x)\, ds}.$$

Let $G_{Th,\gamma}(x) = (Th)^{\frac{1}{2}}(\widehat{\alpha}_{Th,\gamma} - \alpha_\gamma)(x)$, the processes

$$G_{Th,\gamma} - G_{Th,\gamma_0} = (Th)^{\frac{1}{2}}(\widehat{\alpha}_{Th,\gamma} - \widehat{\alpha}_{Th,\gamma_0} - \alpha_\gamma + \alpha_{\gamma_0})$$

and $(Th)^{\frac{1}{2}}(Z_{s,\gamma} - Z_{s,\gamma_0}) = \int_0^t (G_{Th,\gamma_0} - G_{Th,\gamma})(X_s)\, ds$ converge weakly to a Gaussian process on I_X with expectations depending of the variations of the process X on intervals with length $|\gamma-\gamma_0|$, the variance of $(Th)^{\frac{1}{2}}(Z_{s,\gamma}-Z_{s,\gamma_0})$ depends only on the function v_α.

Let u_{Th} in $\mathcal{U}_{Th}^A$, for A sufficiently large, and let $\gamma_{Th,u}=\gamma_0+(Th)^{-1}u_{Th}$. Under P_0, the expectation of $Th(Z_{s,\gamma_{Th,u}}-Z_{s,\gamma_0})$ is

$$ThE_0\int_0^t\{(\alpha_{01}-\alpha_{02})(X_s)1_{\{\gamma_0<X_s\leq\gamma_{Th,u}\}}\,ds$$
$$+(\alpha_{02}-\alpha_{01})(X_s)1_{\{\gamma_{Th,u}<X_s\gamma_0\}}\}\,ds,$$

it is a $O_p(|u_{Th}|)$ and its variance has the same order. As the process $Thl_{Th}(\gamma_{Th,u})$ is tight, it converges weakly on $\mathcal{U}_{Th}^A$ to an uncentered Gaussian process with finite variance function, as A and T tend to infinity. Then, under P_0, the process $Thl_{Th}(\gamma_{Th},X_T)$ converges weakly in $\mathcal{U}_{Th}^A$ to an uncentered Gaussian process L_0 with a finite variance.

The variable $\widehat{u}_{Th}=Th(\widehat{\gamma}_{Th}-\gamma_0)$ maximizes the process $Thl_{Th}(\gamma_{Th,u})$ in $\mathcal{U}_{Th}^A$, for A sufficiently large, and it converges weakly to the location of the minimum of the process L_0. □

9.6 Test of a diffusion with a continuous drift

Let H_0 be the hypothesis of a diffusion process X with a continuous drift function, a test of H_0 against the alternative of a drift with a discontinuity at an unknown threshold γ_0 of the process X is performed with the statistic

$$S_T=Thl_{Th}(\widehat{\gamma}_{Th})=Th\{\widehat{\beta}_{Th}^2-\widehat{\beta}_{Th,\gamma_0}^2\}.$$

The process l_{Th} is the difference of the estimated variances under the alternative of model (9.3) and the hypothesis. We assume that the conditions of Section 9.5 are satisfied. Under H_0, γ_0 is at the end-point of I_X, $\alpha_{01}=\alpha_0$ and $\alpha_{02}=0$, the estimator $\widehat{\alpha}_{2Th}$ is zero as the indicator $1_{\{X>\gamma_0\}}$ is zero, the estimator $\widehat{\alpha}_{0Th}$ is defined like $\widehat{\alpha}_{1Th,\gamma_0}$ and it is a.s. consistent. The process $Z_{0t}=\int_0^t(\alpha_0-\widehat{\alpha}_{0Th})(X_s)\,ds+\int_0^t\beta_0(X_s)\,dB_s$ yields the estimator $\widehat{\beta}_{Th,\gamma_0}^2$.

Proposition 9.5. *Under H_0 and if $\|\beta^2(X)\|_{L^2}$ is finite, the test statistic S_T converges weakly to the maximum S_0 of the process L_0.*

Proof. Under the condition, the diffusion process is bounded in probability under H_0 and γ_0 is bounded. By Theorem 9.7, the process $Thl_{Th}(\gamma_{Th,u})$ converges weakly to an uncentered Gaussian process L_0 and $Thl_{Th}(\widehat{\gamma}_{Th})$ converges weakly to the maximum of this process. □

Proposition 9.6. *Let the conditions of Proposition 9.5 be satisfied. Under a fixed alternative P_θ, the statistic S_T diverges. Under local alternatives $P_{\theta_{Th}}$, the statistic S_T converges weakly to $S_0 + S$, with a non-degenerated variable S.*

Proof. Under an alternative P_θ, with γ distinct from γ_0, the function α_{02} is not zero and $Th(\alpha_0 - \alpha_{02})$ diverges hence S_T diverges.

Under local alternatives $P_n = P_{\theta_{Th}}$, the parameters are $\gamma_{Th} = \gamma_0 - (Th)^{-1}u_{Th}$ and the functions $\alpha_{1,Th} = \alpha_0 + (Th)^{-\frac{1}{2}}v_{1,Th}$ and $\alpha_{2,Th} = \alpha_0 + (Th)^{-\frac{1}{2}}v_{2,Th}$, with sequences $(u_{Th})_T$ and respectively $v_{i,Th}$ converging to non-zero limits u and respectively non-zero functions v_i. Then the function $(Th)^{\frac{1}{2}}(\alpha_{Th} - \alpha_0)(x)$ converges to $v(x) = v_1 1_{\{x \le \gamma_{Th}\}} + v_2 1_{\{x > \gamma_{Th}\}}$. The process

$$G_{Th,\gamma_{Th}} - G_{Th,\gamma_0} = (Th)^{\frac{1}{2}}(\widehat{\alpha}_{Th,\gamma_{Th}} - \widehat{\alpha}_{Th,\gamma_0} - \alpha_{\gamma_{Th}} + \alpha_0)$$

converges weakly under P_n to a Gaussian process sum of its limit under H_0 and v. As in Theorem 9.7, the process $Thl_{Th}(\gamma_{Th}, X_T)$ converges weakly under P_n to an uncentered Gaussian process L_K depending on the function $(v_1 - v_2)$. □

Bibliography

Aalen, O. (1978). Non-parametric inference for a family of counting processes, *Ann. Statist.* **6**, pp. 701–726.

Andersen, P. K. and Gill, R. D. (1982). Cox's regression model for counting processes: a large sample study, *Ann. Statist.* **10**, pp. 1100–1120.

Bai, J. (1993). On the partial sums of residuals in autoregressive and moving average models, *J. Time Series Anal.* **14**, pp. 247–260.

Basseville, M. and Benveniste, B. (1983). Sequential detection of abrupt changes in spectral characteristics of digital signals, *IEEE Trans. Information Theory* **IT-29**, pp. 709–724.

Bhattacharya, P. K. (1994). Some aspects of change-point analysis, *IMS Lecture Notes-Monograph Series* **23**, pp. 28–56.

Billingsley, P. (1968). *Convergence of Probability Measures* (Wiley, New York).

Breslow, N. and Crowley, J. (1974). A large sample study of the life table and product limit estimates under random censorship, *Ann. Statist.* **2**, pp. 437–453.

Chernoff, H. and Zacks, S. (1964). Estimating the current mean of a normal distribution, *Ann. Math. Statist.* **35**, pp. 999–1028.

Cox, D. R. (1969). Some sampling problems in technology, *New Developments in Survey Sampling*, pp. 506–527.

Cox, D. R. (1972). Regression model and life tables (with discussion), *J. Roy. Statist. Soc. Ser. B* **34**, pp. 187–220.

Cox, D. R. (1975). Partial likelihood, *Biometrika* **62**, pp. 269–276.

Csörgo, M. and Horváth, L. (1995). *Limit Theorems in Change-Point Analysis* (Wiley, New York).

Dickey, D. A. and Fuller, W. (1979). Distribution of the estimateurs for autoregressive time series with a unit root, *J. Am. Statist. Assoc.* **74**, pp. 427–431.

Doléans-Dade, C. (1970). Quelques applications de la formule de changement de variable pour les semimartingales, *Zeischrift Wahrscheinlichkeitstheorie verw. Geb.* **16**, pp. 181–194.

Fan, J., Zhang, C. and Zhang, J. (2001). Generalized likelihood ratio statistics and Wilks phenomenon, *Ann. Statist.* **29**, pp. 153–193.

Gill, R. D. (1983). Large sample behaviour of the product-limit estimator on the whole line, *Ann. Statist.* **11**, pp. 49–58.

Haldrup, N. (1994). The asymptotics of single-equation cointegration regressions with I(1) and I(2) variables, *J. Econometrics* **63**, pp. 153–181.

Haldrup, N. and Hylleberg, S. (1995). A note on the distribution of the least squares estimator of a random walk with a drift: Some analytical evidence, *Economics Letters* **48**, pp. 221–228.

Hinkley, D. V. (1970). Inference about the change-point in a sequence of random variables, *Biometrika* **57**, pp. 1–17.

Ibragimov, I. and Has'minskii, R. (1981). *Statistical Estimation: Asymptotic Theory* (Springer, New York).

Jackson, J. E. and Bradley, R. A. (1961). Sequential χ^2 and T^2-tests, *Ann. Math. Statist.* **32**, pp. 1063–1077.

Kaplan, M. and Meier, P. A. (1958). Nonparametric estimator from incomplete observations, *J. Am. Statist. Ass.* **53**, pp. 457–481.

LeCam, L. (1956). On the asymptotic theory of estimation and testing hypotheses, *Proc. 3rd Berkeley Symp.* **1**, pp. 129–156.

Lenglart, E. (1977). Relation de domination entre deux processus, *Ann. Inst. H. Poincaré* **13**, pp. 171–179.

Loader, C. R. (1991). Inference for a hazard rate change point, *Biometrika* **78**, pp. 749–757.

Loader, C. R. (1992). A log-linear model for a Poisson process change point, *Ann. Statist.* **20**, pp. 1391–1411.

Luo, X. (1996). The asymptotic distribution of MLE of treatment lag threshold, *J. Statist. Plann. Inference* **53**, pp. 33–61.

Luo, X., Turnbull, B. and Clark, L. (1997). Likelihood ratio tests for a change point with survival data, *Biometrika* **84**, pp. 555–565.

Matthews, D., Farewell, V. and Pyke, R. (1985). Asymptotic score-statistic processes and tests for a constant hazard against a change-point alternative, *Ann. Statist.* **85**, pp. 583–591.

Nguyen, H. T., Rogers, G. S. and Walker, E. A. (1984). Estimation in change-point hazard rate models, *Biometrika* **71**, pp. 299–304.

Parzen, E. A. (1962). On the estimation of probability density and mode, *Ann. Math. Statist.* **33**, pp. 1065–1076.

Phillips, P. C. B. (1987). Towards a unified asymptotic theory for autoregression, *Biometrika* **74**, pp. 535–547.

Pons, O. (1986). Vitesse de convergence des estimateurs à noyau pour l'intensité d'un processus ponctuel, *Statistics* **17**, pp. 577–584.

Pons, O. (2002). Estimation in a Cox regression model with a change-point at an unknown time, *Statistics* **36**, pp. 101–124.

Pons, O. (2009). *Estimation et tests dans les modèles de mélanges de lois et de ruptures* (Hermès Science Lavoisier, Paris and London).

Pons, O. (2011). *Funtional Estimation for Density, Regression Models and Processes* (World Scientific Publish., Singapore).

Pons, O. (2014). *Statistical Tests of Nonparametric Hypotheses: Asymptotic Theory* (World Scientific Publish., Singapore).

Pons, O. (2017). *Inequalities in Analysis and Probability, 2nd ed.* (World Scientific Publish., Singapore).

Quandt, R. E. (1958). The estimation of the parameters of a linear regression system obeying two separate regimes, *J. Amer. Statist. Assoc.* **53**, pp. 873–880.

Quandt, R. E. (1960). Tests of the hypothesis that a linear regression system obeys two separate regimes, *J. Amer. Statist. Assoc.* **55**, pp. 324–330.

Rebolledo, R. (1980). Central limit theorems for local martingales, *Z. Wahrsch. Verw. Gebiete* **51**, pp. 269–286.

Rosenblatt, M. (1971). Curve estimates, *Ann. Math. Statist.* **42**, pp. 1815–1842.

Sen, P. K. (2002). Shapiro-Wilks type goodness-of-fit tests for normality: asymptotics revisited, In *Goodness-of-fit tests and validity of models*, pp. 73–88.

Shapiro, S. and Wilks, M. B. (1965). An analysis of variance test for normality, *Biometrika* **52**, pp. 591–611.

Tsiatis, A. A. (1981). A large sample study of Cox's regression model, *Ann. Statist.* **9**, pp. 93–108.

Watson, G. S. and Laedbetter, M. (1963). On the estimation of a probability density, *Ann. Math. Statist.* **34**, pp. 480–491.

White, J. S. (1958). The limiting distribution of the serial correlation coefficient in the explosive case, *Ann. Math. Statist.* **29**, pp. 1188–1197.

Whittle, P. (1958). On the smoothing of probability density functions, *J. Roy. Statist. Soc., Ser. B* **20**, pp. 334–343.

Index

Printed in the United States
By Bookmasters